THE GENUS EPIMEDIUM

AND OTHER HERBACEOUS BERBERIDACEAE
INCLUDING THE GENUS PODOPHYLLUM

Professor William T. Stearn (1911–2001)

Botanist, horticulturist, bibliographer, biographer, historian and life-long student of the genus *Epimedium*. Fellow enthusiasts for the genus are thankful that he was able to complete this new account of *Epimedium* and *Vancouveria* shortly before his death in May 2001.

A BOTANICAL MAGAZINE MONOGRAPH

THE GENUS
EPIMEDIUM
AND OTHER HERBACEOUS BERBERIDACEAE

by William T. Stearn

INCLUDING THE GENUS PODOPHYLLUM

by Julian M.H. Shaw

WITH ILLUSTRATIONS BY
Christabel King, Ann Farrer, Mark Fothergill, Pauline Dean,
Joanna Langhorne, Stella Ross-Craig, Matilda Smith, Sydenham Edwards,
Walter Hood Fitch, Kazuko Tajikawa, Ferdinand Bauer

EDITED BY
Peter S. Green and Brian Mathew

SERIES EDITOR
Brian Mathew

Timber Press
Portland, Oregon

© 2002 The Board of Trustees of the Royal Botanic Gardens, Kew

First published 2002

Published in North America in 2002 by
Timber Press, Inc.
The Haseltine Building
133 S.W. Second Avenue, Suite 450
Portland, Oregon 97204, U.S.A.

ISBN 0 88192 543 8

A CIP record for this book is available from the Library of Congress

Production Editor: Ruth Linklater

Design by Jeff Eden, page make up by Christine Beard
Media Resources, Information Services Department
Royal Botanic Gardens, Kew

Origination by Magnet Harlequin

Printed in the European Union by
The Bath Press (CPI Group)

CONTENTS

LIST OF COLOUR PAINTINGS vii

PREFACE AND ACKNOWLEDGMENTS ix

PART I — EPIMEDIUM AND VANCOUVERIA

1. THE FAMILY BERBERIDACEAE 1
    Key to Herbaceous Genera of Berberidaceae 2
2. HISTORICAL INTRODUCTION 3
3. MORPHOLOGY 9
4. CLASSIFICATION 21
5. GEOGRAPHICAL DISTRIBUTION 27
6. CULTIVATION OF *EPIMEDIUM* 35
7. TAXONOMIC TREATMENT OF *EPIMEDIUM* 41
    Generic description 41
    Key to *Epimedium* species 43
        Subgenus I. *Epimedium* 48
            Section i. *Diphyllon* 48
                Series A. *Campanulatae* (species 1–4) 48
                Series B. *Davidianae* (species 5–14) 55
                Series C. *Dolichocerae* (species 15–30) 81
                Series D. *Brachycerae* (species 31–43) 114
            Section ii. *Macroceras* (species 44–49) 137
            Section iii. *Polyphyllon* (species 50) 154
            Section iv. *Epimedium* (species 51–52) 155
        Subgenus II. *Rhizophyllum* (species 53–54) 164
    Appendix 1. *Epimedium* hybrids 174
    Appendix 2. *Epimedium takhtajanii*, the fossil *Epimedium* 196
    Appendix 3. Obscure and excluded taxa 197
    Appendix 4. *Epimedium* cultivars 199
8. TAXONOMIC TREATMENT OF *VANCOUVERIA* 203

PART II — REVIEW OF OTHER HERBACEOUS BERBERIDACEAE

Achlys 211
Caulophyllum 214
Diphylleia 215
Ranzania 219
Jeffersonia 219
Bongardia 223
Gymnospermium 227
Leontice 234
Chromosome counts in Berberidaceae (excluding *Podophyllum*) 237

PART III — THE GENUS PODOPHYLLUM by Julian M.H. Shaw

Morphology of *Podophyllum* 239
Pollination and breeding system in *Podophyllum* 251
Cytology of *Podophyllum* 255
Taxonomic treatment of *Podophyllum* 258
Doubtful and excluded names 314

BIBLIOGRAPHY

A. References and bibliography for *Epimedium* and other genera
 of herbaceous Berberidaceae 315
B. References and bibliography for *Podophyllum* 321

INDEX OF SCIENTIFIC NAMES 337

COLOUR PAINTINGS

Plate 1. *Epimedium davidii* .... ANN FARRER (p. 56)
Plate 2. *Epimedium flavum* .... CHRISTABEL KING (p. 64)
Plate 3. *Epimedium acuminatum* .... CHRISTABEL KING (p. 92)
Plate 4. *Epimedium leptorrhizum* .... PAULINE DEAN (p. 108)
Plate 5. *Epimedium grandiflorum* (three variants) .... CHRISTABEL KING (p. 138)
Plate 6. *Epimedium diphyllum* .... MARK FOTHERGILL (p. 150)
Plate 7. *Epimedium elatum* .... PAULINE DEAN (p. 153)
Plate 8. *Epimedium alpinum* .... FERDINAND BAUER (p. 157)
Plate 9. *Epimedium pubigerum* .... ANN FARRER (p. 160)
Plate 10. *Epimedium pinnatum* subsp. *colchicum* .... ANN FARRER (p. 165)
Plate 11. *Epimedium perralderianum* .... CHRISTABEL KING (p. 172)
Plate 12. *Epimedium* × *rubrum* (upper), *E.* × *cantabrigiense* (lower left), *E. alpinum* (lower centre) .... MARK FOTHERGILL (p. 177)
Plate 13. *Epimedium* × *warleyense* .... MARK FOTHERGILL (p. 181)
Plate 14. *Epimedium* × *versicolor* 'Sulphureum' (left), 'Versicolor' (right) .... ANN FARRER (p. 189)
Plate 15. *Epimedium* × *youngianum* 'Roseum' (upper left), 'Niveum' (right), 'Merlin' (lower left) .... CHRISTABEL KING (p. 192)
Plate 16. *Vancouveria chrysantha* (left), *V. hexandra* (centre), *V. planipetala* (right) .... MARK FOTHERGILL (p. 204)
Plate 17. *Achlys triphylla* .... JOANNA LANGHORNE (p. 212)
Plate 18. *Diphylleia cymosa* .... SYDENHAM EDWARDS (p. 216)
Plate 19. *Ranzania japonica* .... STELLA ROSS-CRAIG (p. 217)
Plate 20. *Jeffersonia dubia* .... STELLA ROSS-CRAIG (p. 220)
Plate 21. *Jeffersonia diphylla* .... SYDENHAM EDWARDS (p. 221)
Plate 22. *Bongardia chrysogonum* .... WALTER HOOD FITCH (p. 225)
Plate 23. *Gymnospermium albertii* .... MATILDA SMITH (p. 228)
Plate 24. *Podophyllum peltatum* .... SYDENHAM EDWARDS (p. 261)
Plate 25. *Podophyllum pleianthum* .... MATILDA SMITH (p. 272)
Plate 26. *Podophyllum versipelle* subsp. *boreale* .... MATILDA SMITH (p. 277)
Plate 27. *Podophyllum hexandrum* .... MATILDA SMITH (p. 308)

The author dedicated this work to

ROBIN and SUSAN WHITE
and
DARRELL PROBST

enthusiastic growers of *Epimedium*;

also

MIKINORI OGISU

for his major contribution to knowledge of the
taxonomy and distribution of *Epimedium* of China

PREFACE

This book on *Epimedium*, *Vancouveria* and other herbaceous *Berberidaceae* (a revised and much enlarged new edition of my monograph published in 1938) remotely owes its origin to the rock-garden enthusiast Reginald John Farrer (1880–1920), of whose many publications I published a bibliography in E.H.M.Cox's *Plant Introductions of Reginald Farrer*: 99–113 (November, 1930). Browsing at that time through Farrer's vividly written encyclopaedia *The English Rock Garden* (1919) I came across an article on *Epimedium*, the name then completely unknown to me. This began: 'Epimedium. The Barren-worts are all much of a muchness except in the colour of their flower-flights plants of extreme but unappreciated value for quiet shady corners of the rock-garden, where they will form wide masses in time, and send up in spring and early summer 10-inch showers, most graceful and lovely of flowers that suggest a flight of wee and monstrous Columbines of waxy texture, and in any colour, from white, through gold, to rose and violet. Then, beginning later than these, appear the leaves, hardly less an adornment in summer than the flowers to spring. For these are of a delicious green, much divided into pointed leaflets, and borne on airy wiry stems.'

Farrer's intriguing account led me to look for 'those delightful things' in Cambridge gardens and to become fascinated by them. The 'dire tangle' of names for Japanese members of the genus mentioned by Farrer seemed to exist for all the cultivated ones.

Accordingly, monographing the genus *Epimedium* promised to be an interesting educational study and indeed proved to be such, because it led not only to elucidation of the taxonomy and nomenclature of the wild and cultivated species and forms, but also to a consideration of past climatic changes responsible for its extraordinary disjunct distribution.

This self-imposed task could not have been achieved without facilities granted by Professor Albert C.Seward (1863–1941) at the Botany School, Cambridge and Mr Frederick Preston at the Cambridge University Botanic Garden in 1941. Not being a university student working for a Ph.D. but a low-paid book-shop assistant working for a living, with time for research confined to lunch hours and evenings, I was grateful to Professor Seward for procuring on loan at my request herbarium material from Berlin-Dahlem, Breslav, Calcutta, Canton, Chicago, Copenhagen, Edinburgh, Kew, Leiden, Manchester, New York, St. Petersburg (Leningrad), Nankin, Paris, Seattle, Shanghai, Uppsala, Vienna and Washington DC. A small grant from the Royal Society of London enabled me to study the types and other specimens in Paris and Geneva. Later, when librarian of the Lindley Library, I was earning enough money to visit and study the material in Bonn, Florence, Munich, Pisa, Prague and Rome. My *Epimedium and Vancouveria*

(*Berberidaceae*), *a Monograph* was, after much delay, eventually published in the *Journal of the Linnean Society, Botany* in December, 1938. Based on the herbarium material in 28 institutions, it was as comprehensive a monograph as was then possible: of 21 species of *Epimedium* and three of *Vancouveria*, *E. leptorrhizum* Stearn and *E. macrosepalum* Stearn were the only new species discovered among the numerous specimens examined. The present work only partly supersedes the 1938 monograph which lists localities for the species then known and provides photographs of types and other specimens.

Meanwhile Mr Preston obtained living plants from 16 botanic and private gardens, which were grown for me by Mr Ralph Thoday (1895–1981; VMH, 1972) in a glasshouse at St. John's College Gardens, Cambridge. They revealed that almost all of the plants portrayed in nineteenth-century horticultural literature still existed in cultivation, although greatly confused as to names. Some had no counterparts in the wild and were undoubtedly hybrids of garden origin. Observation led to the conclusion then that clones of *Epimedium* are self-sterile but interfertile. The horticultural importance of these hybrids led me to devote thirteen pages of the 1938 monograph to their description.

Of the 21 species there distinguished, 13 were Chinese, four of them (*E. acuminatum*, *E. davidii*, *E. fargesii* and *E. sutchuenense*) first described by the Paris botanist Adrien Franchet (1834–1900), the author of an excellent revision of the genus in 1886. Chinese interest in *Epimedium* first became manifest in 1975, when Tsün-Shen Ying published a survey of the Chinese species of *Epimedium*, including two new species, *E. wushanense* and *E. simplicifolium*, remarkable for the leaf having only one leaflet. The Japanese botanist and horticulturist Mikinori Ogisu kindled my latent interest by introducing into cultivation in England in 1989 an *Epimedium* from central China which proved to be a new species (*E. dolichostemon* Stearn). Erroneously recorded from Emi Shan (Mount Omei) but actually from another Sichuan locality, Shizhu. He subsequently introduced 14 more species into cultivation, some completely new and unexpected. No-one else has made so important a contribution to knowledge and understanding of the taxonomy and distribution of Chinese species of *Epimedium*. It is appropriate that *E. ogisui* and *E. mikinorii* are named in his honour. Mention should also be made of the introduction of a few Chinese species by James Compton, Roy Lancaster and Martyn Rix.

A number of new species have been described and, fortunately, illustrated by Chinese botanists; for information about these I have had to rely on their protologues. The general increase in knowledge of the Chinese species, from 12 in 1938 to 42 in 1998, together with a critical survey of the Japanese species in 1990 by Kazuo Suzuki culminating taxonomic and pollination studies, began about 1923 and indicated the desirability of a new account of the genus *Epimedium*. After completion of the monograph of *Epimedium* and *Vancouveria* ultimately published in 1938, I prepared a synopsis of the other genera and species of herbaceous *Berberidaceae*, including *Podophyllum* together with restoration of Spach's neglected genus *Gymnospermium*, but the outbreak of the Second World War in 1939 prevented its completion and publication. It formed, however, the basis of the account of the herbaceous *Berberidaceae* in the *European Garden Flora* 3: 370–371, 389–396 (1989) and the present volume.

Although all clones of *Epimedium* appear to be self-sterile, the species are apparently completely interfertile, thus facilitating the production of garden hybrids, now excessive, especially in Japan, and much to be deplored if it leads to the neglect and loss of original species. Experience has shown that hybridisation between closely related species, such as *E. alpinum* × *pubigerum* (*E.* × *cantabrigiense*) and *E. perralderianum* × *E. pinnatum* subsp. *colchium* (*E.* × *perralchicum*) has produced plants of no greater garden value than their parents, whereas hybridisation between very different species has resulted in plants very different from both parents, e.g. *E.* × *rubrum* and *E.* × *warleyense*, etc.

The taxa of *Epimedium* introduced into cultivation by Mikinori Ogisu and Roy Lancaster have been very successfully cultivated in England by Mr and Mrs Robin White of Blackthorn Nursery, Kilmeston, Hampshire and I am indebted to them for much living material for study. Mr Darrell Probst, Hubbardston, Massachusetts, U.S.A. who grows a comparable collection, has also provided specimens of great interest. To these and other helpful people, notably Mr Tony Hall of the Kew Gardens staff, and members of the Kew Herbarium and Library, I express grateful thanks. It is also a pleasant duty to thank Dr Peter Green for obtaining photocopies of the protologues of taxa described in Chinese journals and Maureen Bradford for typing my extensive manuscript. Brian Mathew, the current Editor of *Curtis's Botanical Magazine*, and Peter Green have carefully edited the manuscript and my thanks go to them for bringing this work to a satisfactory conclusion.

For centuries Chinese pharmacists and drug handlers, and their customers, were content, as they still are, with the cut and dried leaves of Ying Yang Huo plants (*Epimedium sagittatum* and allied species) for their true or imaginary aphrodisiac properties. No plant is known with genuine aphrodisiac properties although these have been attributed to a diversity of plants, among them the tomato or love apple (*poma amori*) and the carrot; it is simply a matter of imagination and self-deception. Traditional cutting of leaves did no great harm to *Epimedium* stocks. Recently however, Chinese peasants have taken to uprooting *Epimedium* rhizomes in quantity and selling them by the roadside for their supposed properties. This can only lead to the elimination of *Epimedium* species, some of which have very limited ranges. The uprooting of *Epimedium* rhizomes for sale, as distinct from cutting leaves, should be condemned by the Chinese Government and Colleges of Traditional Chinese Medicine to save stocks from extinction.

William T. Stearn, 2000.

PART I — EPIMEDIUM AND VANCOUVERIA

1. THE FAMILY BERBERIDACEAE

Berberis vulgaris L. (1753) is the lectotype of the large and widespread genus *Berberis*, thus by implication the type of the family *Berberidaceae*. It is a prickly European shrub with simple leaves and racemes of yellow flowers followed by red few-seeded berries distributed by birds; the petals have two nectarial swellings and the anthers open by up-rolling valves. With this well-known shrub, placed by Linnaeus in his *Hexandria Monogynia* in 1753, Antoine Laurent de Jussieu in his revolutionary *Genera Plantarum secundum Ordines Naturales disposita* (1789: 287) associated the genera *Epimedium* and *Leontice* within his Ordo XVIII *Berberides* (later, by Horantinov and by Lindley, renamed *Berberidaceae*). In this group Jussieu also included two tropical genera *Rinorea* Aubl. and *Conohoria* Aubl. (now classified in the *Violaceae*).

Subsequent addition (by linkage of characters) of *Nandina, Mahonia, Jeffersonia, Vancouveria, Podophyllum, Diphylleia, Caulophyllum, Bongardia* and *Ranzania*, because there seemed no where else in the plant kingdom better to place them, resulted in the *Berberidaceae* becoming a heterogeneous assemblage with no single character common to all the genera. This anomalous situation provided a temptation to fragment the assemblage into small families e.g. *Berberidaceae* (*sensu stricto*), *Nandinaceae, Podophyllaceae, Diphylleiaceae, Leonticaceae* and *Ranzaniaceae*. For this division the *Berberidaceae*, being a very old group with well-defined genera, offers abundant characters. Thus Armen Takhtajan (1997) has divided the order *Berberidales* (which corresponds to the family *Berberidaceae* as generally accepted) into *Berberidaceae, Nandinaceae* (only *Nandina domestica*), *Ranzaniaceae* (only *Ranzania japonica*) and *Podophyllaceae* (including herbaceous genera, such as *Epimedium, Vancouveria, Leontice* etc). The grave objection to this emphasis on differences is that although the segregate families remain associated, the diversity of their names obscures their linkage without adding any new knowledge or understanding. It is better to keep the *Berberidaceae* (*sensu lato*) intact.

The *Berberidaceae* has received much attention from a diversity of viewpoints since Citerne's excellent dissertation on its morphology (1892). However, as observed by Kim & Jansen (1997), in spite of general systematic studies (e.g. by Janchen, 1949) as well as studies of floral anatomy, serology, palynology and cytology, there still remains substantial disagreement regarding phylogenetic relationships between the genera. There is no consensus. The cladistic analysis by Nickol (1995) using 59 morphological characters in PAUP seems as reasonable as any scheme yet proposed.

Key to the herbaceous genera of Berberidaceae

The following key is based on one contributed to the *European Garden Flora* 3: 370–371 (1989) by Stearn, but excluding the woody genera, *Berberis, Mahonia* and *Nandina*.

1a. Flowers stalkless in a many-flowered, slender spike; sepals and petals absent; anthers as broad as long .... **Achlys**
 b. Flowers stalked, solitary or in racemes, corymbs or panicles; sepals and petals present; anthers longer than broad 2

2a. Flowers solitary on a basal, leafless stem; leaves all basal, with 2 stalkless, opposite leaflets or a single lobed leaflet .... **Jeffersonia**
 b. Flowers several or many or if solitary then at apex of stem between 2 leaves 3

3a. Leaves undivided but sometimes lobed almost to stalk, peltate, major veins radiating palmately from top of stalk 4
 b. Leaves divided into leaflets, not peltate, pinnately veined 5

4a. Stem-leaves 2, opposite, with the solitary flower arising between them, or leaf single and flowers several, clustered, drooping; anthers opening by longitudinal slits; fruit a large red or yellow berry .... **Podophyllum**
 b. Stem-leaves 2, well-separated; flowers in erect, many-flowered cymes; anthers opening by up-rolling flaps; fruit a small blue berry .... **Diphylleia**

5a. Leaflets stalkless; rootstock a tuber 6
 b. Leaflets stalked; rootstock a rhizome 8

6a. Leaves all basal, simply pinnate, with 5–8 pairs of leaflets; inner sepals small; petals flat, conspicuous .... **Bongardia**
 b. Leaves basal and on stem, divided into 3 leaflets; inner sepals large, petal-like; petals small, nectar-producing 7

7a. Stems with several, much-divided leaves, bearing terminal and axillary racemes; fruit large, inflated, membranous .... **Leontice**
 b. Stem with 1 leaf, not branched, bearing a terminal raceme only; fruit small, opening before ripening of seeds and exposing them .... **Gymnospermium**

8a. Leaflets deeply 3–5-lobed 9
 b. Leaflets entire or only slightly 3-lobed 10

9a. Flowers greenish, yellowish or purplish, in an erect, terminal raceme
  .... **Caulophyllum**
 b. Flowers lavender-violet on long, drooping stalks between 2 leaves .... **Ranzania**

10a. Petals and stamens 4; petals various but not narrowly oblong .... **Epimedium**
 b. Petals and stamens 6; petals narrowly oblong with a flat or hooded nectar-producing tip .... **Vancouveria**

2. HISTORICAL INTRODUCTION

Epimedium is an Ancient Greek plant name, *epimedion*, of unknown derivation and meaning and of uncertain application, independently perpetuated down to the sixteenth century in the works by the Greek herbalists Dioscorides (1st century A.D.) and his Roman contemporary, the Latin encyclopaedist Pliny the Elder (23–29 A.D.). Dioscorides, in his great work on medicaments, *De Matera Medica* (1st century A.D.), book IV cap. 19, described *Epimedion* as a low-growing plant with 10 or 12 ivy-like leaves, bearing neither flowers nor fruit, with thin black strong-smelling roots unpleasant to the taste, which grew in moist places. A poultice of the leaves beaten fine with oil prevented the breasts from swelling, the root caused barreness and the leaves beaten small and drunk in wine after menstruation prevented conception for 5 days. The *Codex Julianae Aniciae nunc Vindabonensis* (512 AD), 105 verso, portrays this contraceptive plant, which cannot however be identified but is unlike any member of the modern genus *Epimedium* L. Pliny likewise described *Epimedion* in his *Historia Naturalis*, book XXVII, cap. 53, as being a low plant with 10–12 ivy-like leaves, never flowering, with thin black strong-smelling roots; the leaves beaten in wine restrained the breasts of virgins. Neither Dioscorides nor Pliny was acquainted with the works of the other, yet many passages in the one are paralleled in the other, as noted by the French scholar Salmasius in 1629. Dioscorides and Pliny obviously copied these similar passages from an earlier work, according to Max Wellmann the lost herbal of Sextius Niger (fl. 25 BC).

Dioscorides's work provided concise descriptions of more than 600 plants, detailing their real and reputed medicinal properties. Being undoubtedly the most comprehensive work then available and so remaining for some sixteen centuries, the Dioscorides herbal was copied and copied and re-arranged and thus came to exist in many as codices and was translated from Greek into Latin, French, Spanish and German. As stated by E.L.Green in 1909, 'from the year 1516 and for a whole century thereafter the most voluminous and most useful books of botany were in the form of commentaries on Dioscorides'. They introduced into modern botany a vast number of Ancient Greek plant names preserved through the centuries in Dioscorides's work, among them *Epimedion*, and they sought to ascertain their precise application. At least five quite different species seems to have determined the present use.

Botrychium lunaria, *Ornithogalum narbonense*, *Hepatica nobilis* and *Epimedium alpinum*, were regarded by different sixteenth century authors as being the *Epimedion* of the present. Luigi Anguillaria (1512–1570) was the most widely travelled Italian botanist of the sixteenth century and as interested as his contemporaries in the plants known to the Ancients. He roamed not only over most of Italy but through the Balkan Peninsula, including former

Yugoslavia, and inevitably he found many plants hitherto unknown. In his one publication *Semplici* (Venice, 1561), a little volume of great rarity and full of first-hand observations, he described under the heading 'Epimedio' a plant with slender stems which divided into three ivy-like leaves and had a slender root with a heavy smell and astringent taste creeping underground. It grew in woodland near Vicenza. This is the first mention of the species later named *Epimedium alpinum* by Linnaeus. Native to the Colli Euganei east of Padua, the Venetian Republic's University, it was taken into cultivation about this time and quickly passed from Italy into Belgian, German and French gardens. By 1597 John Gerard grew it in London. It long remained the only species of its genus known in Europe. Linnaeus knew it well from plants growing in Holland and Sweden and thus it became the type of the generic name *Epimedium* as established by him in 1753 and 1754.

Fig. 1. ***Epimedium alpinum***. From: Tabernaemontanus, *Eicones Plantarum* (1590).

Although the name *Epimedium* came into widespread use during the seventeenth century for the species, Tournefort was the first to define the genus, in French in 1694 and in Latin in 1700: '*Epimedium* est plantae genus, flore cruciform: quatuor scilicet petalis tubulatis constante. Ex cujus calyce surgit pistillum, quod deinde abit in fructum seu siliquam unicapsulare bival vem et seminibus foetam. *Epimedii* species unica hastenus cognita. *Epimedium* Dod. Pempt. 599. *Epimedion* quorundam J.B. 2, 395'. (Inst. Rei Herb. 232, pl.: 1700). Even today this concisely diagnoses the group except for the Japanese and Chinese species with flat (not tubular or spurred) petals. Tournefort placed *Epimedium* among herbs with cruciform flowers and one-chambered pistils. The same characters led Linnaeus to place *Epimedium* in his artificial group *Tetrandria* (with four stamens) *Monogynia* (with one pistil). Michel Adanson in 1763 and Antoine Laurent de Jussieu in 1783 were the first to associate it with *Berberis* and *Leontice*.

On his journey to the Near East in 1700–1702 Tournefort noted the existence of two taxa which he recorded in 1703 as *Epimedium orientale, flore ex albo flavescente* and *E. orientale, flore albo* (Corollarium: 17), the latter certainly *E. pubigerum*, the former probably also *E. pubigerum*. Of these no specimens exist in his herbarium at Paris.

As Linnaeus knew only the European *E. alpinum*, to which his student Carl Peter Thunberg in 1781 referred 'a poor specimen of a Japanese species', presumably *E. grandiflorum*, the genus remained monotypic until 1821. In that year Augustin Pyramus de Candolle published *E. alpinum* var. *pubigerum*, collected by Olivier and Bruguière near Istanbul (Constantinople), and *E. pinnatum*, from information supplied by F.E.L.Fischer of Moscow who had found it in Pallas's herbarium collected by Carl Luswig Hablizl in northern Iran (Gilan) in 1773. Little was then known about the floristic richness of Eastern Asia. In 1823 a young Bavarian German doctor Philipp Franz von Siebold (1796–1866) arrived at Nagasaki, southern Japan, to be medical officer to the small Dutch trading community there. Everything about Japan fascinated Siebold and engaged his close life-long attention, including its plants, for which he engaged a Japanese artist, Kawahara Keiga, to draw and paint. Keiga's drawings depict *E. grandiflorum* and *E. diphyllum*. Most of the living Japanese plants which he sent to the Leiden Botanic Garden in 1829 died on the long voyage to Europe but among the survivors was *Epimedium diphyllum* which passed from Leiden to the London nursery of Loddiges, and was illustrated in their publication, *The Botanical Cabinet* (19: t.1858) in 1832.

Expelled from Japan in 1830, Siebold came back to Europe with another large collection of living Japanese plants, only 260 out of the original 1200 surviving the six-month voyage. They were planted in the Ghent botanical garden, in what had formerly been the Austrian Netherlands but is now Belgium. A revolution there in August 1830 caused him to remove hastily his books, manuscripts, ethnographical and natural history specimens to Leiden in the Dutch Netherlands, but his plants, now rooted in Ghent, had to say there. Their loss caused Siebold much bitterness, especially as they subsequently contributed so much to the horticultural renown and profit of Ghent. The beauty and interest of the hitherto unknown Japanese epimediums when they flowered in the Ghent botanic garden led two Belgian botanists Charles F.A.Morren

(1807–1858) and Joseph Decaisne (1807–1882) to publish in December 1834 the first monograph of the genus *Epimedium* (Ann. Sci. Nat. II. Bot. 2: 347–361). This they based both on Siebold's living plant and herbarium specimens at the Muséum National d'Histoire Naturelle, Paris.

Morren and Decaisne recognised six species of *Epimedium* proper, namely *E. alpinum* L., *E. pubigerum* (raised from varietal status), *E. elatum* collected by Victor Jacquemont in India, and three, *E. macranthum*, *E. violaceum* and *E. musschianum*, introduced by Siebold from Japan. At the same time they created a new genus *Aceranthus* for a spurless Japanese species *E. diphyllum*, also a Siebold introduction and already figured by Loddiges, and another new genus *Vancouveria* for Hooker's trimerous-flowered *Epimedium* which Menzies, Douglas and Scouler had collected in western North America. Unaware of Morren and Decaisne's 1834 publication, C.S.Rafinesque in early 1837 founded a new genus, *Vindicta*, based on *E. diphyllum* and corresponding to their *Aceranthus*, and *Sculeria* based on *E. hexandrum* (their *Vancouveria*), then in late 1837 *Endoplectris* based on *E. macranthum* (*E. grandiflorum*), although he knew them only from illustrations. The genus *Epimedium* was thus shown to occur in Europe and western and eastern Asia, with a close ally in western North America. Casson's publication of *E. perralderianum* in 1862 extended the generic range to western North Africa. Since its nearest allies are in the Caucasus, this discovery provided another phytogeographical problem not plausibly solved until 1938.

Of the occurrence of numerous species of *Epimedium* in western and central China, little was known until Carl Johann Maximowicz in 1877 described *E. pubescens* from Shanxi (Shensi) province; slightly later Adrien René Franchet, working in Paris on the collections of the French missionaries Pierre Amand David and Paul Hulbert Perny, described *E. davidii* in 1883 from Sichuan (Szechwen) province and *E. acuminatum* in 1886 from Guizhou (Kweichow). Traditional Chinese literature on *Epimedium* is meagre. The name 'Ying yang huo' used by Chinese druggists occurs in ancient Chinese herbals but the celebrated naturalist and encyclopaedist Li Shih-Chen (1518–1593) in his *Pen Tschao Kang Mu* seems to be the first to have described a Chinese *Epimedium* (*E. sagittatum*) recognisably. This species was introduced from China to Japan for its reputed medicinal virtues and the Chinese *E. sagittatum* was described first from Japanese material by Siebold and Zuccarini in 1845 as *Aceranthus sagittatus*.

Meanwhile in Belgium a Dutch gardener Andre Donckelaar (1783–1858), Curator of the Ghent botanic garden from 1835 to 1858, had hybridized the species then in cultivation. According to his biographer D.Spae, 'les charmantes *Epimedium atroroseum, lilacinum, rubrum, versicolor* etc. sont encore les resultats des fecundations artificielles' (Belgique Hort.). This was unknown to the authors of two important papers on the genus, John Gilbert Baker (1834–1920) at the Royal Botanic Gardens, Kew and Adrien René Franchet (1834–1900) at the Muséum National d'Histoire Naturelle, Paris. Since the introduction of *E. grandiflorum* (*E. macranthum*) in 1830 and its subsequent propagation by Belgian nurserymen, the genus had received much attention by cultivators and, while many new taxa had been described and fortunately illustrated in colour, the names in gardens were 'in a state of great confusion' as Baker stated in 1880. This chaos induced him to publish in the *Gardeners' Chronicle* of 1880 (Baker, 1880) a synopsis of the species and forms based on plants at Kew. He recognised 12 species and hybrids and provided excellent descriptions. Franchet overlooked Baker's article when in 1886 he published his paper 'Sur les espèces du genre *Epimedium*' (*Bull. Soc. Bot. France* 33: 38–41, 103–116, 1886) in which he gave an account of the wild species and attempted to put in order the cultivated forms which greatly puzzled him. Neither he nor Baker recognised fully the importance of hybridization within the genus. He wrote, 'C'est toujours une difficulté, parfois insurmontable, de dégager les plantes longuement modifees par la culture et de les ramener d'une façon satisfaite aux types d'ou elles sont issues'. New species from China had aroused his interest in the genus. Following Baillon, he referred both *Aceranthus* and *Vancouveria* back to *Epimedium*. The genus, as he accepted it then based on wild material, consisted of ten species, two with leafless flower stems forming section *Gymnocaulon*, the others with one or two leaves on the flower stem forming section *Phyllocaulon*, and one species in subgenus *Vancouveria*. With the garden plants unknown in a wild state he was less at ease but provided notes on them.

Under *E. hexandrum* he confused the true *E. hexandrum* of Hooker and another species which was carefully distinguished from it the following year (1887) by Silvio Calloni as *Vancouveria planipetala* and later (1890) by E.L.Greene as *V. parviflors*. Greene had already in 1883 described *E. chrysantha*. Thus within a few years the three American species became known. The six supposed 'new species' described by Greene in 1914 are none of them distinct from these three. The paper of 1886 by Franchet is the classic account of *Epimedium*. Collections in Sichuan province, China, by the French missionary Paul Guillaume Farges (1844–1900) provided Franchet in 1884 with material of two new species, *E. sutchuenense* and *E. fargesii*. Franchet thus made known four Chinese species.

The well-known Russian botanist Vladimir L.Komarov (1869–1945) published another revision of *Epimedium* as part of his 'Prolegumena ad floras Chinae necnon Mongoliae' (1908) with suggestive remarks in Russian on the distribution and phylogeny of the sections, at the same time revising *Clematoclethra*, *Codonopsis* and *Nicranus*. He distinguished sixteen species of *Epimedium* proper and three of *Vancouveria*, which he treated as a subgenus of *Epimedium*. Franchet's section *Phyllocaulon* he divided into series, following the example of Bunge and Maximowicz; *Monophylla, Acerantha, Diphylla* and *Polyphylla*. For his new species *E. elongatum* he provided a good illustration but recorded no type, no collector (actually G.N.Potanin) and no locality (actually western China). The consultation of Komarov's work is rendered difficult by the use of three sets of terms for floral parts instead of the standard terminology of 'outer sepals' for the outer-most false-whorl of four segments, 'inner sepals' for the middle false-whorl and 'petals' for the innermost false-whorl.

The bewildering situation in 1880 when J.G.Baker lamented the horticultural confusion of names of *Epimedium* was no better in 1930 when I became interested in the genus, beguiled by Farrer's eulogy because there existed no reliable means of identification and the correct naming of any cultivated plant of *Epimedium*. The previous revisions of the genus by Baker, Franchet and Komarov were quite inadequate, as well as inaccessible, regarding the plants in gardens. Farrer had referred to their 'dire tangle'. I decided that to monograph the whole genus anew, and indeed the only way to end its chaos, would be a challenging and intellectually profitable wide-ranging exercise. Baker based his account solely on the herbarium and living material at Kew, Franchet on that at Paris and Komarov on the herbarium material at St. Petersburg, Paris and Kew. Thanks to the influential co-operation of Professor (later Sir) A.C.Seward, then Professor of Botany in Cambridge, I was able to have on loan for study in the Botany School, Cambridge, herbarium material from Berlin-Dahlem, Breslau, Brno, Calcutta, Chicago, Edinburgh, Hamburg, Cambridge (Mass.), Kew, Leiden, Leningrad, Paris, Seattle, Shanghai, Uppsala, Washington DC. and Vienna, thus being able to determine material by direct reference to type-specimens; later came Canton, Nanking and Peking for data. A Royal Society grant-in-aid enabled me to visit the herbaria at Paris, Geneva and Zurich, and later I saw the herbaria in Weimar, Florence, Pisa, Rome, Heidelberg, Munich and Prague. Together this material of 39 herbaria provided as much direct information as was then available on the characters and distribution of *Epimedium* in the wild. I had also under observation a fairly extensive collection of living plants from 22 botanic and private gardens and grown for me in the Cambridge University Botanic Garden and the garden of St. John's College, Cambridge, thanks to the head gardener, Mr Ralph Thoday. Altogether this revealed not only the varied names often used in gardens for the same plant but that many plants common in gardens had no wild counterparts. No-one then fully appreciated the extent of hybridization within the genus. The resulting monograph published in the *Journal of the Linnean Society of London, Botany* 31: 409–534 (1958) distinguished 3 species of *Vancouveria* and 23 species of *Epimedium*, with 10 subordinate categories and 11 horticulturally distinct plants of hybrid origin. In addition to keys and taxonomic descriptions, it included a detailed account of morphology and phytogeographic history. The living plants were planted out in St. John's College's Wilderness garden. Those which survived the Second World War included a hitherto unknown hybrid between *E. alpinum* and *E. pubigerum* now known as *E.* × *cantabragiense*.

Botanical investigation of the Chinese floras by Chinese botanists after the devastating and sterilizing Cultural Revolution led to a paper in China by Tsün-shen Ying in 1975 on the Chinese species of *Epimedium*, in which he distinguished 13 species, including the new *E. simplicifolium* and *E. wushanense*. Since then the description of new species by Chinese authors and by me, using material collected by Mikinori Ogisu, has brought the number of known Chinese species to 29 and the world total to 54.

China, where the evolution of *Epimedium* has continued without interruption probably since the origin of the genus, is not only the richest in species of *Epimedium*, it is the only region where new species may yet be found.

Pollination, hybridization and cytological studies of Japanese taxa of *Epimedium* in the wild by Kazuo Suzuki over many years have culminated in a biological and taxonomic monograph, *Nippon no Ikariso* (Tokyo, 1990).

3. MORPHOLOGY

Epimedium (*sensu lato*) is a group of perennial woodland herbs – more often hemicryptophytes than geophytes – spreading vegetatively by woody, irregularly branched, interlacing rhizomes which creep horizontally a little below the surface, with their numerous, fine, moderately branched roots occupying the upper 10–30 cm of soil. Rhizomes and roots are admirably portrayed by Tatemi Shimizu and Masayoshi Umbayashi (1995: 70–73).

RHIZOME

The rhizome is generally cylindric or slightly nodose and covered with brown membranous bracts, these being comparable to ligulate stipules with the petiole reduced to a minute prominence; the tip of the rhizome turns upward each year, elongating into a leafy aerial shoot which perishes usually before the next season; growth is continued from buds arising at irregular intervals in bract-axils behind the tip. The form of the rhizome, i.e. the degree of elongation and thickness, and also the average size of the terminal winter-bud, is fairly constant for each species, and sometimes offers contrasts of taxonomic value (*cf. E. alpinum* and *E. pubigerum*; *E. leptorrhizum* and *E. brachyrrhizum*; *E. fangii* and *E. davidii*, *E. rhizomatosum* and *E. membranaceum*. It is almost thread-like in certain Chinese species, e.g. *E. pauciflorum*, *E. leptorrhizum*, and *E. platypetalum*, although stouter and more compact in others, e.g. *E. sagittatum*, and may be more than 1 cm thick in *E. pubigerum* of Europe and Asia Minor. Gardeners, in general, being used to propagating herbaceous perennials by division of the root, possess a more first-hand knowledge of the underground organs than taxonomists working from dried material.

Despite Himmelbaur's suggestion (1913: 743) that slender rhizomes characterise the extremes of the generic area and stouter rhizomes the centre, there is no correlation between the form of the rhizome and geographical distribution.

STEM

The stem is terete, wiry, and shows little variation apart from pubescence (Stearn, 1938: 415). The young growths are coiled crosier-like, so that the bend of the stem or petiole emerges first and then draws into the light the downward directed young buds and leaflets; the latter are involutely rolled and covered with long hairs which usually fall early. The leaves present at flowering time are smaller in the length of their sheaths and the area of leaflet than those formed later. The rudiments of the next season's leaves and flower-buds may be found well-formed on dissecting the terminal bud of the rhizome in autumn.

FOLIAGE

The foliage of *Epimedium* is so characteristic that the group may be recognised by this alone. The compound leaves vary much in degree of subdivision; the leaflets are cordate with slender terete petiolules and are normally in threes. The most widely distributed type of leaf is biternate, occurring throughout the range and in all groups except *Epimedium diphyllum* and series *Dolichocerae*. A compound type seems to be the primitive leaf-form of the *Berberidaceae*, from which those which are simple have been derived by reduction and fusion. The extreme of leaf-division is found in *E. elatum*, the leaves of which sometimes consist of more than forty leaflets. The biternate leaf of nine leaflets may be easily modified to an imparipinnate leaf of five leaflets, as in *E. pinnatum* subsp. *pinnatum*, by the two lateral secondary petioles bearing only one leaflet each and the median one remaining trifoliolate. This imparipinnate type may be further reduced to a trifoliolate type, as in *E. pinnatum* subsp. *colchicum*, *E. platypetalum* and *E. davidii* which, from the same rhizome, produce both three- and five-foliolate leaves. Leaves never with more than three leaflets characterise *E. perralderianum*, *E. pubescens*, *E. acuminatum*, and most species of section *Diphyllon*; occasionally these produce basal leaves reduced to only one leaflet. An unusual modification of the trifoliolate leaf is found in *E. diphyllum*, the median leaflet being suppressed so that the leaf consists of two very oblique-based leaflets. There is thus a tendency to vary from bi- or even tri-ternate leaves through pinnate to trifoliolate or even bi- or uni-foliolate leaves as in *E. simplicifolium*, so that the species of the systematic groups may be arranged in parallel series according to the amount of leaf-reduction.

For diagnostic purposes it is convenient to distinguish leaves arising direct from the rhizome (*folia radicalia*) and those on the flowering stem (*folia caulina*); morphologically they differ little, except that the petiole of a basal leaf broadens at the base into a thin ligulate stipular sheath. In *Epimedium* subgenus *Rhizophyllum*, as in *Vancouveria*, all leaves are basal; in subgenus *Epimedium* the stem bears leaves, although basal leaves are usually also present. The actual number of stem-leaves is important: Franchet founded his classification upon it, remarking that it was a feature he had never known to vary in either the wild or cultivated plants. It is not, however, so unvarying as he supposed: in *E. sagittatum*, and others which normally have two apparently opposite stem-leaves, a third leaf sometimes arises at almost the same level as these; while in *Vancouveria* and in *Epimedium* subgenus *Rhizophyllum* a leaf is occasionally carried on the flowering stem: such abnormalities do not, however, seriously diminish the practical diagnostic value of the character. There are also plants (*E.* × *versicolor*, *E.* × *warleyense*), here considered hybrids between *E. pinnatum* (with no stem-leaves) and *E. grandiflorum* or *E. alpinum* (with one stem-leaf), which produce leafless and leafy inflorescences in almost equal numbers from the same rhizome; these plants are known only in gardens. However, the Chinese species *E. epsteinii* may have either one leaf or two leaves on the flower stem. By the normal number of stem-leaves subgenus *Epimedium* could be divided into three informal groups: "Monophyllon" having one stem-leaf (but in the classification proposed here, this is split between section *Macroceras* and section *Epimedium*), "Diphyllon" (here treated as section *Diphyllon*) having two stem-

Fig. 2. ***Epimedium pubescens***, to show general morphology of an *Epimedium* plant. A, whole plant showing inflorescence, basal and cauline leaves; B, non-flowering plant with rhizome and basal foliage; C, bud, showing outer and inner sepals; D, whole flower, showing large inner sepals, small saccate petals and stamens; E, flower with sepals removed showing petals (detached), stamens and gynoecium; F, fruit. From: Maximowicz, *Diagnoses Plantarum Novarum* in *Bulletin de l'Académie Impériale des Sciences de St. Pétersbourg* 29: t.1 (1883).

leaves and "Polyphyllon" (here treated as section *Polyphyllon*) having several stem-leaves; further division of the large section *Diphyllon* according to floral characters results in four quite natural series (see Classification, page 24). In sect. *Diphyllon* the two leaves are usually opposite but occasionally alternate, e.g. in *E. davidii*; *E. leptorrhizum* and *E. brachyrrhizum* are anomalous species — by the sum of characters near to *E. sutchuenense* (*Diphyllon*, series *Dolichocerae*) — in which the two leaves are unequally developed or one even suppressed, so that in this state they can resemble a species of "Monophyllon"; convergence is indicated by dissimilarity in other characters. Another anomalous species is *E. elongatum*, which usually bears several alternate stem-leaves like *E. elatum* (section *Polyphyllon*), but may occasionally have only two opposite stem-leaves as in *Diphyllon*, although its leaflets are different; the position of this species on floral characters is in section *Diphyllon*. It is significant that the comparatively unstable species occur in western China, where the genus is best represented and where its evolution may still be proceeding.

The petioles and petiolules are slender, stiff, terete, and have the same general structure as the stem; the scattered bundles often tend to form two rings; the nodes, as in *Nandina*, are swollen and frequently pilose.

LEAFLETS

The leaflets vary considerably between different species and within a species. They are normally in threes, of which the middle leaflet alone is symmetric; the lateral leaflets have their outer basal lobe larger than the inner lobe, and the outer 'half' of the leaflet is broader than the inner 'half'; this asymmetry is conspicuous in *E. diphyllum* and *E. sagittatum*. The terminal or middle leaflet is somewhat larger than the others. Owing to this variation it is possible to indicate only the general or prevailing form of the leaflets, which may be nearly orbicular, broadly ovate (i.e. length: breadth: 6:5), ovate (length: breadth: 3:2), narrowly ovate (length: breadth: 2:1), or lanceolate (length: breadth: 3:1) according to the species, the broadest part being below the middle; obovate leaflets are unusual. The base is nearly always cordate, with a sinus formed by the two lobes, which may be rounded and overlap or diverge at a wide or narrow angle; in *E. sagittatum* the larger lobe is acute or acuminate. The margin is stiffened by a fibrous bundle and usually furnished in *Epimedium* with short spines (usually 1–2 mm long), although in *E. diphyllum* and *E. pinnatum* subsp. *colchicum* these are often few or even absent; *Vancouveria* always has spineless leaflets. The tip may be mucronate, as in *E. elatum*, or acuminate, as in *E. acuminatum* and most Chinese and Japanese species, being usually terminated by a spine in *Epimedium* but indented in *Vancouveria* and most Chinese and Japanese species. The young leaflets of *V. hexandra* are peculiar for the hairs they bear on the upper surface, these disappearing with age; in other species of *Vancouveria* and *Epimedium* the upper surface is always glabrous. The cells of the epidermis are irregular in outline, faintly impressing a jigsaw pattern on the cuticle; the stomata, of the usual ranunculaceous type, occur in the lower epidermis. Like the petioles and flowering stem, the lower leaf-surface is at first covered with long white or reddish hairs, usually shed by maturity. They are multicellular, being composed of several superposed more or less equal cells, in most species, but in *E. acuminatum* and *E. sagittatum* there occurs another type of hair with the base formed of several very small cells and the upper part of one or two comparatively long cells. Under a lens, these hairs of *E. acuminatum* and *E. sagittatum* appear as appressed bristles. Occasionally they are suppressed, so that the leaflet is glabrous. Their development was first studied by Citerne (1892); one of the lower epidermal cells grows into a small conical projection which divides parallel to the leaf-surface; the upper half elongates into the long cell, while the lower divides again into three or four basal cells. The lower epidermis is often glaucescent; it may be smooth or papillose, as in certain Chinese species, with rounded projecting cells. The internal structure of the leaf is simple; there are about four or five layers of rounded cells not very well differentiated into palisade and spongy mesophyll. The primary veins have a fibrous sheath most developed on the lower side and they rise as a network, more or less prominent according to the species, from the surrounding tissue.

INFLORESCENCE

The inflorescence may be a simple raceme, e.g. in subgenus *Rhizophyllum*, or a compound raceme having the upper pedicels one-flowered and the lower peduncles cymosely several-flowered, e.g. in *E. acuminatum* and *E. elongatum*, or all peduncles several- (usually 3- or 5-) flowered as in *E. sagittatum*, *V. planipetala*, etc.; *E. elatum* has a very large loose panicle of branching peduncles. The pedicels and axis may be glabrous or furnished conspicuously or sparsely with multicellular gland-tipped hairs. Their presence seems a feature liable to much fluctuation within a species and is of little taxonomic value, although some species are always glandular and others nearly always glabrous (*cf. Vancouveria*).

FLOWER COLOUR

Epimedium has a greater range of flower colour★ than other berberidaceous genera, varying from white (e.g. *E. diphyllum*) and yellow (e.g. *E. pinnatum*) to rose, crimson, and violet (e.g. *E. grandiflorum*); in *Vancouveria* the flowers are white, sometimes lavender-tinged, or pale yellow. Usually the colour is constant for a species, but *E. grandiflorum*, *E. acuminatum*,

★Note: By the courtesy of the Director of the John Innes Horticultural Institution in 1934, Dr Rose Scot-Moncrieff kindly made a preliminary examination of the flower-pigments of *Epimedium*. She reported that some ivory anthoxanthin (flavone or flavonol), in varying quantity, occurs in the cell-sap of all forms examined; this is present alone in the pure white *E.* × *youngianum* nothosubsp. *niveum*. The other sap-pigment is a 3-pentose-glycosidic or 3-monoglycosidic anthocyanin of the delphinidin or petunidin type, this being responsible for the red coloration of *E. alpinum*, *E.* × *rubrum*, *E.* × *versicolor* 'Versicolor', and *E.* × *warleyense*. Yellow results from the presence of plastid pigment; no yellow (as distinct from ivory) anthoxanthin has been observed in any form; *E. pinnatum* subsp. *colchicum* and *E. perralderianum* have much plastid yellow with a trace of ivory anthoxanthin; *E.* × *versicolor* 'Sulphureum' (*E. pinnatum* × *E. grandiflorum*), plastid yellow likewise but less than in these species; *E. alpinum*, much anthocyanin in its sepals, associated with anthoxanthin and a trace of plastid yellow, and no anthocyanin but much anthoxanthin and a trace of plastid yellow in its petals; *E.* × *rubrum* (*E. alpinum* × *E. grandiflorum*), less anthocyanin and less anthoxanthin than *E.* × *alpinum* in its sepals and less anthoxanthin but a trace of anthocyanin in its petals; *E.* × *youngianum* nothosubsp. *niveum*, much ivory anthoxanthin, but no anthocyanin, and no plastid yellow; *E.* × *youngianum* nothosubsp. *roseum*, anthoxanthin, with a little anthocyanin but no plastid yellow; *E.* × *warleyense* (*E. pinnatum* × *E. alpinum*), anthocyanin, plastid yellow in greater quantity than in *E. alpinum* and much ivory anthoxanthin. These results, though necessarily provisional owing to limited material, support the view that these cultivated forms described are hybrids.

and *E. membranaceum* are heterochromic. The flowers are of frail texture and soon fall. They are protogynous and visited by bees for the nectar concealed within the usually saccate or spurred petals. Loew observed *Osmia rufa* visiting cultivated *E. pinnatum* (probably subsp. *colchicum*) and *Bombus agrorum* visiting E. × *rubrum*; Kunth describes the honey-bee (*Apis mellifera*) as methodically sucking the nectaries of *E. alpinum* and touching and dusting the stigma with pollen brought from another plant; I have watched it at work on *E. pinnatum* subsp. *colchicum*. In Japan Kazuo Suzuki has observed *Bombus niger*, *B. ardens*, *Tetralonis nipponensis*, *Lasioglossum* spp. and *Andraea* spp. as pollinators of *Epimedium* (Suzuki, 1990).

FLOWERS

The flowers of *Epimedium* and *Vancouveria* are regular and gamophyllous. Surrounding the androecium and gynoecium are three false-whorls of floral segments, for which various terms have been used in systematic literature. In this paper the outer segments are called 'outer sepals' or '*sepala exteriora*' ('*sepala*' Morren & Decaisne; or '*bracteae*' Franchet); the middle segments 'inner sepals' or '*sepala interiora*' ('*petala*' Morren & Decaisne; '*sepala*' of Franchet); the inner segments 'petals' or '*petala*' ('*nectaria*' of Morren & Decaisne; '*petala*' of Franchet). Komarov made the consultation of his work difficult by using three sets of terms indiscriminately, the outer sepals being called '*bracteae*' in some descriptions, in others '*sepala exteriora*' or '*sepala*'; the inner sepals '*sepala*', '*sepala interiora*' or '*petala*'; the petals '*petala*' or '*nectaria*', so that the 'petals' of one description correspond to the 'sepals' of another or the 'petals' of one to the 'nectaries' of another, which is a little confusing. The segments of one false-whorl are opposite those of the next, this arrangement resulting from the compression of two whorls into one false-whorl. The aestivation is imbricate.

In *Epimedium* (a dimerous group) two pairs of small membranous scales or outer sepals compose the outer false-whorl; there may also be two additional, usually deltoid, bracts appressed to the outside of the bud. In *Vancouveria* there are six to nine outer sepals. These fall as the flower opens and so have little diagnostic utility: nevertheless, their form when mature is fairly constant and distinctive. The inner are larger than the outer. They are frequently narrowly ovate, but may be broadly obovate or oblong, blunt or subacute, from 1–5 mm long, 1–2 mm broad, according to the species. In *Vancouveria hexandra* and *V. chrysantha* they are beset with very short glandular hairs, but they are glabrous in *V. planipetala* and nearly all species of *Epimedium*.

The two inner false-whorls form the conspicuous part of the flower and exhibit great variation in size and shape. In *Epimedium* there are normally four segments to a false-whorl, but in *Vancouveria* six, this being characteristic of most *Berberidaceae*. *Epimedium campanulatum* and *E. platypetalum* approximate most to the general type of the family. The relation in size between the petals and inner sepals is important. In the subgenus *Rhizophyllum* (*E. pinnatum* and *E. perralderianum*), where the petals are reduced to small nectariferous sacs, the yellow, broadly ovate, elliptic or obovate, rounded inner sepals are large and conspicuous. The inner sepals also much exceed the petals in the group *Diphyllon*, series *Brachycerae*, unless, as in *E. sagittatum*, both alike have been greatly reduced. Remarkably long, acuminate, narrowly ovate inner sepals, about 1.7 cm long, 3 mm broad, occur in *E. fargesii* and *E. sutchuenense*. Other distinct kinds are the narrowly ovate, dorsally concave, and boat-like inner sepals of section *Epimedium* (*E. alpinum* and *E. pubigerum*), the obovate clawed inner sepals of *Vancouveria*, and the minute triangular inner sepals of *E. platypetalum*. At anthesis they spread widely or even reflex (e.g. in *E. fargesii* and *Vancouveria*).

The six (*Vancouveria*) or four (*Epimedium*) segments constituting the innermost false-whorl, the nectaries of some authors, are here called petals; as Citerne remarks, there is no reason to consider them as nectariferous glands which have become petaloid or as staminodes; they are true petals with nectariferous tissue on their inner face. Their form and size are of great taxonomic importance. In *E. diphyllum* and *E. platypetalum* they are obovate, rounded at the tip, and more or less flat with a slight median nectariferous furrow. *Vancouveria planipetala* also has flat petals, but they are minute and of quite different shape, having the tip divided into three deltoid lobes of which the outer are nectariferous. In *V. hexandra* and *V. chrysantha* the petal consists of a narrowly oblong or cultrate stalk expanded and bent over at the tip so as to enclose the nectariferous tissue in a pocket. In other species of *Epimedium* the petal bulges outwards above or at the base as a blunt nectariferous sac or an acute spur (*calcar*), the form and size being characteristic of the species. The area of nectariferous tissue bears no relation to the length of the spur and is confined to the inside of the tip. It consists of several layers of close small many-sided cells supplied by a vein running straight from the base of the petal, the lateral veins not reaching so far, and covered by an epidermis of more cubical cells with a thin cuticle through which the liquid nectar diffuses into the cavity of the petal.

Accepting the flat petals of *E. platypetalum* and *E. campanulatum* as nearest to the ancestral type, one can imagine the development of the other petal-types from this by the protrusion outwards of the basal portion to form a pouch, small and blunt in some species, elongated and tapering in others, with nectariferous tissue on the inside of the tip, this process being sometimes accompanied by the reduction of the flat petal proper (lamina) to a mere fringe around the mouth of the cavity. Thus one can distinguish six main types of *Epimedium*:

1 Lamina flat, obovate, with a slight median nectariferous furrow but no spur,
   e.g. *E. diphyllum*, *E. platypetalum* and *E. campanulatum*.
2 Lamina rounded and conspicuous, but expanded outwards at base into an elongated
   spur usually longer than the inner sepal, e.g. *E. grandiflorum*, *E. davidii* and *E. hunanense*.
3 Lamina reduced to a rim around the mouth of the elongated basally-swollen spur which
   is longer than the inner sepal, e.g. *E. acuminatum*, *E. membranaceum* and *E. elongatum*.
4 Lamina reduced to a rim around the mouth of the slipper-like cylindric spur, which
   is shorter than the inner sepal, e.g. *E. alpinum*, *E. pubigerum*, *E. fargesii* and *E. elatum*.
5 Lamina reduced to a lacerated quadrate fringe at the mouth of the minute spur,
   e.g. *E. pinnatum* and *E. perralderianum*
6 Lamina and spur alike reduced, forming a minute pouch, e.g. *E. sagittatum* and
   E. pubescens.

Fig. 3. The spurless flowers of **Epimedium ecalcaratum.** Photograph: John Fielding
Fig. 4. **Epimedium flavum**, showing the long spur and very raised lamina of the petals, enclosing the stamens. Photograph: John Fielding.
Fig. 5. **Epimedium fargesii** showing the large sepals, the smaller, slipper-shaped petals and exserted stamens. Photograph: John Fielding.
Fig. 6. **Epimedium sagittatum.** The tiny flowers have much-reduced, pouch-like petals. Photograph: Phillip Cribb.

A classification based on the form and size of the petal is thus fairly natural provided other matters such as the number of stem-leaves, or their absence, geographical distribution and C-banding are also taken into consideration. The same process of petal reduction has apparently taken place with *Leontice* (*sensu lato*, Prantl), the petals being large and flat (*cf.* no. 1) in *Bongardia*, much smaller in *Gymnospermium*, and represented by scale-like nectaries (*cf.* no. 6) in *Caulophyllum* and *Leontice* (*sensu stricto* — *Euleontice* Prantl).

STAMENS

The stamens are four in *Epimedium*, six in *Vancouveria*, and are opposite the perianth segments. They stand erect, connivent, and more or less appressed to the ovary, and are usually about 5 mm long; *E. fargesii* and *E. dolichostemon* with stamens up to 10 mm long are a striking exception. They protrude in some species, but are enclosed by the lamina or the petals in others. The filament varies in length from species to species, being extremely short with the anther almost sessile as in *E. diphyllum*, about as long as the anther as in *E. pinnatum*, and many times longer than the anther in *E. fargesii* and *E. dolichostemon*. The relation of anther to filament is constant and supplies specific characters. In *Epimedium*, and in *Vancouveria planipetala*, the stamens are glabrous, but in *V. hexandra* and *V. chrysantha* they are dotted with very short glandular hairs. The anther is basifixed, as in other *Berberidaceae*, with two pairs of pollen-sacs and opens outwards by two oblong valves which separate from the anther along their base and sides, but remain attached at the top, curling up and crowning the stamen. The pollen is usually yellow, but is green in *E. chlorandrum*, *E.* × *warleyense* and *E. pinnatum* subsp. *circinnatum*. The grains are ellipsoid, almost smooth, slightly rugulose or striate, as in *E. planipetala*, with three longitudinal furrows (colpi); they vary in length from 26 μm–36.4 μm in *Epimedium* 32 μm–41 μm in *Vancouveria* (*cf.* Nowicke & Skvarla (1981)).

FRUITS

The fruit is a thin-walled capsule splitting longitudinally and completely into two valves, with seeds having a large oily caruncle (elaisome). They are about 3–4 mm long and weigh about 0.003–0.004 g. Ants, together with rain-wash, disperse the seeds (Berg, 1972). In America, wasps (yellow-jacket) have been observed to pick up the seeds of *Vancouveria* (Pellmyr, 1984).

Epimedium and *Vancouveria* have no close allies and the genera associated with them in the tribe *Epimediinae* vary from author to author (*cf.* Kim & Jansen, 1995). Meacham (1986) associated them with the one-flowered *Jeffersonia*, having a peculiar laterally and obliquely slit capsule, and wind-pollinated *Achlys,* having a spicate inflorescence without perianth. Loconte & Estes (1989a & b) group together *Epimedium, Vancouveria, Jeffersonia, Bongardia* (often associated with *Leontice*), *Podophyllum* and *Diphylleia*, all very distinct genera of great antiquity.

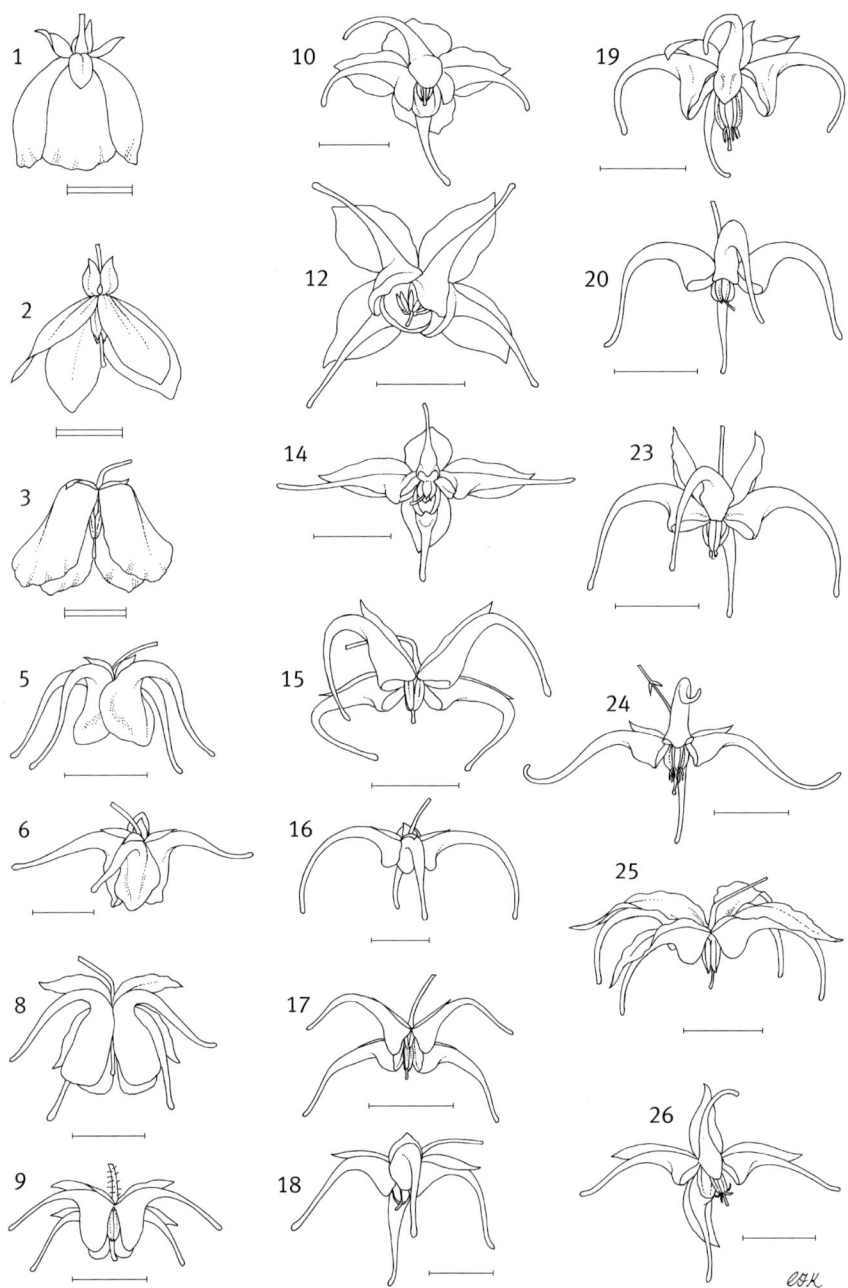

Fig. 7. **Epimedium flowers 1**. (numbered in sequence of classification in monograph). 1, *E. campanulatum*; 2, *E. platypetalum*; 3, *E. ecalcaratum*; 5, *E. davidii*; 6, *E. fangii*; 8, *E. flavum*; 9, *E. ilicifolium*; 10, *E. epsteinii*; 12, *E. ogisui*; 14, *E. mikinorii*; 15, *E. elongatum*; 16, *E. membranaceum*; 17, *E. rhizomatosum*; 18, *E. lishihchenii*; 19, *E. acuminatum*; 20, *E. franchetii*; 23, *E. chlorandrum*; 24, *E. wushanense*; 25, *E. leptorrhizum*; 26, *E. brachyrrhizum*. Single bar scale = 10 mm; double bar scale = 5 mm. Drawn by Christabel King.

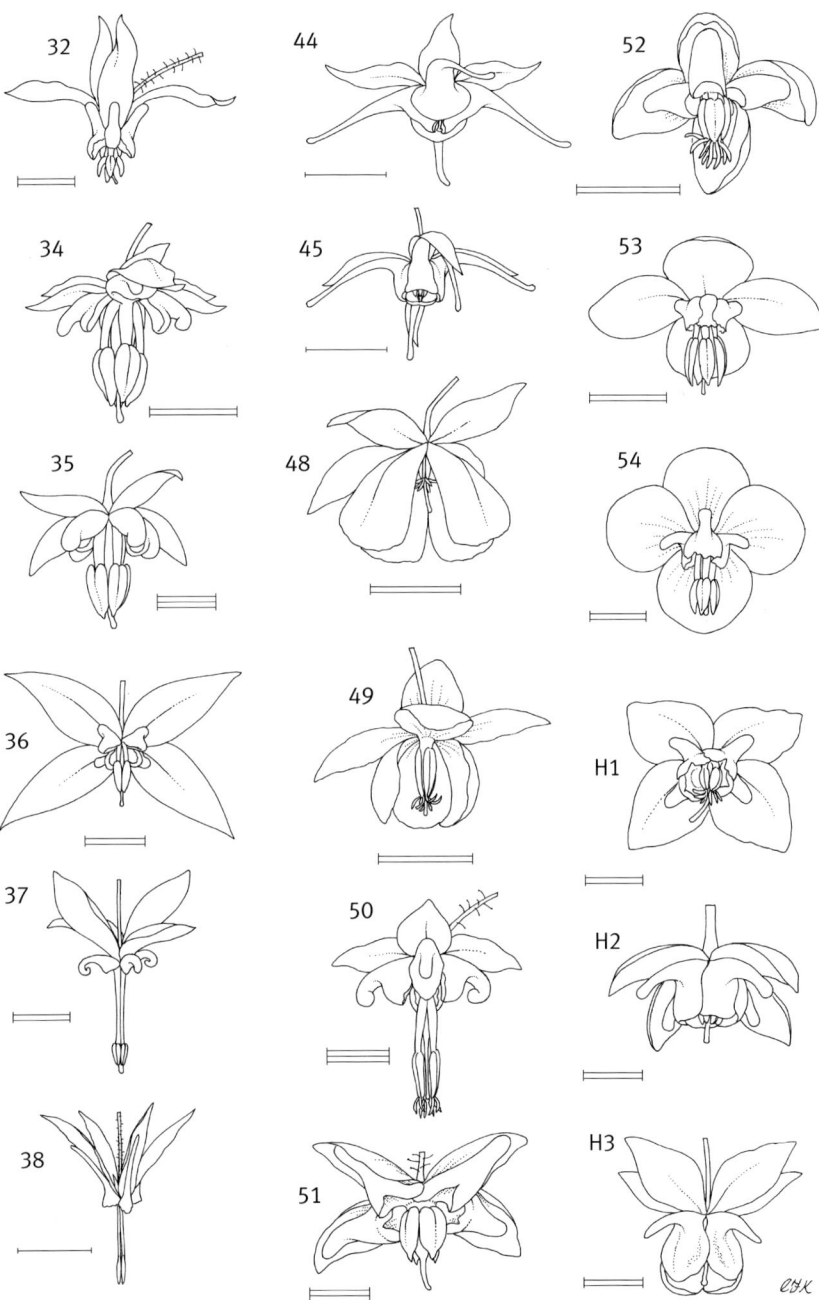

Fig. 8. **Epimedium flowers 2**. (numbered in sequence of classification in monograph). 32, *E. brevicornu*; 34, *E. sagittatum*; 35, *E. myrianthum*; 36, *E. stellulatum*; 37, *E. dolichostemon*; 38, *E. fargesii*; 44, *E. grandiflorum*; 45, *E. sempervirens*; 48, *E. trifoliolatobinatum*; 49, *E. diphyllum*; 50, *E. elatum*; 51, *E. alpinum*; 52, *E. pubigerum*; 53, *E. pinnatum*; 54, *E. perralderianum*; H1, *E.* × *warleyense*; H2, *E.* × *rubrum*; H3, *E.* × *youngianum* 'Niveum'. Single bar scale = 10 mm; double bar scale = 5 mm; triple bar scale = 2.5 mm. Drawn by Christabel King.

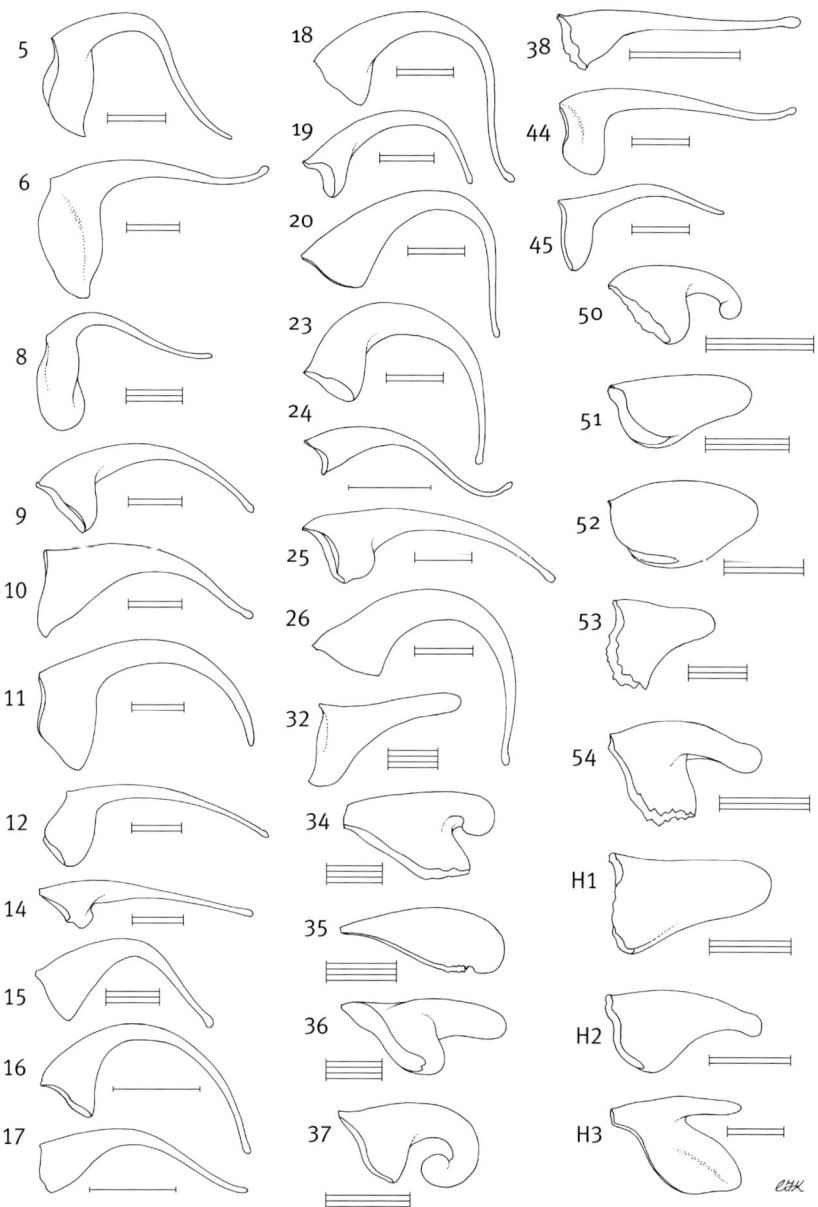

Fig. 9. **Petals of Epimedium flowers** (numbered in sequence of classification in monograph). 5, *E. davidii*; 6, *E. fangii*; 8, *E. flavum*; 9, *E. ilicifolium*; 10, *E. epsteinii*; 11, *E. latisepalum*; 12, *E. ogisui*; 14, *E. mikinorii*; 15, *E. elongatum*; 16, *E. membranaceum*; 17, *E. rhizomatosum*; 18, *E. lishihchenii*; 19, *E. acuminatum*; 20, *E. franchetii*; 23, *E. chlorandrum*; 24, *E. wushanense*; 25, *E. leptorrhizum*; 26, *E. brachyrrhizum*; 32, *E. brevicornu*; 34, *E. sagittatum*; 35, *E. myrianthum*; 36, *E. stellulatum*; 37, *E. dolichostemon*; 38, *E. fargesii*; 44, *E. grandiflorum*; 45, *E. sempervirens*; 50, *E. elatum*; 51, *E. alpinum*; 52, *E. pubigerum*; 53, *E. pinnatum*; 54, *E. perralderianum*; H1, *E.* × *warleyense*; H2, *E.* × *rubrum*; H3, *E.* × *youngianum* 'Niveum'. Single bar scale = 10 mm; double bar scale = 5 mm; triple bar scale = 2.5 mm; quadruple bar scale = 1 mm. Drawn by Christabel King.

4. CLASSIFICATION

Before W.J.Hooker's description and illustration in 1829 of *Epimedium hexandrum* (now *Vancouveria hexandra*), *Epimedium* was considered a monotypic genus, the only known species being *E. alpinum*, so named by Linnaeus in 1753. Philipp Franz von Siebold's introduction in 1830 of living plants from Japan into cultivation at the Ghent botanic garden stimulated two Belgian botanists, Charles Morren (1807–1858) and Joseph Decaisne (1807–1882), to produce in 1834 the first monograph of the genus. Here they recognised six species of *Epimedium* proper (i.e. with spurred dimerous flowers) and two monotypic genera, *Aceranthus* and *Vancouveria*, of debatable autonomy. Accepted by Bentham and Hooker in 1862, these have been fused with *Epimedium* by all others who have studied the whole group, e.g. by Baillon, Franchet, Prantl, Citerne, Tischler, Komarov and Himmelbaur. A middle course is adopted here. The type of *Aceranthus*, as also of *Vindicta* Raf., is *E. diphyllum* Graham, a Japanese species with white, flat, obovate, spurless petals and peculiar bifoliolate leaves, the median leaflet of the usual *Epimedium* triad being suppressed. On these characters Bentham maintained the genus, but, as similar spurless petals occur also in three Chinese species (*E. platypetalum*, *E. campanulatum* and *E. ecalcaratum*) with three- or five-foliolate leaves of the usual generic pattern, and as there exist plants (*cf. E.* × *youngianum* = × *Bonstedtia* H.R.Wehrh.) which produce spurred and spurless flowers in the same raceme and are intermediate in leaf-habit between *E. diphyllum* and *E. grandiflorum*, the maintenance of *Aceranthus* as a genus is artificial and impracticable.

For *Vancouveria* there is a better case, it being a North American group with flower-parts in threes; the Old-World epimediums have flower-parts in pairs. The three species form a homogeneous natural group, recognisable at a glance by their spineless and often trilobed leaflets, scapes of small trimerous flowers, reflexing obovate inner sepals and clawed petals, as well as by their semi-orbicular seeds. Their facies is thus distinct from that of the Old-World epimediums, among which nearly all these characters occur individually. Both groups have six pairs of chromosomes with $2n = 12$ (Langlet, 1928; Mayaji, 1930; Takahashi, 1989), except for the tetraploid *V. planipetala* with $2n = 24$. The number of flower-parts, the form of the seeds (to a certain extent), and the geographical distribution furnish the only constant distinctions, and it is accordingly a matter of taste whether they should be treated as genera or subgenera. They are certainly 'natural genera' as defined by Otto Stapf (*Botanical Magazine* t.8985, 1923) in the sense of being 'genera with a tolerably uniform facies as well as floral structure and at the same time a range suggestive of a continuous common history of their members' and, since *Vancouveria* has been used in all American floras and most horticultural works since 1834, this traditional treatment is accepted here.

The six remaining species were placed by Morren & Decaisne in two sections, the large-flowered Japanese species (*E. grandiflorum* and forms) in their section *Macroceras*, the others (*E. alpinum*, *E. pubigerum* and *E. elatum*) with smaller flowers and blunt spurs in their section *Microceras*; this is a fairly good arrangement when confined to these, but becomes inadequate and artificial if extended to the many other species now known. The following note of Fischer & Meyer (1846) initiated the present system:

'Genus *Epimedium* commode in sectiones tres, valde naturales, dividi potest.
I. *Macroceras*. Nectaria lamina explanata erecto-conniventia, basi calcarata vel cucullata. Caulis foliosus. Flores albidi vel violacei. Species omnes japonicae.
                                         A. Calcaria elongata
  1. *E. macranthum* Morren et Decaisne
  2. *E. violaceum* M. et Dcn.
  3. *E. Musschianum* M. et Dcn.
                                  B. Calcaria abbreviata cuculliformia
  4. *E. Youngianum* Fisch. et Mey. *E. Musschianum* Botanical Magazine tab. 3745 (non Decaisne). Species nobis nota solum ex icone citata, sed ab omnibus descriptis speciebus nectarii lamina explanata petaloidea instructis, calcare abbreviato differt; ab *E. musschiano* Decaisne praeterea distat foliis biternatis.
II. *Microceras*. Nectaria brevia, depressa, cucullata; lamina nulla. Caulis foliosus. Petala sanguinea. – Species europaeae, unica Indiae orientalis.
  5. *E. alpinum* Linn.
  6. *E. pubigerum* M. et Dcn.
  7. *E. elatum* M. et Dcn.
III. *Rhizophyllum*. Nectaria brevia, depressa, cucullata; lamina nulla. Folio omnia radicalia. Scapus aphyllus. Petala flava.– Species caucasica.
  8. *E. pinnatum* Fisch.'

Epimedium pinnatum, which Morren & Decaisne (1834) referred to the section *Microceras*, is here made the type of a new section, differing from the others by its leafless inflorescence. Sixteen years later Baillon proposed a similar arrangement, his section *Dimorphophyllum* (1862) corresponding to Fischer & Meyer's *Rhizophyllum* (1846) and Franchet's *Gymnocaulon* (1886); he also retained *Microceras* and *Macroceras* as sections, but added *Aceranthus* and, later, *Vancouveria* (1871) as other sections. The next attempt was made by Franchet in 1886 as a result of acquaintance with two new Chinese species, *E. davidii* and *E. acuminatum*, which led him to revise the entries. He proposed the subgenera *Euepimedium* (i.e. *Epimedium sensu stricto*) and *Vancouveria*, and emphasised a hitherto little-appreciated character, the number of leaves borne on the flowering stem, which is constantly one for species of Europe and Asia Minor, Manchuria and Japan but varying from three to one in Chinese species, a few of which may have one or two stem leaves in the same species.

Komarov accepted Franchet's classification in 1968, as I did, with some modifications, in 1938. It began with the most advanced group *Rhizophyllum* (*Gymnocaulon*). Using also differences in the relative size and form of the inner sepals and petals, the 21 species then known could be put into eight small groups, some distinguished as series, following the procedure of Bunge and Maximowicz in 1872 and 1873, which were geographically significant. More than twice as many species are now known. Attention to some evolutionary tendencies manifest within the *Berberidaceae* as a whole and some manifest only in *Epimedium* have suggested a new arrangement of species.

It is customary to distinguish as 'primitive' the character states assumed to have existed in a remote ancestor and as 'advanced' those diverging most from these. Provided development in one part of an organism is not dependent, through correlation, with development in another part for the efficient functioning of the individual organism, primitive characters can exist alongside advanced characters in the same species. Obviously, because of this, species of the same age cannot be satisfactorily arranged in phylogenetic order, but on the other hand the modification of an organ in a presumed evolutionary progression from a primitive to an advanced state can often be traced from species to species or from genus to genus. Such a sequence can be postulated as regards the form of corolla and leaf in *Epimedium* and suggest a thematic and phonetic not necessarily phylogenic arrangement.

Since genera of *Berberidaceae* as diverse as *Nandina*, *Ranzania*, *Mahonia*, *Berberis*, *Bongardia*, *Jeffersonia*, *Diphylleia* and *Podophyllum* all have flat unspurred petals, the far distant ancestral stock, from which these, together with *Caulophyllum*, *Epimedium* and *Vancouveria*, have diverged, can be postulated likewise to have had free flat petals. Such flat unspurred petals exist within the genus *Epimedium*: the flowers of the Chinese species *E. campanulatum* and *E. platypetalum*, and the Japanese species *E. diphyllum* obviously represent the original state of corolla of the genus. The beginning of a nectarial projection or spur just above the base of the petal is evident in the shouldered flowers of the allied *E. ecalcaratum*. The development of this projection into an elongated spur, with the original petal surviving as its basal lamina and the stamens enclosed, has resulted in the long-spurred corollas of *E. davidii*, *E. fangii*, *E. hunanense*, *E. latisepalum*, *E. ogisui*, *E. pauciflorum*, *E. grandiflorum*, *E. sempervirens* and *E. macrosepalum*). The stages by which a long-spurred corolla might have developed can be seen in *E.* × *youngianum* with variable spur length on the same plant. The virtual disappearance of the flat petal, by its reduction to a spine around the mouth of a horn-shaped spur so characterises *E. acuminatum*, *E. sutchuenense*, *E. elongatum*, *E. franchetii* and *E. membranaceum*. In these species the spur has become the conspicuous attractant for insect visitors, mostly, perhaps exclusively, bees. The transfer of this function to conspicuous inner sepals and accompanying reduction of the petal spur to a small nectariferous pouch, straight, curved or coiled, is evident in *E. brevicornu*, *E. pubescens*, *E. fargesii*, *E. dolichostemon*, *E. stellulatum* and *E. pubigerum*. Broadening of the inner sepals and extreme reduction of the petal reaches its ultimate state in *E. pinnatum* and *E. perralderianum* which, with their leafless flower stems, form the subgenus *Rhizophyllum*. This group is so distinct as to suggest the former existence of an evolution centre south of the Caucasus. The only known fossil leaflet, *E. takhtajanii*, from Sarmatian deposits, is unlike that of any western species and indicates lost diversity (see Appendix 2, page 196).

Parallel reduction of leaf in complexity within the genus *Epimedium* and from genus to genus in other *Berberidaceae* has occurred. The elaborate pinnately and ternately divided leaves of *Epimedium elatum* and *Nandina*, both with very numerous leaflets, are linked by the biternate leaves of *E. pubigerum*, *E. elongatum*, *E. grandiflorum*, *Leontice leontopetalum*, and the pinnate leaves of *E. pinnatum*, *Bongardia* and *Mahonia*, to the trifoliolate leaves of *Achlys*, *E. perralderianum* and most Chinese species of *Epimedium* and the bifoliolate leaves of *E. diphyllum* and *Jeffersonia diphylla*, these providing a link to the simple (unifoliolate) leaves of *E. simplifolium*, *E. baojingense*, *E. glandulosopilosum* and *Jeffersonia dubia*.

A further trend in the herbaceous *Berberidaceae* has been the reduction of leaves on the flowering stem from many (3–8) in *E. elatum* to two in most Chinese species, *E. diphyllum* and *Ranzania* to one in *E. grandiflorum*, *E. alpinum* and allied species and from three to none in *E. pinnatum*, *E. perralderianum* and *Jeffersonia*.

On the basis of these tendencies, notably modifications of the corolla, C-banding of chromosomes and geographical distribution, the genus may be divided into two subgenera: (I) subgenus *Epimedium* (= section *Phyllocaulon* Franch.), typified by *E. alpinum*, with 4 sections and (II) subgenus *Rhizophyllum*, typified by *E. pinnatum*.

(I) Subgenus *Epimedium*

1. Section *Diphyllon* (Kom.) Stearn, comprising all the Chinese species mostly with two stem leaves or sometimes three (*E. elongatum*) and sometimes one (*E. leptorrhizum*, *E. brachyrrhizum*). This is the largest and most diverse assemblage, with C-banding Type A (Takahashi, 1989). It comprises series *Campanulatae* Stearn, typified by *E. campanulatum*, series *Davidianae* Stearn, typified by *E. davidii*, series *Dolichocerae* Stearn, typified by *E. membranaceum*, and series *Brachycerae* Stearn, typified by *E. sagittatum*.

2. Section *Macroceras* C.Morren & Decne., comprising *E. grandiflorum* of Japan, Korea and Manchuria and its allies, and *E. diphyllum*, with one stem leaf and C-banding Type B (Takahashi, 1989).

3. Section *Polyphyllon* (Kom.) Stearn, comprising only *E. elatum* of the western Himalaya, with 3–8 ternately compound stem leaves, diffuse many-flowered inflorescences and flowers as in section *Epimedium*.

4. Section *Epimedium*, comprising *E. alpinum* and *E. pubigerum* of southern Europe and northern Asia Minor with one biternate stem leaf, flowers 8–13 mm in diam. and C-banding Type C (Takahashi, 1989).

(II) Subgenus *Rhizophyllum* (Fisch. & C.A.Mey.) Stearn, comprising *E. pinnatum* of the Caucasus and *E. perralderianum* of Algeria, with no stem leaves, broad yellow inner sepals, and minute petals.

The following is the arrangement of infrageneric taxa and species in the present work:

Genus EPIMEDIUM

I. Subgenus EPIMEDIUM

i. Section DIPHYLLON

A. Series Campanulatae [all from China]
1. *E. campanulatum*
2. *E. platypetalum*
3. *E. ecalcaratum*
4. *E. shuichengense*

B. Series Davidianae [all from China]
5. *E. davidii*
6. *E. fangii*
7. *E. hunanense*
8. *E. flavum*
9. *E. ilicifolium*
10. *E. epsteinii*
11. *E. latisepalum*
12. *E. ogisui*
13. *E. pauciflorum*
14. *E. mikinorii*

C. Series Dolichocerae [all from China]
15. *E. elongatum*
16. *E. membranaceum*
17. *E. rhizomatosum*
18. *E. lishihchenii*
19. *E. acuminatum*
20. *E. franchetii*
21. *E. enshiense*
22. *E. sutchuenense*
23. *E. chlorandrum*
24. *E. wushanense*
25. *E. leptorrhizum*
26. *E. brachyrrhizum*
27. *E. simplicifolium*
28. *E. zhushanense*
29. *E. baojingense*
30. *E. glandulosopilosum*

D. Series *Brachycerae* [all from China]

31. *E. pubescens*
32. *E. brevicornu*
33. *E. reticulatum*
34. *E. sagittatum*
35. *E. myrianthum*
36. *E. stellulatum*
37. *E. dolichostemon*
38. *E. fargesii*
39. *E. elachyphyllum*
40. *E. truncatum*
41. *E. coactum*
42. *E. borealiguizhouense*
43. *E. lobophyllum*

ii. Section *MACROCERAS* [from Japan, Korea, Manchuria, Far Eastern Russia]

44. *E. grandiflorum*
45. *E. sempervirens*
46. *E. koreanum*
47. *E. macrosepalum*
48. *E. trifoliolatobinatum*
49. *E. diphyllum*

iii. Section *POLYPHYLLON* [from western Himalaya]

50. *E. elatum*

iv. Section *EPIMEDIUM* [from Europe, Caucasus and northern Turkey]

51. *E. alpinum*
52. *E. pubigerum*

II. Subgenus *RHIZOPHYLLUM* [from Caucasus and North Africa]

53. *E. pinnatum*
54. *E. perralderianum*

5. GEOGRAPHICAL DISTRIBUTION

The present distribution of plants cannot be understood without considering their history and environmental requirements. The requirements of *Epimedium* and its herbaceous allies — such as *Achlys, Jeffersonia, Ranzania, Caulophyllum*, all woodland plants — are clear, but the fossil record yields no direct information about their former range and wanderings. One can only surmise how their existing very disjunct range has been attained from what is known of various other groups sharing their habitats and presenting the same peculiarities of distribution.

Epimedium is an Old World assemblage, dispersed between Japan and Algeria but with enormous gaps, its major area of diversity existing in China. In North America its place is taken by *Vancouveria*, a group of three species which extends within the Humid Transition Zone from north Washington to middle California, ranging not more than 130 miles inland; like *Achlys*, an allied berberid occurring also in eastern Asia, it is absent from Atlantic North America. Essentially woodland plants, the vancouverias often grow in the shade of redwoods (*Sequoia sempervirens*), a suggestive association since, although the living species of *Sequoia* and *Sequoiadendron* are confined to California and southern Oregon, the generic range was formerly much greater. *Vancouveria hexandra* (with thin leaflets) is a species found in Washington, Oregon and north California. *Vancouveria planipetala* (with coriaceous leaflets) ranges between middle California and the Rogue River in south Oregon, which marks the northern limit of many Californian plants. Thus their ranges overlap in north California and south Oregon. *Vancouveria chrysantha* (with coriaceous leaflets) is confined to a small area on the Oregon-California boundary within the overlap of its two wide-ranging allies and, though unique in possessing yellow flowers, it otherwise combines their characters. In its trimerous flowers *Vancouveria* furnishes a transition from other *Berberidaceae* to *Epimedium* proper, and may be regarded as preserving in this respect a primitive condition of the family. In its leafless flowering stems it resembles the Caucasian and North African *Epimedium* subgenus *Rhizophyllum*.

Epimedium occupies two widely separated regions, in eastern Asia and the Mediterranean lands; within each region there are extensive areas from which the genus is entirely absent; no one species of the genus has a very wide range and many are very local.

The Eastern Asiatic group is centred on western and central China and has outlying members in Kashmir and Japan. Section *Epimedium* is represented by two species in the Mediterranean region. The corresponding section *Macroceras*, likewise with one stem leaf, in eastern Asia consists of a variable species, *E. grandiflorum*, common to Japan and Korea, and three close allies, *E. sempervirens* apparently endemic to Japan, *E. koreanum* to northern Korea and *E. macrosepalum* to Far Eastern Russia, together with the aberrant *E. diphyllum* and the hybridogen *E. trifoliolatobinatum*. Since *Epimedium* is a group confined ecologically to woodland and scrub, and seems able to spread only by the slow process of rhizome growth and the transport of its seeds by woodland ants, this range was most probably attained before the sinking of south-east Asia had parted Japan from the mainland (i.e. probably during the Pliocene period; *cf.* Arldt, 1919: 422) but, judging from herbarium material, no morphological characters distinguish the long-isolated island populations from the mainland populations of *E. grandiflorum*. The Japanese endemic *E. diphyllum* is primitive only in its spurless petals. Three species with spurless petals occur in China but are not at all closely related to this. Section *Diphyllon* — the largest group within the genus — is confined to western and central China. There remains section *Polyphyllon*, with one species, *E. elatum*, in Kashmir; *E. elatum* is geographically isolated between the eastern Asiatic and Mediterranean groups and stands morphologically apart from both. In its much reduced floral structure it resembles species of section *Epimedium*, but, by virtue of its numerous and much divided leaves, it suggests a type from which, by reduction, the other leaf-forms of the genus have been derived.

The Chinese species of *Epimedium* occur within four of the eight 'floral provinces' (including Korea) proposed by Handel-Mazzetti (1927a, 1927b, 1931): (1) the north-east Sino-Korean province of mixed woods; (2) the mountains on the southern boundary of the north China loess-steppe province; (3) the temperate or warm temperate mountain zone to the west of the middle Sino-Japanese laurel province; (4) the temperate or warm temperate zone of the high mountain province of Yunnan and west Sichuan. Thus the genus is absent from tropical China, the Gobi desert, the western Yunnan monsoon province, and the eastern Tibetan grassland province, and is poorly represented in the northern mixed forest and loess provinces. It avoids cold and dry as well as low, hot, and very moist regions, and grows normally in temperate forest and scrub in mountainous regions. This temperate forest vegetation is now largely restricted to western China, but fossil evidence testifies to its former existence within regions where it no longer occurs or is poorly represented. The relationship of this former forest extension to the wide and discontinuous range of *Epimedium* will be considered later.

Of the four Mediterranean species, two belong to section *Epimedium*; *E. alpinum* (the generitype) is a west Balkan species extending across northern Italy along the southern foot-hills of the Alps; its close ally, *E. pubigerum*, ranges from the south-east of the Balkan peninsula, in moist woods of *Fagus orientalis*, *Populus tremula*, *Prunus laurocerasus*, *Rhododendron ponticum*, etc., along the northern coastal fringe of Asia Minor to the west Caucasus. Its associates are of interest because fossil remains show them to have been important elements in Tertiary forests. The other two western species, *E. pinnatum* and *E. perralderianum*, make up the subgenus *Rhizophyllum*. They resemble one another closely, but are remarkably distinct from other members of the genus. Two subspecies of *E. pinnatum* (subsp. *colchicum* and subsp. *circinnatum*) inhabit the western Caucasus, while a third (subsp. *pinnatum*) grows in mountain woods in the eastern Caucasus and northern

Iran. Both areas, one by the Black Sea, the other by the Caspian, have a moist temperate climate, approximating to that of the temperate zone of the middle Sino-Japanese laurel province, and are noted for their luxuriant vegetation. The intervening region of the central Caucasus is drier, and *Epimedium*, in common with certain Caucasian genera having Sino-Japanese affinities, e.g. *Pterocarya* and *Zelkova*, is absent from it. The only remaining member of subgenus *Rhizophyllum*, *E. perralderianum*, lives in North Africa, in the mountains of Kabylia in Algeria, fully 3200 km (2,000 miles) distant from its nearest allies and is the only African representative of the genus. The distinctness of these Caucasian-Mediterranean species from those of China suggests that there was a secondary centre of evolution in western Asia, almost obliterated by climatic change.

Despite the wide and discontinuous distribution outlined above, the ecological requirements of the species appear to be much alike. The frequency of such notes in the literature and on collectors' labels as 'woodland', 'margins of thickets', 'on mossy banks', etc., together with details of altitude and latitude and meteorological data from stations nearest to habitats of *Epimedium*, all point to a prevailing type of habitat for the group as a whole. Wherever *Epimedium* is found, woodland and scrub in temperate hilly or montane regions provide the essential conditions for its existence and, since the range of species halts abruptly where these ecological and climatic conditions cease, without them the genus would not survive.

This ecological uniformity of the genus is confirmed by its behaviour in gardens. Species from Algeria, southern Europe, Asia Minor, the Caucasus, China and Japan, as well as the vancouverias from the Pacific Coast of North America, thrive in British gardens under the same cultural conditions, delighting especially in the shelter and dappled shade of light woodland, nor are they of difficult cultivation elsewhere in northern Europe. Though indigenous only south of the Alps, *E. alpinum* has become naturalized in Belgium, Czechoslovakia, France, Germany, Great Britain, Sweden and Switzerland. The absence of the genus from north-western Europe today cannot be attributed to unsuitable climate; the reason must be sought in past climatic changes.

If the species of *Epimedium* have had a common origin, which there is no cause to doubt, obviously the generic area cannot always have been so broken. From the ecological and morphological similarity of species now geographically remote, it is reasonable to suppose that their common ancestors had like requirements; in other words, *Epimedium* has always been a group needing the shade of temperate woodland for its survival and for its spread under natural conditions. Hence its wide range at the present day can only be explained by the former existence of temperate woodland in what are now treeless or subtropical regions

As a result, primarily, of the work of C. and E.M. Reid (1915; *cf.* also Reid, E.M., 1920) on the Pliocene flora of north-western Europe, the close resemblance between this and the present mountain forest flora of west China has become almost a palaeobotanic commonplace. This resemblance is marked by the presence of Pliocene deposits on the Dutch-Prussian border of 'various species which, although now extinct in Europe, cannot be separated from living Chinese plants', while others have their closest allies in species inhabiting the temperate belt of the Chinese mountains. It is significant that the distribution of *Epimedium* coincides in many respects with the modern distribution of some of these Tertiary genera. Thus *Pterocarya* consists of one Caucasian species, one Japanese, and several Chinese; *Zelkova* has today one species in the eastern and western Caucasus and northern Iran, with a range very like that of *E. pinnatum*, as well as one in Crete and the others in China. By analogy with these, as also from the existence of herbaceous genera such as *Incarvillea* in Europe during the Oligocene period of the Tertiary era (Bembridge flora), it is reasonable to suppose that *Epimedium* had likewise acquired its present form and had spread widely long before the end of the Pliocene period. Its marked distinctness from other *Berberidaceae* suggests that the differentiation of the group must have taken place in a remote period of geological time.

During the early part of the Tertiary, according to palaeobotanists, an evergreen flora of subtropical aspect flourished in Europe and as far north as western Greenland. In northern Asia there grew mixed but primarily summer-green forests built of essentially temperate genera such as *Sequoia*, *Taxodium*, *Fagus*, *Alnus*, *Betula*, *Corylus*, *Juglans*, etc. Their composition was probably not unlike that of the Sino-Japanese mountain forests wherein, as already noted, *Epimedium* is now best represented. This forest-belt is considered to have existed in Asia since the last stages of the Cretaceous. Before the end of the Tertiary a temperate forest flora had spread westward and had become characteristic of Europe. The mild conditions prevailing over the northern hemisphere during most of the Tertiary gave way subsequently to the colder conditions of the Quaternary (Pleistocene or Diluvial) ice-ages. At the periods of maximum glaciation a vast ice-sheet covered Scandinavia, most of Great Britain, and northern Germany, while the glaciers of the Alps, the Caucasus, eastern Asia, and elsewhere were greatly increased (*cf.* Penck, 1906; Wright, 1914; Antevs, 1929). The Pyrenees, Alps and Himalaya barred the southward migration of temperate plants retreating from the cold of advancing ice-sheets, and between the Scandinavian and Alpine ice sheets and north of the Himalaya probably the greater part of Tertiary forest vegetation perished. No such barrier entirely cut off the Balkan peninsula from central Europe; accordingly it appears to have become an area of refuge and a centre from which, during the long interglacial periods and the postglacial period, the more vigorous types recolonized lost territory; here *E. alpinum* has its maximum area. Other remnants of temperate Tertiary forest vegetation held out on the southern slopes of the Caucasian and Himalayan ranges; these are the areas inhabited by *E. pinnatum* and *E. elatum* and these plants have evidently here enjoyed uninterrupted evolution. The glaciation of western and central Asia was less severe than that of Europe, but the deserts and steppes of these regions do not support forest growth, and *Epimedium* is absent from them. The greatest area of refuge was in western China; here the mountains trend obliquely north and south, rather than east and west, and upon their high and feebly glaciated slopes plants were able, by climbing or descending as the climate changed from cold glacial to warmer interglacial or vice versa, to keep within the conditions necessary for their survival; here *Epimedium* reaches its maximum development.

Epimedium may be regarded as having originated in the north-eastern Asiatic forest-belt and shared its history. Such a view would account for the diversity of forms in eastern Asia, where the present mountain forest flora is held to be a lineal descendant of the Tertiary forest-belt. It also accounts for the occurrence of the genus in less abundance in the other regions (Kashmir, Caucasus, Balkan peninsula), where remnants of this forest have survived, and the absence of the genus from regions where — owing to glaciation and climatic deterioration with increasing aridity — this forest is believed to have suffered destruction or never existed.

The surprising presence in north-west Africa of *E. perralderianum*, a plant closely related to the Caucasian *E. pinnatum*, can likewise be explained after consideration of former climatic conditions and the ranges of various of its associates.

Between the existing stations of *E. pinnatum* and *E. perralderianum* stretch over three thousand kilometres (two thousand miles) of sea and land, most of the latter being desert and unsuited to the growth of temperate forest vegetation. The area in the mountains of Constantine, Algeria, which *E. perralderianum* inhabits, is one of the coolest and wettest along the arid north African coast; snow covers it for several months in winter, but little rain falls in summer (Barbey, 1934). Transferred from the mountains to Algiers in the lowlands, *E. perralderianum* has failed to survive (R.Maire, in litt.); transferred to British gardens the plant waxes to a size unknown under natural conditions. The species is clearly a relict lingering under conditions which approximate sufficiently closely to the usual requirements of the genus for its survival, but apparently not enough for its optimum development.

Of the trees under whose shade *E. perralderianum* persists upon the Babor Massif, the three most abundant — *Quercus afares, Abies numidica, Cedrus atlantica* – are all likewise noteworthy in being isolated and endemic north-west African species, so little differentiated from western Asiatic species as to leave no doubt of their derivation from Asiatic stocks. Thus *Quercus afares* is almost conspecific with *Q. castaneifolia*, an oak confined, as is *E. pinnatum* subsp. *pinnatum*, to mountain woods by the Caspian Sea and forming with another Caspian oak, *Q. sintenisii*, a distinct group within the genus, the distribution of which (*Quercus* ser. *Castaneifoliae*) thus parallels significantly the distribution of *Epimedium* subgenus *Rhizophyllum*. Of *Cedrus*, besides *C. atlantica*, there exist only three species or geographical races, *C. libani* in Lebanon and Asia Minor, *C. brevifolia* in Cyprus, *C. deodara* in the Himalaya, kept apart by sea and desert areas, but differing little in morphological characters. *Abies numidica* is more closely allied to *A. cilicica*, an associate of *Cedrus libani* in Asia Minor, than it is to the geographically nearer *A. pinsapo* of Spain and Morocco. *Populus tremula*, which grows sparsely on the Babor Massif, has here (*fide* R.Maire) its one station in Africa. These plants can only be regarded as isolated remnants of a flora which was more widely distributed in Tertiary times. Of this more extensive range, indications have been given by Stojanoff and Stefanoff (1929), and by Stefanoff and Jordanoff (1935), in identifying with *Cedrus libani* (*sensu lato*) the fossil cone-scales of a conifer found in Pliocence sediments of the plain of Sofia (for fossil records of *Cedrus* in France, Germany and Russia during the Tertiary, see Studt, 1926:

257) and with *Quercus castaneifolia* (*sensu lato*) the leaf-impressions, fossil cupules, and acorns of the oak formerly distinguished as *Quercus drymeja* occurring in the same Bulgarian deposits (as well as being recorded from Austria, Switzerland, Yugoslavia, Italy, Asia Minor and the Caucasus). The abundant fossil material shows these plants to have been common in the forest vegetation of south-eastern Europe in late Tertiary times. Today neither this cedar nor the oak grows wild in Europe. From the same Bulgarian deposits other species found in the Babor forest, i.e. *Populus tremula*, *Ilex aquifolium*, *Taxus baccata* and *Acer campestre*, have also been recorded. Possibly these plants reached their present North African habitat by different routes and at different times, but their present restricted association in the Babor forest and their inability to grow in the arid Mediterranean lowlands suggest that they came as a community at a time when the Mediterranean had a higher rainfall and a more temperate climate than it now possesses. In other words, the Babor Massif may be regarded as sheltering a fragment of the Tertiary mixed forest-belt. The clue to its presence in north-west Africa is supplied by the geological evidence indicating periods of higher rainfall in Syria, Palestine, Egypt and Cyrenaica in former times. These pluvial periods have been correlated with periods of glaciation in Europe and attributed to the turning southward, by the high pressure areas over the great Scandinavian and Alpine ice-sheets, of the storms from the Atlantic which now water Europe north of the Alps. During such a period a temperate forest-flora could extend westward along the north African coast from western Asia. At the close of such a period this flora could only survive by ascending the mountains in the west, the region to which *Cedrus*, *Abies*, *Quercus* and *Epimedium* are now confined in Africa. With regard to the length of time these north African plants have been cut off from their west Asiatic representatives, it may be mentioned that the deposit by the Nile of the alluvial muds south of Cairo, which is regarded as beginning towards the end of the last glacial period (*cf.* Brooks, 1922: 72–74), is calculated to have taken about 14,000 years. The isolation of *Cedrus*, *Abies*, *Quercus* and *Epimedium* probably came about much earlier than this, for no gravel terraces such as indicate pluvial conditions in Egypt during the Günz, Mindel, and Riss glaciations of the Alps are known for the Würm interglacial, i.e. for about the last 100,000 years. If so, the feeble differences achieved illustrate the slowness of evolutionary change in these groups.

As already noted, the species and subspecies of *Epimedium* are usually both geographic and morphological entities; in other words, a species of more or less continuous range is fairly homogeneous, but when the range is broken the isolated populations often, but by no means always, differ to an extent sufficient for their recognition as subspecies. Floral differences are usually correlated with differences in rhizome, leaf, or inflorescence; most of the species are well defined and occupy areas separated from those of their nearest allies. A consideration of differences between species (*cf.* contrasts in key to species, page 43) reveals few of such utility and direct survival value as to have been developed through natural selection. Behaviour of different pollinators may have influenced modifications of flower form, although about pollinators of plants in the wild virtually nothing is known. In general life form all the species are much alike and grow under fairly similar ecological conditions; characters helpful for specific distinction, such as length of marginal spines on leaflets, form of hairs on the lower leaflet surface, space between pedicels on flower stem, etc. appear to be of neutral value. *Epimedium acuminatum* and allies and *E. sagittatum* and allies are very distinct in flower but vegetatively some members of the one group resemble some members of the other as to be not well distinguishable out of flower. Accordingly their differences appear uncorrelated with the environment. Most botanical taxonomists are probably in agreement with the zoologists Robson and Richards (1936: 274): 'The most striking impressions that a taxonomic survey of any large group conveys to one's mind are the manifold diversities of species, the distinctness of the majority of these groups, the fact that they usually differ in several associated characters and the apparent triviality of these distinctions.'

How isolated fragments of a variable population can become morphologically and physiologically distinct, independently of any kind of environmental selection, has been made clear by A.L. and A.C.Hagedoorn (quoted in Du Rietz, 1930). When a population colonizes an area and becomes geographically divided, even though the parts are isolated under similar conditions, the reduction of variability in these will probably lead to different results; the original composition of the parts is not likely to be exactly the same, and even if a certain gene is at first equally represented in all parts its fate in these will be largely determined by chance, possibly becoming eliminated from one population, but present throughout another. This automatic differentiation may, as Du Rietz pointed out, be increased and directed by isolation under different environmental conditions. The species of subgenus *Rhizophyllum* seem to have been differentiated in this way, *E. perralderianum* of Algeria differing from *E. pinnatum* of the Caucasus principally by the constant lack of a character, the power of forming leaves with more than three leaflets, which together with their firmer texture may be related to its drier habitat. The extent, however, to which isolation alone can influence specific differentiation must depend upon the stage of 'down-grade evolution' or stability already reached by the population becoming divided. Obviously a population may attain a state so uniform that, however long thereafter parts may be isolated, its potentialities of variation will be too slight for any appreciable divergence to take place. Probably all species of *Epimedium* outside China are now in this state.

One example must suffice. No species has a range exceeding 1600 km (1000 miles) east to west or 800 km (500 miles) north to south; *E. pubigerum*, spread from eastern Bulgaria to the western Caucasus, a distance of about 1400 km (900 miles), is accordingly one of the most widely ranging. It has been introduced into cultivation from extreme stations (Strandja, Istanbul, Trabzon) between which there has long ceased to be interchange of genes; nevertheless these plants appear indistinguishable apart from a slight colour difference in the colour of the inner sepals. If, then, isolation be taken as a factor important in the differentiation of species from populations blessed with high variability, i.e. a rich gene pool, and in a youthful stage of development, its feeble effect upon the populations of *Epimedium* subgenus *Rhizophyllum* and associated *Abies*, *Cedrus* and *Quercus* mentioned above suggest that, before their populations were parted, these had already reached almost the last stage in their development. The same applies to the Asiatic and North American genera *Achlys* and *Caulophyllum*.

Although the formation of species and subspecies from an originally variable stock (coenospecies), believed to have been located in north-eastern Asia for *Epimedium* and allies, may thus be postulated, still there remains the cardinal problem of how they would have arisen. Whatever the inner cause of the inbuilt tendency of living matter to vary, there can be little doubt that certain features that are confined to one or a few species, and sharply distinguishing them from near allies, have arisen suddenly, i.e. as mutations, making morphological innovations for the group. To this category belong, for example, the remarkable long stamens of *E. fargesii* and *E. dolichostemon*, species which are vegetatively hard to distinguish from some other Chinese species, and the glandular ovary of *Vancouveria hexandra* and *V. chrysantha*, this being glabrous in *V. planipetala* and *Epimedium* proper. If such characters arose by mutation, then more widely distributed characters probably originated likewise and survived because they were neutral as regards natural selection.

Thus, on general evidence, *Epimedium* (*sensu lato*) appears to belong to a very old group which probably originated in the early Tertiary (or earlier) mixed temperate forest belt of north-eastern Asia; by the end of the Tertiary period they had acquired a wide range across the northern hemisphere, despite their reliance on woodland ants for seed dispersal, remaining everywhere restricted to woodland conditions. That range was disastrously broken by Quaternary (Pleistocene) glaciation, very severe in Europe but seemingly of little effect in China, and the drying up of central Asia which would have obliterated any secondary centre of evolution for *Epimedium* south of the Caucasus. China has remained its major centre of specific differentiation, with the greatest number of species in Sichuan province.

The haploid chromosome number (n) is 6 for *Epimedium* and *Vancouveria* (Takahashi, 1989), as also for *Jeffersonia*, *Achlys* and *Diphylleia* (Langlet, 1928), *Bongardia* and *Podophyllum*, although 7 in *Ranzania* and 10 in *Nandina* (Miyaji, 1930); no cytological irregularities have been observed in the hybrids of *Epimedium*, due probably to the cytological uniformity of the parent species. Dermen (1931) and Giffen (1937) found a like uniformity in *Berberis* and *Mahonia*; in all forms studied, except the tetraploid *B. turcomanica* var. *integerrima*, the haploid number was 14; the numerous and widespread species of these genera exhibit no important differences either in chromosome number or size. Dermen concludes that 'species differentiation in *Berberis* is not due to changes in chromosome number or to any fundamental change in chromosome structure or genetic constitution. Most of the differences between species are those which might be attributed to mutation association with geographic isolation..... the production of polyploid types or fundamental change in the chromosome complex, produced by wide species hybridization, has evidently not played an important part in the formation of species in *Berberis*'. This conclusion seems equally applicable to *Epimedium* and other herbaceous *Berberidaceae*.

6. CULTIVATION OF *EPIMEDIUM*

Cultivation creates experimental environments for plants, and their behaviour in gardens may indicate the bounds of climatic tolerance and other factors limiting their range. On the other hand, knowledge of their habitats in the wild may be a guide to their successful cultivation. Epimediums throughout their distribution inhabit deciduous or mixed temperate woodland, usually on slopes providing light dappled shade. Being woodland plants, they thrive best under moderately cool and half-shady conditions, preferably in moist but well-drained humus-rich soil. Their rhizomes creep shallowly below the surface and in the wild are mulched by fallen leaves. The fine roots occupy the upper 10–30 cm (4–12 inches) of soil. Accordingly, dry shallow soils should be enriched with abundant humus, preferably leaf-mould, to which a little slow-acting fertiliser may be added. An ideal situation is one where they have such a good leafy soil and are not exposed to frost and harsh drying winds, which damage their first leaves, and to summer drought.

Epimediums are attractive garden plants and by no means difficult to grow. Most of the better-known species, together with their hybrids, appear to be winter-hardy throughout Britain, and many will grow even further north without protection, for example at Uppsala, Sweden. *Epimedium alpinum*, naturalised at Linne's Hammarby, near Uppsala for two and a half centuries, is recorded as being grown as far north in Sweden as Hernosand (63°37'N); presumably the plants are given spruce covering in winter. It is the custom in some gardens to remove all the leaves of evergreen epimediums in late autumn, for tidiness, leaving the soil bare all winter instead of being covered with pleasing foliage, which may become decoratively bronzed or even reddish. If foliage is so removed, a few leaves should be left untouched to identify the position of the clumps. In southern England the leaves need not be cut away before February, after which new growth emerges from the soil; further north they can be left longer. Plants benefit by a top-dressing of leaf-mould or composted bark.

When planting it is a wise precaution against labels being lost or misplaced to write the name and other particulars on a piece of paper, seal this within a discarded plastic photographic film container and bury that shallowly by the plants.

Those kinds that have creeping rhizomes, such as *E. alpinum*, *E. perralderianum*, *E.* × *perralchicum*, *E. pinnatum*, *E. sempervirens* and *E.* × *warleyense*, can be left to make their own way into fresh soil. Clump-forming kinds with short annual rhizome growth, such as *E. acuminatum*, *E. davidii*, *E. diphyllum*, *E. franchetii*, *E. grandiflorum* and *E.* × *youngianum*, need to be divided every three or four years, using a sharp stout knife or pair of secateurs to cut through their woody rhizomes. The divided pieces should be planted in fresh soil and

Fig. 10. **_Epimedium_ × _rubrum_** with _Onoclea sensibilis_ and hardy geraniums in a London garden. Photograph: John Fielding.

Fig. 11. ***Epimedium* x *versicolor* 'Neosulphureum'** with *Dicentra eximia* at The Garden House, Devon. Photograph: John Fielding.

Fig. 12. ***Epimedium acuminatum***. Epimediums are excellent for mixed woodland plantings. Photographed by John Fielding in a garden near London.

kept moist while re-establishing themselves. In southern England this is best done from mid August to late September.

Epimediums can be used for cool glasshouse decoration from March to May by lifting small clumps from the outdoor garden, unless already grown in outdoor containers, and potting them in a mixture of loam, leaf-mould or composted bark and grit or coarse sand, with the rhizome about 2.5 cm (1 inch) below the surface; they can be put below the staging until shoots begin to unroll. On the staging, their elegant young leaves, sometimes brightly coloured, and their exquisite flowers, come nearer to eye-level for appreciation than in the outdoor garden and repay examination with a ×8 hand lens. They should be moved back to the outdoor garden after flowering (containers should be sunk to the brim in soil or sand).

For the outdoor garden the most robust one is *E. pinnatum* subsp. *colchicum*, spectacular with its profuse yellow flowers. It thrives in any good garden soil, and it has endowed with vigour its hybrid progeny, *E.* × *perralchicum*, *E.* × *warleyense* and *E.* × *versicolor*, notably the clone 'Sulphureum'. These can be used at the front of a herbaceous border or as ground cover among deciduous shrubs. Non-aggressive plants like the Japanese *E. diphyllum*, *E. grandiflorum* and *E.* × *youngianum*, as well as Chinese species such as *E. acuminatum*, *E. brevicornu*, *E. davidii* and *E. stellulatum*, are admirable for shady ledges of the rock garden. Evergreen kinds include *E. acuminatum*, *E. dolichostemon*, *E. fangii*, *E. franchetii*, *E.* × *perralchicum*, *E. perralderianum*, *E. pinnatum*, *E. pubigerum*, *E. sempervirens*, *E.* × *warleyense*, *E.* × *versicolor* 'Sulphureum' and 'Neosulphureum' and most Chinese species. Non-evergreen (deciduous) kinds include *E. alpinum*, *E. elongatum*, *E. grandiflorum* and *E.* × *versicolor* 'Versicolor'.

Until recent years epimediums could be reproduced only by division of rhizomes, because the single clones by which species were then represented gave no seed unless bee-pollinated from a clone of another species; *E.* × *warleyense*, *E.* × *cantabrigiense*, *E.* × 'Kew Hybrid' and *E.* × 'Kaguyahime' arose thus. Growing recently introduced Chinese species side by side has resulted in abundant seed, thanks to nectar-seeking bumble-bees which, unlike hive bees, have no flower constancy and go to and fro between the different species. Such promiscuously produced seed generates only hybrids. To raise a species like *E. acuminatum* true from seed, different clones of it should be placed together, well away from other species. *Epimedium* seed is best sown immediately after it drops to the ground from the open capsules.

Epimediums suffer from no special pests but may be attacked by slugs, snails, rabbits and vine weevils. Pheasants will eat the young leaves and leaf-cutter bees (*Megachile* sp.: see *Gartenflora* 80: 297, 1938) cut neat slices from mature leaves, as they do the leaves of other plants, to line their nests. Some fungi are recorded on wild plants (see Saccardo, *Sylloge Fungorum*; 1882–1931).

A detailed, well-illustrated account of the cultivation of *Epimedium* in southern England by Robin White has been published in *The Garden* 121: 208–214 (1996).

7. TAXONOMIC TREATMENT OF *EPIMEDIUM*

Epimedium [Tourn., Elemens Bot. 1: 198 (1694); Inst. Rei Herb.: 232, t.117 (1700)]: Linn., Sp. Pl. 1: 117 (1753); Gen. Pl., ed. 1, 30, no. 81 (1737), & ed. 5, 53, no. 138 (1754); Benth. & Hook. f., Gen. Pl. 1: 44 (1862); Franch. in Bull. Soc. Bot. France 33: 103, p.p. (1886); Prantl in Engl. & Prantl, Nat. Pflanzenfam., 3 (2): 75, p.p. (1888); N.Busch in Fl. Cauc. Crit. 3 (3): 207, p.p. (1903); Moss, Cambr. Brit. Fl. 3: 154 (1920); Lemée, Dict. Gen. Phan. 2: 887 (1930); Stearn in J. Linn. Soc., Bot. 30: 436 (1938) & in Europ. Garden Fl. 3: 389 (1989). Type species: *Epimedium alpinum* Linn., based on Dodoens' *Epimedium*, not on Dioscorides'.

Aceranthus C.Morren & Decne. in Ann. Sci. Nat. Bot. II, 2: 349 (1834); Benth & Hook. f., *op. cit.*: 44 (1862). Type species: *Aceranthus diphyllus* C.Morren & Decne.

Vindicta Raf., Fl. Tellur. 2: 52, no. 187 (1837). Type species: *Vindicta begonifolia* Raf., based on *Epimedium diphyllum* Lodd.

Endoplectris Raf., Fl. Tellur. 3, 56, no. 636 (1837). Type species: *Endoplectris tricolor* Raf., based on *Epimedium macranthum* C.Morren & Decne.

Epimedium [sect. *vel* subgen.] *Euepimedium* Franch. in Bull. Soc. Bot. France 33: 41 (1886); Prantl in Engl. & Prantl, *op. cit.* 76 (1888); N.Busch, *loc. cit.* 208 (1903); Kom. in Acta Horti Petrop. 29: 128 (1908); Himmelbaur in Denkschr. Kaiserl. Akad. Wiss. Math.-Naturwiss. Kl. 89: 744 (1914).

× *Bonstedtia** H.R.Wehrh., Gartenstauden 1: 455 (1930); Wehrhahn in Bonstedt, Pareys Blumeng. 1: 621 (1930); Silva Tarouca & Schneider, Unsere Freiland-Stauden, ed. 5: 99 & 476 (1934). Type: × *Bonstedtia* H.R.Wehrh. is a hybrid group, comprising the hybrids between *Epimedium*, as represented by *E. macranthum*, and *Aceranthus* as represented by *E. diphyllum*. The hybrid × *B. youngiana* (Fisch. & C.A.Mey.) H.R.Wehrh., synonymous with *Epimedium youngianum* Fisch. & C.A.Mey., *sensu stricto*, may be taken as the lectotype.

GENERIC DESCRIPTION. *Perennial herbs. Rhizome* sympodial, irregularly branched, horizontally creeping, furnished with brown membranous leaves. *Leaves* basal or cauline, biternate, or rarely many times ternately divided (e.g. *E. elatum*), imparipinnate,

---

* Dedicated to Herr Carl Bonstedt (b. 1866), editor of *Pareys Blumengärtnerei*, etc., from 1900–1931, Gartenoberinspektor in Göttingen.

trifoliolate, bifoliolate (e.g. *E. diphyllum*), or abnormally unifoliolate; stipules dimorphic, those of basal leaves expanded and united in front of the petiole into a membranous ligulate sheath to 1.5 cm long, 1.3 cm broad, those of cauline leaves distinct, oblong, inconspicuous, scarcely 2 mm long, 1 mm broad; petioles terete but swollen and usually furnished with multicellular hairs at their junction and at insertion of leaflet; leaflets glabrous above, frequently pubescent below, cordate at base, the lateral leaflets asymmetric, with the outer basal lobe longer than the inner, acute or acuminate at the tip, usually spiny at the margin but occasionally entire, involutely rolled before expansion. *Flowering stem* leafless or bearing 1 to 6 leaves. *Inflorescence* simple or compound, with cymose lateral peduncles, few- or many-flowered, glabrous or glandular. *Pedicels* subtended by and occasionally bearing a small membranous bract. *Flowers* dimerous, the terminal one occasionally pentamerous, regular, glabrous, white, rose, yellow, crimson, or violet, with imbricate aestivation; parts free, opposite. *Outer sepals* ('*sepala*' Morren & Decne.) 4, unequal, inner pair larger than the outer pair and up to 0.5 cm long, usually scarious and quickly falling. *Inner sepals* ('*petala*' Morren & Decne.) 4, petaloid, spreading horizontally or rarely reflexing at anthesis. *Petals* ('*nectaria*' Morren & Decne.) 4, flat and petaloid or saccate and produced outwards into nectariferous pouches or spurs shorter or longer than the inner sepals. *Stamens* 4, connivent and appressed to the ovary but free from one another; filaments glabrous, without appendages; anthers dehiscing by two oblong valves which separate from the connective along their base and sides but remain attached at the top, curling upwards; pollen grains elliptic, smooth, with three longitudinal furrows, *c.* 25–40 μ long. *Ovary* superior, glabrous, with parietal placentation; ovules several, in two series, anatropous, almost horizontal; style slender, with slightly dilated stigma. *Capsule* splitting to the base as two independently veined valves, the larger and dorsal valve bearing the seeds and persistent style. *Seeds* smooth, almost black, rounded at both ends, allantoid, about 4 mm long, with a conspicuous aril.

DISTRIBUTION. Southern Europe, North Africa, Asia Minor, Caucasus, India (western Himalaya), China, Manchuria, Far Eastern Russia, Korea, Japan.

Features unusual in the genus, and thus useful aids to determination, are: the 2-foliolate leaves of *E. diphyllum* and *E.* × *youngianum*; the suborbicular and coriaceous leaflets of *E. platypetalum*; the appressed bristle-like hairs on the lower side of the leaflets in *E. sagittatum* and *E. acuminatum*; the leafless flowering stem of subg. *Rhizophyllum*; the numerous stem-leaves and leaflets of *E. elatum*, its tall growth and diffuse inflorescence; the minute inner sepals of *E. platypetalum*; the conspicuous yellow inner sepals of subg. *Rhizophyllum*; the spurless petals of *E. diphyllum* and a few other species; the 1 cm long stamens of *E. dolichostemon* and *E. fargesii*, in other species *c.* 5 mm long.

A Key to the Species of *Epimedium*

If a cultivated plant fits none of the species below, try the 'Key to named Hybrids' (p. 176). An essential tool is a ×8 lens

1.	Flower stem leafless; all leaves basal; inner sepals yellow; petals brown, 2–3 mm long	2
1.	Flower stem with 1 to 8 leaves	3
2(1).	Leaves with 3, 5, 9 or 11 leaflets; marginal spines 0.05–0.9 mm; veins on upper leaflet surface same green as rest of surface. Caucasus	53. ***E. pinnatum***
2.	Leaves with 1–3 undulate-margined leaflets; marginal spines 0.5–1.9 mm; veins on upper leaflet surface lighter green than rest of leaflet. Algeria	54. ***E. perralderianum***
3(1).	Flower stem with one leaf	4
3.	Flower stem with 2 or more leaves	17
4(3).	Leaves with only 2 leaflets or with 2 sets of 3 leaflets; flowers small, white, petals flat and spurless or with spur 10–15 mm long. Japan	5
4.	Leaves with 3–9 or more leaflets	6
5(4).	Petals spurless	49. ***E. diphyllum***
5.	Petals with short spurs	48. ***E. trifoliolatobinatum***
6(4).	Petals long-spurred, the spur tapering, subulate, 12–30 mm long, usually much longer than the flat inner sepal	7
6.	Petals short-spurred, the spur 3.5–4 mm long, slipper-shaped, or flat and spurless	15
7(6).	Leaves with 9 or more leaflets. Japan, Korea	8
7.	Leaves with 3 or 5 (rarely 7) leaflets. East Asiatic mainland	10
8(7).	Leaves perishing in winter; hairs, if present beneath, erect; rhizome compact. Japan	44. ***E. grandiflorum***
8.	Leaves evergreen; rhizome elongated	9
9(8).	Flowers white or purple. Japan	45. ***E. sempervirens***
9.	Flowers yellow. Korea	46. ***E. koreanum***
10(7).	Leaflets long acuminate; petals without basal lamina, pale rose, lilac or purple	11
10.	Leaflets acute, obtuse or rounded; petals with basal lamina 6–8 mm high, white, yellowish or yellow	13
11(10).	Leaflets with sparse short bristle-like hairs beneath; inner sepals ovate	10. ***E. epsteinii***
11.	Leaflets almost glabrous or with scattered, long, spreading multicellular hairs or short, curled reddish hairs beneath, especially on the veins; inner sepals lanceolate or narrowly elliptic	12

12(11). Rhizome elongated, slender (1–2 mm thick); inner sepals narrowly elliptic or lanceolate .... 25. ***E. leptorrhizum***
12. Rhizome compact; inner sepals lanceolate .... 26. ***E. brachyrrhizum***

13(10). Inner sepal 20 mm long, 8–11 mm broad, about as long as spur of petal .... 47. ***E. macrosepalum***
13. Inner sepal 8–11 mm long, 2.5–4 mm broad, much shorter than spur of petal 14

14(13). Flowers white. Inner sepal 2.5 mm broad .... 13. ***E. pauciflorum***
14. Flowers yellow; inner sepal 4 mm broad .... 8. ***E. flavum***

15(6). Petal flat and spurless; leaves with 3 or 5 leaflets. China 24
15. Petal slipper-shaped, saccate, shorter than the concave inner sepal; leaves usually with 9 leaflets. Europe and Asia Minor 16

16(15). Rhizome long-creeping; yearly branch growth 4 cm or more; leaves perishing in winter; inflorescence shorter than stem leaf, hence flowers somewhat hidden beneath leaves .... 51. ***E. alpinum***
16. Rhizome compact: yearly branch growth 2–3 cm; leaves evergreen; inflorescence longer than stem leaf, hence flowers held well above leaves .... 52. ***E. pubigerum***

17(3). Flower stem with 3–8 leaves, with 50 or more leaflets on larger leaves, 50–140 cm high. Kashmir .... 50. ***E. elatum***
17. Flower stem with 2 or 3 leaves, with 1–9 leaflets, not more than 70 cm high. China 18

18(17) Leaves with 1 leaflet 19
18. Leaves with 3–9 leaflets 22

19(18). Inflorescence simple 20
19. Inflorescence compound .... 27. ***E. simplicifolium***

20(19). Leaflets with rounded basal lobes. Hunan 21
20. Leaflets with acute basal lobes. Sichuan .... 30. ***E. glandulosopilosum***

21(20). Leaflets sparsely pubescent with dark yellow hairs beneath, ovate, 7–13 × 4.5–8 cm .... 29. ***E. baojingense***
21. Leaflets densely tomentose beneath, almost orbicular, 10–14 × 7–11 cm .... 28. ***E. zhushanense***

22(18). Petals spurless, flat or slightly saccate at base 23
22. Petals with spur or definitely saccate 30

23(22). Petals 6–9 mm long (but see 4. *E. shuichengense*: 3–4 mm long) 24
23. Petals 1–4 mm long 26

24(23). Leaflets broadly ovate to almost orbicular; inflorescence simple; petals 3.5–5 mm broad .... 2. ***E. platypetalum***

24.	Leaflets narrowly ovate to ovate; inflorescence simple or compound; petals 5–7 mm broad	25
25(24).	Leaves with 3 leaflets; petals flat; base of corolla rounded .... 1. ***E. campanulatum***	
25.	Leaves 3, 5 or 7 leaflets; petals with small nectarial projection at base, hence base of corolla slightly truncate or shouldered .... 3. ***E. ecalcaratum***	
26(23).	Inflorescence few-flowered (with 8–12 flowers), simple; inner sepals yellow; petals purple .... 39. ***E. elachyphyllum***	
26.	Inflorescence many-flowered (with 25–100 flowers), compound	27
27(26).	Apex of terminal leaflet divided into 3 (or 2, 4 or 5) acuminate lobes .... 43. ***E. lobophyllum***	
27.	Apex of terminal leaflet not divided	28
28(27).	Leaflets glabrous beneath, the two lateral leaflets not cordate at base but obliquely truncate; petals almost orbicular .... 40. ***E. truncatum***	
28.	Leaflets hairy beneath, all cordate at base	29
29(28).	Leaflets ovate; petals elliptic .... 41. ***E. coactum***	
29.	Leaflets narrowly lanceolate; petals broadly obovate .... 42. ***E. borealiguizhouense***	
30(22).	Flowers 1 cm across, yellow; petal with lamina 3–4 mm high and erect slender spurs 4–5 mm long almost parallel with it .... 4. ***E. shuichengense***	
30.	Flowers not with this combination of characters	31
31(30).	Flowers large, (13–)20–60 mm across; petal spur elongated, subulate, (7–)10–30 mm long (if spur only *c.* 7 mm, see 21. *E. enshiense*)	32
31.	Flowers small, 8–20 mm across; petal saccate, obtuse, 2–9.5 mm long	52
32(31).	Petal almost straight, horizontal, purple .... 14. ***E. mikinorii***	
32.	Petal distinctly curved (if almost straight and yellow, orange or light pink, see 6. *E. fangii* or 24. *E. wushanense*)	33
33(32).	Petals with basal laminae enclosing stamens as if within a cup	34
33.	Petals without basal laminae, horn-shaped, stamens partly exposed or enclosed	41
34(33).	Petals white	35
34.	Petals yellow	37
35(34).	Spur of petal *c.* 12 mm long; inner sepal 8–10 mm long; leaflets broadly ovate or almost orbicular .... 13. ***E. pauciflorum***	
35.	Spur of petal 15–30 mm long; inner sepal 16–19 mm long; leaflets ovate or narrowly ovate	36
36(35).	Basal lobes of leaflet usually touching or overlapping; spur of petal much longer than inner sepal .... 11. ***E. latisepalum***	
36.	Basal lobes of leaflet diverging; spur of petal as long as or only slightly longer than inner sepal .... 12. ***E. ogisui***	
37(34).	Inner sepals 10–13 mm long	38

37.	Inner sepals 4–6 mm long	39
38(37).	Leaflets ovate, rounded at apex, glabrous beneath	8. ***E. flavum***
38.	Leaflets lanceolate, long acuminate at apex, pubescent beneath...	9. ***E. ilicifolium***
39(37).	Leaflets 10–13 cm long, with basal sinus deep and lobes tending to overlap	7. ***E. hunanense***
39.	Leaflets to 8 cm long, with basal sinus shallow and lobes not overlapping but diverging; inner sepals red	40
40(39).	Rhizome compact; leaflets with minute erect hairs beneath; inflorescence glandular-hairy; spur of petal curved downwards	5. ***E. davidii***
40.	Rhizome elongated, creeping; leaflets glabrous; inflorescence glabrous; spur of petal almost straight and horizontal	6. ***E. fangii***
41(33).	Leaves of flower stem mostly with 9 or more (but sometimes 2 or 5) ovate to almost orbicular leaflets; flower stem with 2–3 alternate leaves; leaves perishing in early winter	15. ***E. elongatum***
41.	Leaves of flower stem with 3 or 5 leaflets; flower stem with 2 opposite or 3 whorled leaves; leaves evergreen, those of previous year present at flowering time	42
42(41).	Leaflets narrowly lanceolate or lanceolate, with almost straight sides, 7–23 cm long, rigid, strongly nerved, the basal lobes of a lateral leaflet very unequal	24. ***E. wushanense***
42.	Leaflets lanceolate to broadly ovate, with curved sides, mostly less than 14 cm long, the basal lobes of a lateral leaflet unequal or almost equal	43
43(42).	Leaflets with a few scattered, erect grey hairs beneath; rhizome elongated, 1–3 mm thick; inner sepals 15–17 mm	22. ***E. sutchuenense***
43.	Leaflets glabrous or with very short, appressed unicellular hairs or bristles, or long multicellular hairs, beneath; rhizome compact or elongated	44
44(43).	Leaflets with a few spreading or almost erect hairs, or glabrous; petal almost straight, horizontal, purple	14. ***E. mikinorii***
44.	Leaflet with minute appressed hairs or bristles beneath, or glabrous; petals curved	45
45(44).	Leaflets elliptic or broadly ovate, glabrous beneath but pilose near base	21. ***E. enshiense***
45.	Leaflets narrowly ovate to lanceolate, with hairs beneath	46
46(45).	Inflorescence compact, simple or compound, the space between the pedicels *c.* 0.5–2 cm	47
46.	Inflorescence usually loose, elongated and with 2–5-flowered peduncles in lower part, the space between them *c.* 1.5–6 cm	49
47(46).	Basal sinus of leaflets with widely diverging lobes; inner sepals white; petals pale purple. Hunan	10. ***E. epsteinii***

47.	Basal sinus of leaflets narrow, the lobes almost touching; inner sepals pale yellow or yellowish; petal pale yellow	48
48(47).	Hairs on underside of leaflet appressed, bristle-like	20. ***E. franchetii***
48.	Hairs on underside of leaflet loosely spreading, multicellular	18. ***E. lishihchenii***
49(46).	Inner sepals slightly ascending, not appressed to petals; anthers green	23. ***E. chlorandrum***
49.	Inner sepals appressed to petals; anthers yellow	50
50(49).	Leaflets with minute appressed bristles on underside; inner sepals 3–7 mm broad, much broader than base of petal; petal, in cultivated Mount Emei form, deep purple	19. ***E. acuminatum***
50.	Leaflets with long flaccid or spreading multicellular hairs on underside; inner sepals 2.5–3 mm broad, not broader than base of petal; petals yellow	51
51(50).	Rhizome compact; leaflet margins with about 9–11 spines to 3 cm of margin	16. ***E. membranaceum***
51.	Rhizome elongated; leaflet margin with about 15–17 spines to 3 cm of margin	17. ***E. rhizomatosum***
52(31).	Leaves on flower stem with 9 or 5 leaflets	53
52.	Leaves on flower stem with 3 (rarely 5) leaflets	54
53(52).	Leaflets ovate to broadly ovate; inner sepals *c.* 10 mm long, much longer than petals	32. ***E. brevicornu***
53.	Leaflets narrowly ovate to lanceolate; inner sepals *c.* 4 mm long, only slightly longer than petals	34. ***E. sagittatum***
54(52).	Stamens 7–10 mm long, the filament much longer than the anther	55
54.	Stamens 2–6 mm long, the filament equalling or much shorter than the anther	56
55(54).	Inner sepals 15–18 mm long, narrowly lanceolate, ascending or reflexing; petals almost straight, 6.5–9.5 mm long	38. ***E. fargesii***
55.	Inner sepals 8–9 mm long, narrowly elliptic, spreading horizontally; petals coiled, to 3 mm long	37. ***E. dolichostemon***
56(54).	Inner sepals *c.* 12 mm long	36. ***E. stellulatum***
56.	Inner sepals 2–7 mm long	57
57(56).	Inner sepals only slightly longer than petals	34. ***E. sagittatum***
57.	Inner sepals much longer than petals	58
58(57).	Leaflets persistently pubescent to tomentose beneath with few–many multicellular spreading or curled grey hairs, especially on the veins	31. ***E. pubescens***
58.	Leaflets glabrous or with very minute appressed hairs beneath	59
59(58).	Leaflets glabrous except for pilose veins beneath	33. ***E. reticulatum***
59.	Leaflets with very minute appressed hairs beneath	35. ***E. myrianthum***

Subgenus I. EPIMEDIUM

(Species number 1–52)

Section *Phyllocaulon* Franch. in Bull. Soc. Bot. France 33: 40 (1886). Type species: *E. alpinum* L.

This subgenus comprises all the species that have cauline leaves, between one and eight, usually accompanied by basal leaves. The other subgenus, *Rhizophyllum*, normally has leafless flower stems, all the leaves being basal. The species of subgenus *Epimedium* are distributed widely throughout the range of the genus, except for North Africa.

Section i. DIPHYLLON

(Species number 1–43)
Section *Diphyllon* (Kom.) Stearn **stat. nov.** Type species: *E. sagittatum* (Sieb. & Zucc.) Maxim.
Series *Diphyllon* Kom. in Acta Horti Petrop. 29: 134 (1908).
Subsection *Diphyllon* (Kom.) Stearn in J. Linn. Soc., Bot. 51: 490 (1938).

This heterogeneous group includes all the Chinese species. The flowering stem in most species bears two opposite leaves, occasionally three verticillate, but in a few species one leaf is suppressed and then the stem bears only one (as in *E. flavum* and *E. leptorrhizum*), or the stem may bear three or four leaves (as in *E. elongatum*). The leaves of most species have only three leaflets but in *E. brevicornu* they have nine leaflets.

Series A. CAMPANULATAE

(Species number 1–4)
Series *Campanulatae* Stearn **ser. nov.** Flores campanulati, parvi (6–8 mm longa). Petala plana vel basi leviter gibbosa. Type species: *E. campanulatum* Ogisu.
Flowers campanulate, small (6–8 mm long), yellow. Petals, flat or with a slight nectarial swelling at base.

1. EPIMEDIUM CAMPANULATUM

The third species of *Epimedium* with spurless yellow petals to be discovered, this, like its close allies *E. platypetalum* and *E. ecalcaratum*, is an elegant plant with numerous pendulous small yellow flowers. It differs from *E. platypetalum* in having narrowly ovate leaflets, those of *E. platypetalum* being broadly ovate or almost orbicular. The flowers in *E. ecalcaratum* differ in the petals being slightly pouched at base.

Epimedium campanulatum Ogisu in Kew Bull. 51: 401, fig.1A–H (1996). Type: China, Sichuan, Dujiangyan Shi, May 1994, *Ogisu* 94305 (holotype K).

ILLUSTRATION. Kew Bull. 51: 403, fig.1A–H (1996).

DESCRIPTION. *Flowering stem c.* 15–40 cm long, with one leaf, or two leaves which are usually alternate, rarely opposite. *Rhizome* compact or short-creeping. *Basal and cauline leaves c.* 7–14 cm long, with 3 leaflets, the petioles pilose with long reddish hairs forming conspicuous tufts at the nodes; leaflets narrowly ovate, *c.* 1.5–5 × 1–3 cm, the tip mucronate, rounded or obtuse, the margin spinous with spines 0.5–1 mm long, the base

Fig. 13. **Epimedium campanulatum**. Photographed by John Fielding at Blackthorn Nursery, Hampshire.

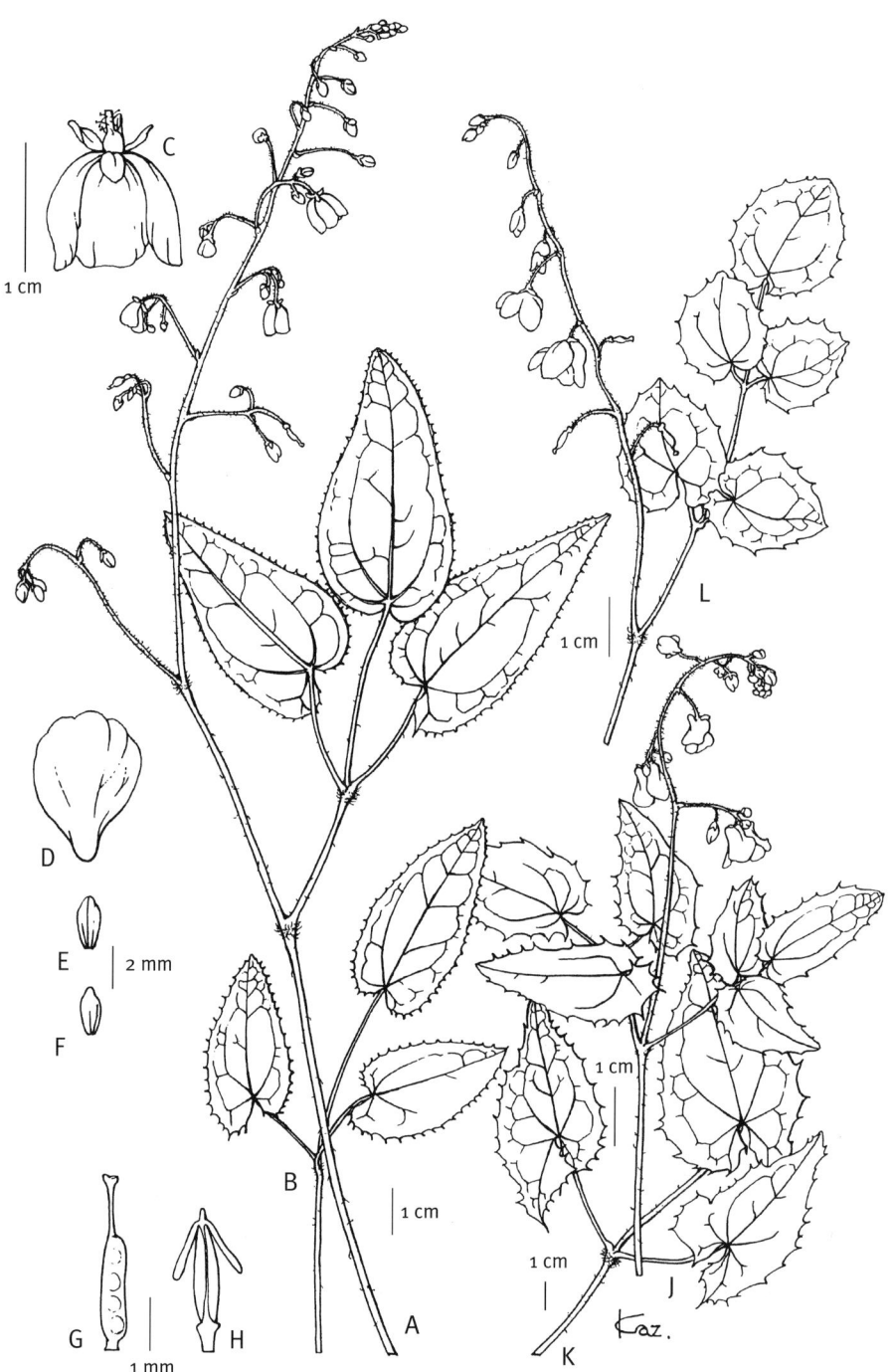

Fig. 14. ***Epimedium campanulatum*** (*M. Ogisu* 94305) A, flowering stem; B, basal leaf; C, flower; D, petal; E, inner sepal; F, outer sepal; G, gynoecium; H, stamen. ***Epimedium ecalcaratum*** (*M. Ogisu* 93082) J, flowering stem; K, basal leaf. ***Epimedium platypetalum*** (*M. Ogisu* 93085) L, inflorescence. Drawn by Kazuko Tajikawa.

shallowly cordate with small usually rounded but sometimes subacute lobes, persisting green all winter, subcoriaceous when mature, glossy above, the underside glaucous sprinkled with very minute erect hairs, and only the whitish midrib evident. *Inflorescence* compound or rarely simple, *c.* 7–21 cm long, many-flowered (with 12–37 flowers), pilose or almost glabrous when mature; pedicels 1.2–1.8 cm long, with short glandular hairs. *Flowers* cup-shaped, pendulous, *c.* 6–8 mm long, with the mouth (in a living state) 6–8 mm across. *Outer sepals* 2–2.5 mm long, green, soon falling. *Inner sepals* oblong, obtuse, 2.5–3 × 1–1.5 mm, pale sulphur-yellow or red-tinged, slightly ascending. *Petals* obovate, flat and spurless, 6–8 × 5–7 mm, pale sulphur-yellow or yellow, the tip rounded, almost truncate. *Stamens* included, 2–3 mm long; filaments 1–2 mm long; anthers yellow 0.8–1 mm long.

DISTRIBUTION. Central China, northwestern Sichuan (Szechwan) province, Dujiangyan Shi (31°10'N, 103°35'E), 2000 m.

2. EPIMEDIUM PLATYPETALUM

Epimedium platypetalum, when described and named in 1922 by K.Meyer, was unique among known Chinese species of *Epimedium* in having small campanulate flowers with flat spurless petals. It remained known only from the type-collection made in 1914 by Hans Wolfgang Limpricht in western Sichuan until introduced into cultivation by Mikinori Ogisu (no. 93885). The flowers are yellow like those of the closely related *E. campanulatum*, not 'probably white' as stated in 1938.

The epithet *platypetalum* is from Greek Πγατυs (platys, broad or wide).

Epimedium platypetalum K.I.Mey. in Repert. Spec. Nov. Regni Veg. Beih. 12: 380 (1922); Stearn in J. Linn. Soc., Bot. 51: 489 (1938); Ogisu in Kew Bull. 51: 404 (1996). Type: China, Sichuan, *Limpricht* 1386 (syntypes WRSLE, WU).

ILLUSTRATION. J. Linn. Soc., Bot. 51: t.25 (1938); Kew Bull. 51: 403, fig.1, L (1996).

DESCRIPTION. *Flowering stem* 18–30 cm long, bearing one leaf. *Rhizome* long-creeping, very slender. *Basal and cauline leaves* with 3 or 5 leaflets; leaflets broadly ovate to almost orbicular, the apex rather round but acute, the margin very spinous with spines *c.* 0.8 mm long, the base deeply cordate with the subequal lobes rounded and nearly touching, ultimately almost coriaceous, papillose and sparingly pilose beneath, 1–5 cm long, 1–5 cm broad. *Inflorescence* simple, 14–18 cm long, very glandular-hairy, few-flowered (with 6–14 flowers); pedicels 0.5–1 cm long, glandular hairy. *Flowers* small, campanulate. *Outer sepals* broadly ovate, blunt, 1.5–3 mm long. *Inner sepals* triangular, 2–2.5 mm long. *Petals* flat, spurless, obovate, yellow, 7–8 mm long, 3.5–5 mm broad. *Stamens* included, 2 mm long, yellow, filament 0.5 mm

DISTRIBUTION. China, Sichuan (Szechuan).

Fig. 15. ***Epimedium platypetalum*** (*M.Ogisu* 93085). Photographed at Blackthorn Nursery, Hampshire, by John Fielding.

Fig. 16. ***Epimedium platypetalum*** growing in Sichuan Province, China, at Wenchuan, 1700 m. Photograph: Mikinori Ogisu.

THE GENUS EPIMEDIUM
3. EPIMEDIUM ECALCARATUM

Fig. 17. ***Epimedium ecalcaratum*** (*M.Ogisu* 83082). Photographed by John Fielding at Blackthorn Nursery, Hampshire.

Fig. 18. ***Epimedium ecalcaratum***, showing variation in the wild. The petals are typically without spurs, but may have small spurs developed to varying degrees. Photographed at Baoxing, Sichuan Province, China, by Mikinori Ogisu.

3. EPIMEDIUM ECALCARATUM

Although having much in common with *E. platypetalum* and *E. campanulatum*, likewise yellow-flowered, *E. ecalcaratum* differs in having the base of the petals slightly saccate, so that the flower has a slightly shouldered base and can be regarded as beginning the development of a nectar-producing spur. All have the same pleasing and distinctive appearance. It was introduced into cultivation from western Sichuan (Szechwan) by Mikinori Ogisu (no. 93082).

Epimedium ecalcaratum G.Y.Zhong in Acta Phytotax. Sin., 29: 89 (1991); Ogisu in Kew Bull. 51: 402 (1996). Type: China, province Sichuan (Szechwan), Baoxing, G.Y.Zhong 87–02 (holotype SM, *n.v.*).

ILLUSTRATION. Acta Phytotax. Sin. 29: 90 (1991), Kew Bull. 51: 403, fig.1, J & K (1996).

DESCRIPTION. *Flowering stem* 30–65 cm long, bearing one leaf, or two or three leaves which are opposite or alternate. *Rhizome* compact, short-creeping. *Leaves* basal and cauline with 3, 5 or 7 leaflets; leaflets narrowly ovate to ovate, the apex subacute to acute, the margin very spinous with spines *c*. 1 mm long, the base cordate, the lobes almost equal and acute and separated by a narrow or wide sinus, minutely pilose beneath, 1.5–4 cm long, 1.5–3 cm broad. *Inflorescence* simple or compound, glandular, hairy, 15–23 cm long, many-flowered (with 20–30 flowers); pedicels 1–2 cm long, glandular hairy. *Flowers* small, campanulate. *Outer sepals* 2–3 mm long. *Inner sepals* lanceolate, acute, red, 2.3–4 mm long, 1–1.5 mm broad. *Petals* flat except for slightly saccate base, obovate, yellow, 8–9 mm long, 4–5 mm broad. *Stamens* 2.5 mm long, yellow; filaments 0.3 mm long.

DISTRIBUTION. China, Sichuan (Szechwan) province.

4. EPIMEDIUM SHUICHENGENSE

Epimedium shuichengense is an anomalous small-flowered species pollinated by short-bodied bees and apparently represents a stage in floral evolution transitional from the spurless small-flowered species of series *Campanulatae* to the long-spurred series *Davidianae*. It is a native of Guizhou (Kweichow) province, southern China, from which nine species of *Epimedium* have been recorded. The author, He Shun-Zhi, states that it resembles *E. sutchuenense* but differs in having spinulose bracts, smaller flowers, with the spur of the petals 4–5 mm long.

Epimedium shuichengense S.Z.He in Acta Bot. Yunnan. 18: 209, fig.1 (1996). Type: Guizhou, Shuicheng, *He Shun-Zhi* 94058 (holotype in Herb. Guizhou Inst. Chinese Mat. Med., *n.v.*).

ILLUSTRATION. Acta Bot. Yunnan. 18: 210, fig.1 (1996).
DESCRIPTION. *Flowering stem* 15–30 cm long bearing two opposite leaves. *Rhizome* long and creeping, elongated, slender, 1.5–2.5 mm thick. *Leaves* basal and cauline 3-foliolate; leaflets ovate, the tip acuminate, the margin spinous, the base cordate with a narrow sinus, 3–17 cm long, 2–2.5 cm or more broad, the underside with white appressed hairs. *Inflorescence* simple, loose, few-flowered (with up to 9 flowers), pedicels 1–1.5 cm long, glandular-pilose or almost glabrous. *Flowers* small, 1 cm across. *Outer sepals* purple, 5 mm long. *Inner sepals* broadly ovate, yellowish, 4 mm long, 3 mm broad. *Petals* with lamina 3–4 mm long, the spur strongly upcurved not spreading, 4–5 mm long. *Stamens* enclosed, 4 mm long, the filaments 2 mm long.
DISTRIBUTION. China, Guizhou province.

Series B. DAVIDIANAE

(Species number 5–14)

Series *Davidianae* Stearn **ser. nov.** Flores magni. Lamina petali expansa calcari elongato curvato munita. Type species: *E. davidii* Franch.

Flowers large. Lamina of petal expanded, with an elongated curved spur. Stamens enclosed.

5. EPIMEDIUM DAVIDII

Many botanical and zoological names such as *Davidia involucrata* (the DOVE or HANDKERCHIEF TREE), *Clematis armandii*, *Lilium davidii*, *Elaphurus davidianus* (PÈRE DAVID'S DEER) and *Sciurus davidi* indicate the esteem in which Père Armand David (1826–1902) was held by scientists at the Paris Muséum d'Histoire Naturelle as a natural history collector, notably by the botanist Adrian Franchet and the zoologist Henri Milne-Edwards. David went to China in 1862 as a Lazarus missionary. Here he collected mammals, birds, insects and plants and discovered the giant panda. His travels in China amounted to 3000 miles. In 1869 he visited the Mission Station at Mupin (Moupine — now Baoxing — in western Sichuan, 30°22'N, 102°49'E). Among the many new plants he collected was an *Epimedium* which Franchet named *E. davidii*. Martyn Rix visited the type region in 1985 and introduced this species into cultivation. No Chinese species except *E. sagittatum* was in cultivation when I prepared my 1938 monograph and I identified a plant collected on Mount Omei (Emei Shan) as *E. davidii*. This plant is now in cultivation and its characters of rhizome etc. reveal it to be a different species which has been named *E. fangii* and is apparently an endemic of Mount Omei; *E. fangii* has an elongated creeping rhizome and nearly straight, almost horizontal, petal-spurs. *Epimedium davidii* is a pleasing species with a compact rhizome and strongly curved petal-spurs.

Epimedium davidii

ANN FARRER

5. EPIMEDIUM DAVIDII

Fig. 19. **Epimedium davidii** (*E.M.Rix* 4125). Photographed at Blackthorn Nursery, Hampshire by John Fielding.

Epimedium davidii Franch. in Nouv. Arch. Mus. Hist. Nat. II, 8: 195, t.6 (1885), & in Bull. Soc. Bot., France 33: 109 (1886); Stearn in J. Linn. Soc., Bot. 51: 490 (1938) *pro parte*, excl. pl. Mount Omei, & in Bot. Mag. 12: 19, fig.A & B (1995). Type: China, Sichuan, 'ad mentes lapidoris prope Moupine', *David* (holotype P, isotype K).

ILLUSTRATIONS. Nouv. Arch. Mus. Hist. Nat. II, 8: t.6 (1885); Curtis's Bot. Mag. 12: 19, fig.A & B (1995).

DESCRIPTION. *Plant* 30–50 cm high when in flower. *Rhizome* compact. *Leaves* basal and cauline, 5- or 3-foliolate, with tufts of reddish hairs at the nodes; leaflets narrowly to broadly ovate, the tip usually rounded and mucronate, the margin very spinous-serrate, the base deeply or shallowly cordate with the usually rounded lobes almost touching or diverging up to 150°, subcoriaceous, both sides distinctly reticulated with veins, beneath glaucescent, papillose and sparingly pubescent with short erect hairs, usually less than 6 cm long, 4.5 cm broad. *Flowering stem* normally bearing two, opposite, 3- or 5-foliolate leaves, rarely alternate or three. *Inflorescence* usually compound (with usually 3-flowered peduncles) below, simple above, loose, very glandular, 6–24-flowered; pedicels 1.5 cm (in flower) to 3 cm long (in fruit). *Flowers* 2–3 cm across, yellowish. *Outer sepals* blunt, ovate, 2–4 mm long. *Inner sepals* narrowly ovate, subacute, reddish, 4 mm long, 1 mm broad. *Petals* much longer than inner sepals, with distinct petaloid, rounded laminae

58 THE GENUS EPIMEDIUM
5. EPIMEDIUM DAVIDII

Fig. 20. Baoxing (Moupin), China, where Armand David first collected **Epimedium davidii** in 1869. Photograph: Phillip Cribb.

Fig. 21. **Epimedium davidii** (*E.M.Rix* 4125). A, inflorescence; B, flower. **Epimedium fangii** (*M. Ogisu* 81001). C, fruiting stem and cauline leaf; D, inflorescence; E, flower. Bar scales = 1 cm. Drawn by Christabel King.

forming a cup 7–13 mm deep, and slender downward-curved subulate spurs 1–1.5 cm long. *Stamens* included, about 4 mm long; anthers 3 mm long.

DISTRIBUTION. Western China, in mountain woods of west Sichuan province.

6. EPIMEDIUM FANGII

Epimedium fangii commemorates the Chinese botanical professor Wen-pei Fang (1899–1983), of Chengdu University, Sichuan, author of *Icones Plantarum Omeiensium* (2 parts, 1943 and 1944), who had a special interest in the astonishingly rich flora of the sacred Chinese mountain Emei Shan (Mount Omei). According to Roy Lancaster, *Travels in China*: 85 (1989), 'for fifty-five years until his death in 1983 he studied its flora, visiting it on many occasions at different seasons. No-one knew it better or loved it more than he.' Fang estimated its total flora to exceed 3000 species. His friend Mikinori Ogisu introduced *E. fangii* into cultivation and found hybrids between it and *E. acuminatum*.

My 1938 account of *Epimedium davidii* (Stearn, 1938) was based entirely on herbarium specimens, including not only the type material collected by David at Mupin (Baoxing) but also specimens collected by Yü & Tü at 1850 m and 1900 m on Emei Shan. Now that plants from Mupin and Emei Shan are both in cultivation, it is evident that they represent two species, *E. davidii* and *E. fangii*, as distinguished in the

Fig. 22. ***Epimedium fangii*** (*M.Ogisu* 81001). Photographed at Blackthorn Nursery, Hampshire by John Fielding.

Fig. 23. ***Epimedium fangii*** growing on Mt. Omei (Emei Shan) in western Sichuan, China at 1660 m. Photograph: Mikinori Ogisu.

Fig. 24. Mt. Omei (Emei Shan) in Sichuan, China, where ***Epimedium fangii***, ***E. acuminatum*** and their hybrid ***E. × omeiense*** occur. Photograph: Phillip Cribb.

key above (couplet 40), the fundamental differences being in their rhizomes and leaf hairs. *Epimedium davidii*, introduced from its type locality, has a compact rhizome and leaflets with minute erect hairs on the underside. *Epimedium fangii*, although seemingly glabrous, has minute appressed hairs on the underside. In this they parallel other closely allied species much alike in the form and size of their flowers. The leaflets of *E. fangii* are more glossy and a lighter green than those of *E. davidii* grown under the same conditions.

Epimedium fangii Stearn in Curtis's Bot. Mag. 12: 18, fig.C, D, E, & p. 20 (1995). Type: China, Sichuan, Emei Shan, *M.Ogisu* 81001 (citation of 81007 is incorrect), cultivated at Blackthorn Nursery, Hampshire, England, 14 April 1994 (holotype K).

ILLUSTRATION. Curtis's Bot. Mag. 12: 19, fig.C, D, E, & p. 20 (1995).

DESCRIPTION. *Flowering stem* c. 25 cm long, bearing 2 opposite leaves. *Rhizome* long-creeping, slender (1.5–2 mm thick). *Basal and cauline leaves* with 3 leaflets; leaflets narrowly ovate, 4–8 cm long, 2.5–5.5 cm broad, the tip obtuse or almost acute, the margin spinous with spines 0.5–1.2 mm long, the base moderately cordate with the

Fig. 25. ***Epimedium fangii***, leaf and rhizome. Bar scale = 1 cm. Drawn from *M.Ogisu* 81001 by Christabel King.

lobes rounded and separated by a narrow sinus, those of the lateral leaflets moderately unequal, coriaceous when mature, persisting green all winter, the underside glaucous and almost glabrous except for scattered appressed minute hairs. *Inflorescence* simple, loose, few-flowered (with 6–10 flowers), to 13 cm long; pedicels glabrous, *c.* 2 cm long. *Flowers* large, *c.* 4.5 cm across. *Outer sepals* soon falling, *c.* 2.5–3.5 mm long. *Inner sepals* boat-shaped, reddish, obtuse, *c.* 6 mm long, 2.5 mm broad, spreading horizontally. *Petals* much longer than the inner sepals, pale yellow; spur elongated, subulate, almost straight and horizontally spreading, *c.* 2.2 cm long, expanded at base into a lamina *c.* 1 cm high. *Stamens* included, *c.* 3.5 mm long; filaments *c.* 1 mm long; anthers *c.* 2.5 mm long, pale yellow.

DISTRIBUTION. Central China, Sichuan, on Emei Shan (Mount Omei)

7. EPIMEDIUM HUNANENSE

Epimedium hunanense would seem to be a species of limited distribution in the Chinese province of Hunan, seemingly known only from the type specimens collected in 1919 by a Chinese collector for Heinrich von Handel-Mazzetti. It is related to *E. davidii* of central Sichuan but distinguished by narrow ovate much larger leaves and slightly larger flowers with thicker horizontally spreading spurs.

Plants at one time grown as '*E. hunanense*' belong to another species, *E. franchetii*.

Epimedium hunanense (Hand.-Mazz.) Hand.-Mazz., Symb. Sinicae 7: 324 (1931); Stearn in J. Linn. Soc., Bot. 51: 492 (1938). Type: China, Hunan, *WangT.-H.* in Pl. Sin. 43 (holotype WU).

E. davidii var. *hunanense* Hand.-Mazz. in Anz. Akad. Wiss. Wien, Math.-Naturf. Kl. 62: 131 (1926).

ILLUSTRATION. J. Linn. Soc., Bot. 51: t.27 (1938).

DESCRIPTION. Flowering stem 40 cm high bearing two opposite leaves. *Rhizome* short creeping, 3 mm thick. *Leaves* basal and cauline, 3-foliolate; *leaflets* (of mature basal leaf) narrowly ovate, the apex acuminate, the margin very spinous-serrate, the base deeply cordate with the lobes tending to overlap, coriaceous in texture, beneath glaucescent, papillose and sparingly pubescent or almost glabrous, 10–13 cm long, 6 cm broad. *Inflorescence* compound with the lower peduncles 2- or 3-flowered, loose, almost glabrous, 10–16-flowered; *pedicels* 1–2 cm long. *Flowers* 3.5 cm across, yellow. *Outer sepals* oblong-elliptic, blunt, 4 mm long, 2 mm broad. *Inner sepals* broadly elliptic, blunt, 5–6 mm long, 3–4 mm broad. *Petals* much longer than the inner sepals, with distinct rounded laminae forming a cup about 8 mm deep and with fairly stout nearly cylindric blunt spurs 1.5–1.8 cm long. Stamens 5 mm long; anthers 4 mm long.

DISTRIBUTION. China, Hunan province.

Epimedium flavum

CHRISTABEL KING

8. EPIMEDIUM FLAVUM

Epimedium flavum is a pleasing, small, few-flowered species from Sichuan province related to *E. davidii* with likewise yellow-spurred flowers. Its inner sepals are pale yellow about 11 mm long whereas *E. davidii* has reddish inner sepals 4–6 mm long. In the form and colour of its flowers *E. flavum* is rather like the garden hybrids *E.* × *versicolor* 'Sulphureum' and 'Neosulphureum' but their leaves are quite different. Unlike most Chinese species, *E. flavum* may have either one leaf or two leaves on the flowering stem. Not being a vigorous plant, *E. flavum* is more suitable for pot-culture in a cool glasshouse than in the open ground.

Epimedium flavum Stearn in Curtis's Bot. Mag 12: 21, pl.263 (1995). Type: China, Sichuan: Erlang Shan, Tianquan Xian (29°52'N, 102°19'E), 2000 m, 1992, *M.Ogisu* 92036; cultivated at Blackthorn Nursery by R.White, April 1994 (holotype K). Other specimens: 4 May 1992, *M.Ogisu* 92032 (K), 92033 (K).

ILLUSTRATION. Curtis's Bot. Mag. 12: pl.263 (1995).

DESCRIPTION. *Flowering stem* 13–30 cm long, with one leaf or two leaves at the same height or at different heights. *Basal leaves* mostly with 5 leaflets, sometimes 3. *Cauline leaves* with 5 or 3 leaflets; leaflets ovate, to 4 cm long, 3 cm broad, the tip rounded, the margin spinous with spines 0.5–1.0 mm long, the base cordate with the lobes rounded and slightly diverging or separated by a narrow sinus, those of the lateral leaflets moderately unequal, almost

Fig. 26. **Epimedium flavum** in the wild at Tianquan, western Sichuan, China, 2090 m. Photograph: Mikinori Ogisu.

coriaceous when mature, persisting all winter, the underside glabrous. *Inflorescence* simple, loosely few-flowered (with 3–10 flowers), *c.* 6 cm long; pedicels with short glandular hairs, to 2 cm long. *Flowers* large, *c.* 3 cm across. *Outer sepals* soon falling. *Inner sepals* lanceolate, pale sulphur yellow, obtuse, spreading horizontally, 11 mm long, 4 mm broad. *Petals* a little longer than inner sepals, pale sulphur yellow; spur elongated, subulate, slightly curved, 1.3 cm long, expanded at base into a lamina 8 mm high. *Stamens* included, *c.* 3 mm long; filaments pale sulphur yellow, 0.5 mm long; anthers pale sulphur yellow, 2.5 mm long.
DISTRIBUTION. Central China, Sichuan province, Erlang Shan.

9. EPIMEDIUM ILICIFOLIUM

A member of series *Davidianae*, this species is nearest in flower to *E. flavum*, both having yellow inner sepals, but the two are very distinct. *Epimedium flavum* has ovate leaflets rounded at the apex and glabrous beneath, the inflorescence simple with 3–10 flowers, while *E. ilicifolium* has lanceolate acuminate leaflets, pubescent beneath, and the inflorescence compound with 25–32 flowers. Geographically they are distinct, *E. flavum* being a native of Sichuan, *E. ilicifolium* of Shanxi.

In a living state the glossy, firm, leaflets with strongly undulate spiny margins resemble the leaves of some Asian species of holly (*Ilex*). The Latin name *Ilex*, genitive *ilicis*, was applied in classical times to the holm oak (*Quercus ilex* L.) or to the chermes oak (*Quercus coccifera* L.) but in the eighteenth century Linneaus used it as the generic name for holly (*Ilex aquifolium* L.), discarding Tournefort's *Aquifolium*. In modern botanical Latin usage, *ilicifolius* accordingly means 'holly-leaved'.

Epimedium ilicifolium Stearn in Kew Bull. 53: 213 (1998). Type: China, Shanxi (Shansi) province, Jinxinling, Zhenping, 1650 m, *Ogisu* 93020, cultivated at Blackthorn Nursery, Kilmeston, Hampshire (holotype K).

ILLUSTRATION. Kew Bull. 53: 215, fig.1 (1998).
DESCRIPTION. *Flower stem c.* 28–42 cm long, bearing two opposite leaves. *Rhizome* compact. *Basal and cauline leaves* with 3 leaflets; leaflets lanceolate, 5–7 cm long, 1.2–3 cm broad, the apex long acuminate, the margin undulate sinuate and spinous, with spines 1–2 mm long, the base cordate with an open sinus and acute lobes, coriaceous, persisting green all winter, the upper side glossy; the underside covered with minute erect hairs. *Inflorescence* compound, loose, many-flowered (with 25–32 flowers), 18–23 cm long, the lower peduncles 3-flowered, 2–3 cm long; pedicels glandular, *c.* 2 cm long. *Flowers* large. *Outer sepals* soon falling, almost black, white-margined. *Inner sepals* elliptic or narrowly ovate, acute, pale yellow, 10–12 mm long, 5–6 mm broad. *Petals* much longer than inner sepals, pale yellow, about 2 cm long; spur elongated, subulate, strongly curved with basal lamina about 7 mm high. *Stamens* enclosed, 4 mm long, yellow; filament 1 mm long; anthers yellow.
DISTRIBUTION. China, Shanxi province.

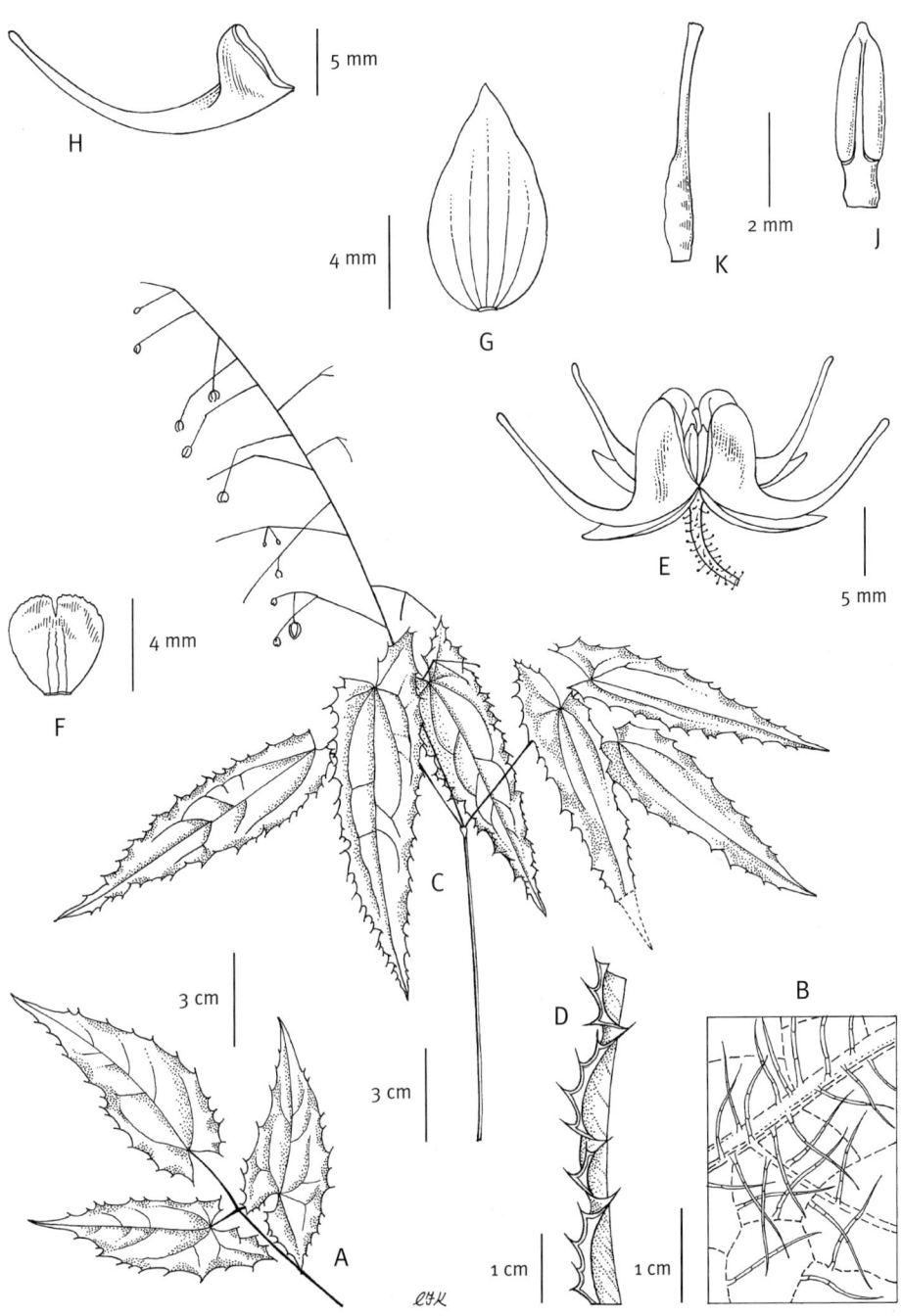

Fig. 27. ***Epimedium ilicifolium***. A, basal leaf; B, underside of basal leaf; C, inflorescence; D, margin of cauline leaf; E, flower; F, outer sepal; G, inner sepal; H, petal; J, stamen; K, gynoecium. Drawn from *M.Ogisu* 93020 by Christabel King.

10. EPIMEDIUM EPSTEINII

Notable, among Chinese species, for the minute bristle-like hairs on the underside of the leaflets, in combination with white inner sepals and purplish long-spurred petals, *Epimedium epsteinii* was unknown until 1994 when plants were collected in Hunan for the Beijing Botanic Garden. Mr Darrell Probst noted, on its flowering with him in 1995, an extraordinary difference in size between the outer pair and the inner pair of the four inner sepals on the middle flower of a three-flowered peduncle. The outer pair were about 7 mm long, 8 mm broad, and obtuse, the inner pair 13 mm long, 9 mm broad, and acute. This species has been named in honour of the late Harold Epstein (1897–1997) of Larchmont, New York, renowned as a grower and collector of *Epimedium*, for which he has been an enthusiast over 45 years, and esteemed as President Emeritus of the North American Rock Garden Society.

Epimedium epsteinii Stearn in Kew Bull. 52: 662 (1997). Type: China, Hunan province, Tianpingshan mountains, Beijing Botanical Garden Exped., cult. 1995 at Hubbardston, Mass., U.S.A. by *Darrell Probst* CPC 94.0255 (holotype K).

ILLUSTRATION. Kew Bull. 52: 665, fig.3 (1997).

DESCRIPTION. *Flower stem* 12 cm or more long, bearing two opposite leaves or one leaf. *Rhizome* long-creeping, slender, 2–3 mm thick. *Basal and cauline leaves* with 3 leaflets; leaflets narrowly ovate, 4–5 × 2–3 cm, the apex acuminate, the margin spinous with

Fig. 28. ***Epimedium epsteinii*** (*D.Probst* CPC 94.0347). Photographed at Blackthorn Nursery, Hampshire by John Fielding.

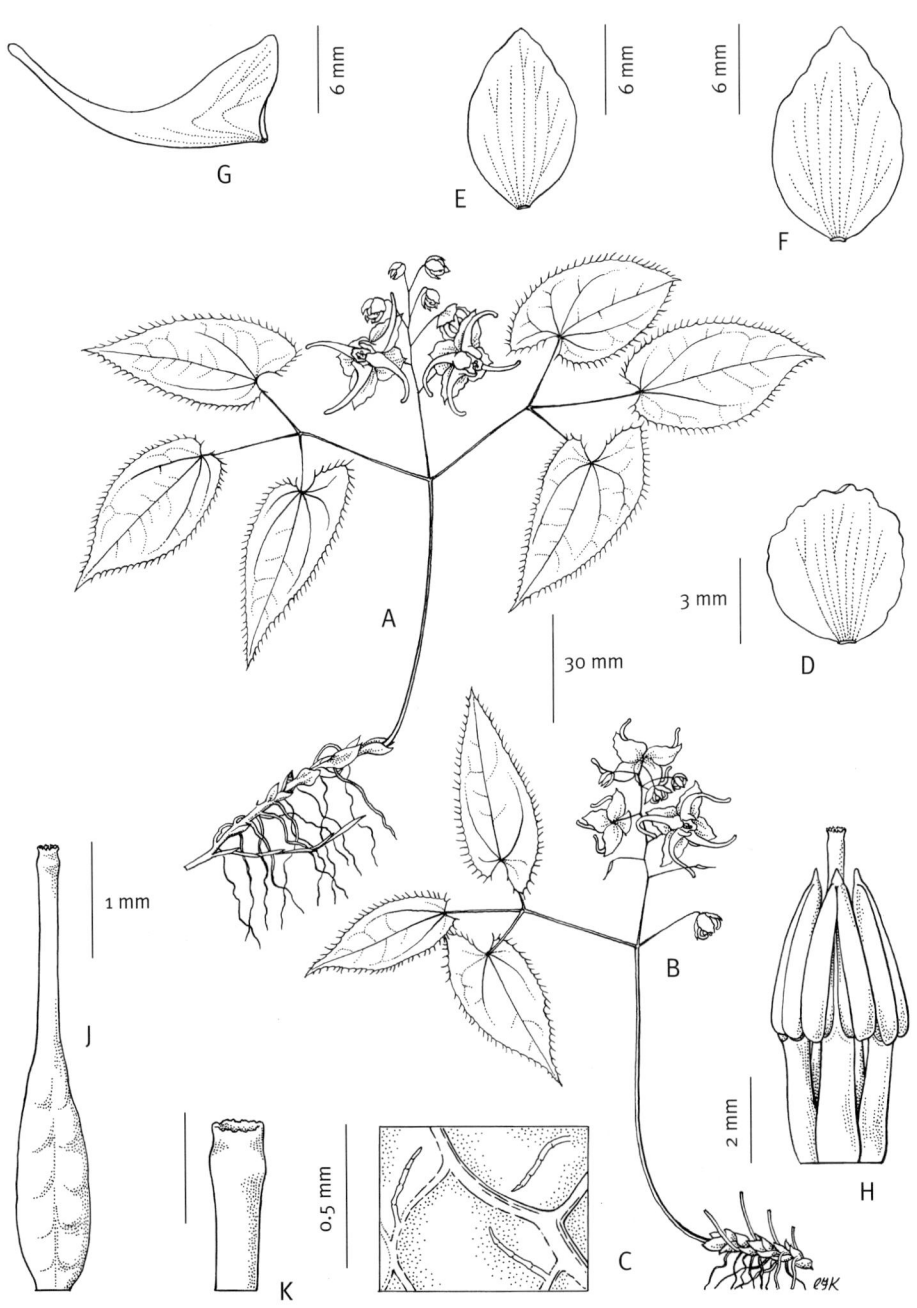

Fig. 29. ***Epimedium epsteinii***. A & B, habit; C, detail of underside of leaf; D, outer sepal; E, F, inner sepals; G, petal; H, stamens & gynoecium; J, gynoecium; K, apex of style. Drawn from *D.Probst* CPC 94.0255 by Christabel King.

numerous spines 1.5–2 mm long, the base shallowly cordate with the lobes separated by a moderately wide sinus, those of the terminal leaflet rounded, of the lateral leaflets the smaller lobe rounded, the larger lobe acute, the underside glaucous and with sparse minute appressed bristle-like hairs, persisting green during winter. *Inflorescence* simple or compound with lower peduncles 3-flowered, compact, with the pedicels about 1 cm apart, few-flowered (with 6–15 flowers), 6 cm or more long; pedicels about 3 cm long, glabrous. *Flowers* large, 3 cm across. *Outer sepals* soon falling, 2.5–3.5 mm long, greenish. *Inner sepals* ovate, white, 13 × 9 mm, acute (but outer pair of terminal flower 7 × 8 mm, and obtuse). *Petals* a little longer than inner sepals, pale purple; spur elongated, subulate, slightly curved, about 15–16 mm long, deeper purple and swollen at the base to 5 mm high, but without a well-developed basal lamina. *Stamens* enclosed, 4 mm long; filaments 0.5 mm long.

DISTRIBUTION. Central China, Hunan province, in Tianpingshan mountains, in half shade along the edges of woods and near streams, 400–1000 m.

11. EPIMEDIUM LATISEPALUM

Epimedium latisepalum is one of the most beautiful species of the genus, with large, well-spaced, pendulous white flowers, rivalled only by *E. ogisui* with slightly smaller flowers. It was introduced into cultivation from Baoxing Xian, Sichuan by Mikinori Ogisu.

The epithet *latisepalum* is from *latus* (broad, wide) and *sepalum* (sepal) referring to the broad inner sepals.

Fig. 30. ***Epimedium latisepalum*** (*M.Ogisu* 91002). Photographed by John Fielding at Blackthorn Nursery, Hampshire.

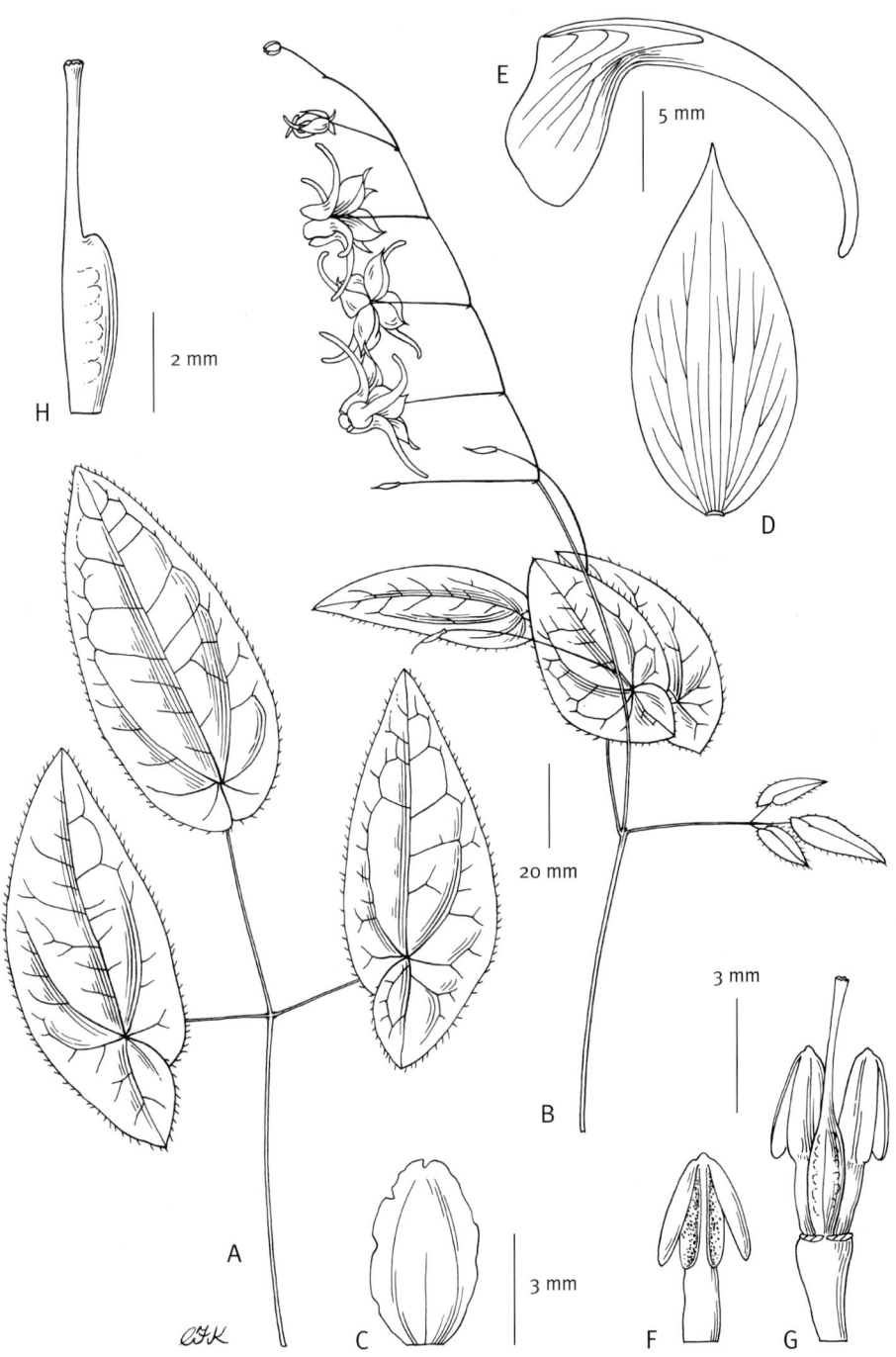

Fig. 31. ***Epimedium latisepalum***. A, basal leaf; B, flowering stem; C, outer sepal; D, inner sepal; E, petal; F, stamen; G, gynoecium with two stamens present and two removed; H, gynoecium. Drawn from *M.Ogisu* 91002 by Christabel King.

11. EPIMEDIUM LATISEPALUM

Epimedium latisepalum Stearn in Kew Mag. 10: 180 (1993). Type: cult. Blackthorn Nursery, origin China, Sichuan province, Xiaoguanzi, Baoxing Xian, *Ogisu* 91002 (holotype K).

ILLUSTRATION. Kew Mag. 10: 181 (1993).

DESCRIPTION. *Flowering stem* c. 30 cm long, leaning, 2-leaved. *Basal and cauline leaves* with 3 leaflets; petioles up to 10 cm long; leaflets narrowly ovate, 6–9 cm long, 2.5–4 cm broad, the tip shortly acuminate or acute, the margin spinous with spines 0.5–1.2 mm long, the base deeply cordate with the lobes acute or rounded and touching or slightly overlapping, those of the lateral leaflets very unequal, coriaceous when mature, persisting green all winter, the underside very glaucous and with short erect scattered hairs. *Inflorescence* simple or sometimes compound below (with usually 2-flowered peduncles), loose, few-flowered (with *c*. 8 flowers), *c*. 20 cm long; pedicels glabrous, 2.5–5 cm long. *Flowers* large, *c*. 4–5 cm across, pendulous. *Outer sepals* soon falling, green, 3–5 mm long. *Inner sepals* elliptic, white, shortly acuminate, *c*. 16 mm long, 8–9 mm broad, spreading horizontally. *Petals* much longer than inner sepals, white but slightly yellowish or purple-tinged at base; spur elongated, subulate, ± straight to slightly curved, *c*. 2.5 cm long, expanded at base into a lamina *c*. 7 mm high. *Stamens* included, *c*. 4 mm long; filaments white, *c*. 1.5 mm long; anthers yellow, *c*. 2.5 mm long.

DISTRIBUTION. China, Sichuan (Szechwan) province.

12. EPIMEDIUM OGISUI

No one man has made a greater contribution to the knowledge of the genus *Epimedium* in China, notably in the province of Sichuan (Szechwan), by introducing live plants for study in cultivation, by collecting excellent herbarium specimens and photographing plants in their habitats, together with their floral parts, than Mikinori Ogisu. This special attention to *Epimedium* has led to the discovery and description of several previously unknown species, i.e. *E. campanulatum*, *E. dolichostemon*, *E. fangii*, *E. flavum*, *E. franchetii*, *E. latisepalum*, *E. mikinorii* and *E. ogisui*, as well as the introduction of species previously known only from herbarium specimens, i.e. *E. brevicornu*, *E. ecalcaratum*, *E. elongatum*, *E. fargesii*, *E. lishihchenii*, *E. pauciflorum* and *E. platypetalum*. The name *E. ogisui*, for a beautiful white-flowered species, is a tribute to this fruitful activity in the mountainous wilds of China.

Epimedium ogisui, closely akin to the white-flowered *E. latisepalum*, differs from that species in the basal lobes of its shallowly cordate leaflets being separated by a short triangular sinus (not overlapping as in the deeply cordate leaflets of *E. latisepalum*), narrower sepals and shorter petal spurs, about as long as the inner sepals. It was introduced by Mikinori Ogisu (no. 92002) from Lushan Xian, Sichuan, where it grows at about 950 m.

Fig. 32. **Epimedium ogisui** growing near Lushan, western Sichuan, China, at 980 m. Photograph: Mikinori Ogisu.

Fig. 33. **Epimedium ogisui** (*M.Ogisu* 92002). Photographed in a garden near London by John Fielding.

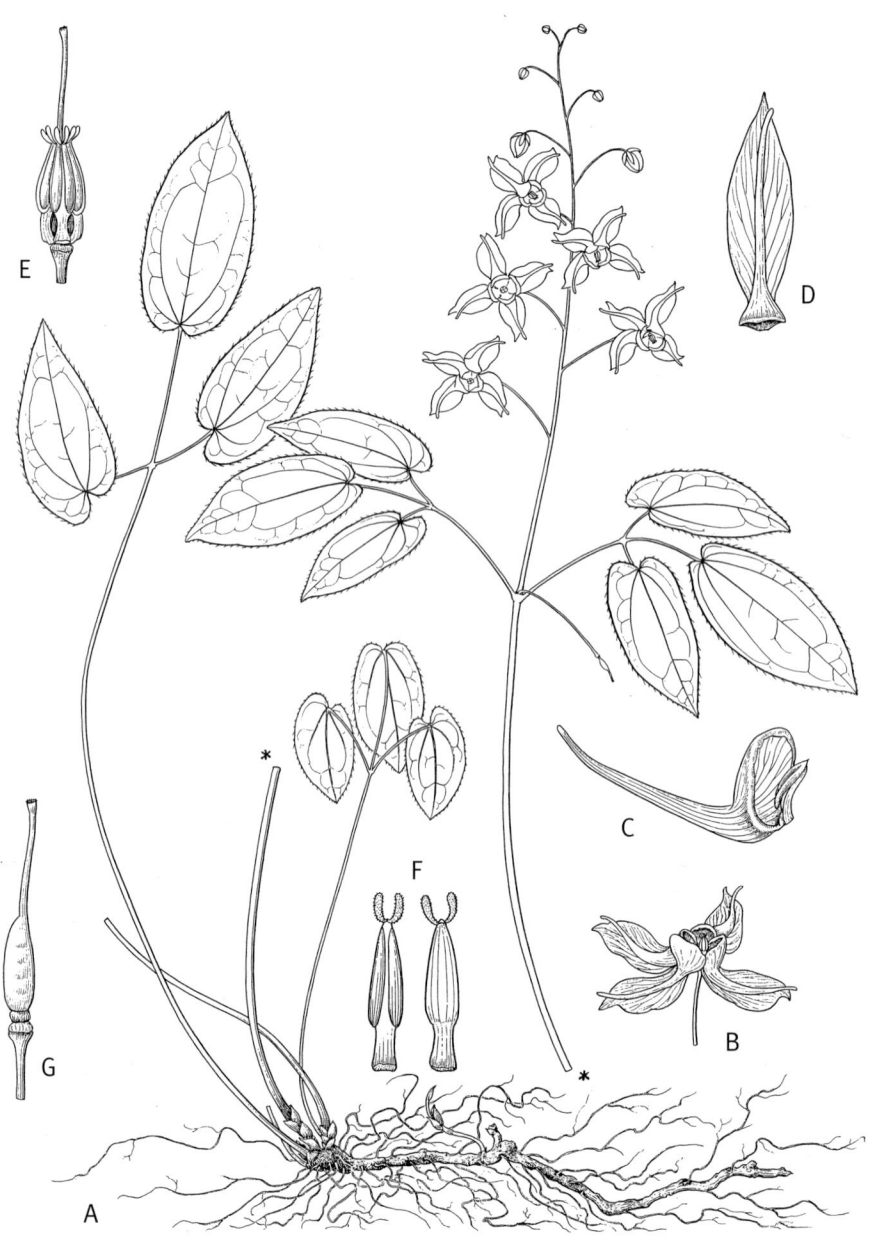

Fig. 34. ***Epimedium ogisui.*** A, flowering plant, × 1; B, flower (inverted), × 1; C, petal with stamen, × 2; D, inner sepal and petal, × 2; E, stamens with dehisced anthers and projecting style, × 4; F, stamen, front and back views, × 5; G, ovary with style, × 4. Drawn from *M.Ogisu* 92002 by Mikinori Ogisu.

Epimedium ogisui Stearn in Kew Mag. 10: 182 (1993). Type: China, Sichuam province, Shuangsi, Lushan Xian, *Ogisu* 92002 (holotype K).

ILLUSTRATION. Kew Mag. 10: 183 (1993).
DESCRIPTION. *Flowering stem* 25–35 cm long, bearing 2 opposite leaves. *Rhizome* long-creeping, slender (*c*. 1 mm thick). *Basal leaves* with 1 or 3 leaflets, the petioles 5–13 cm long. *Cauline leaves* with 3 leaflets, the petioles 2–5 cm long; leaflets ovate or narrowly ovate, 3–6 cm long, 1–3 cm broad, the tip acute, the margin spinous with spines 0.7–1 mm long, the base moderately cordate with the lobes rounded and separated by a narrow sinus, those of the lateral leaflets moderately unequal, coriaceous when mature, the underside glaucous and almost glabrous except for scattered erect short hairs. *Inflorescence* simple, loose, few-flowered (with 3–12 flowers), 12–14 cm long; pedicels glabrous, 2–3 cm long. *Flowers* large, *c*. 2.5 cm across. *Outer sepals* soon falling, 2–3.5 mm long. *Inner sepals* lanceolate, white, acuminate, spreading horizontally, *c*. 16–19 mm long, 7–9 mm broad. *Petals* about as long as the inner sepals, white; spur elongated, subulate, slightly curved, *c*. 1.5–1.8 cm long, expanded at base into a lamina 7–8 mm high. *Stamens* included, *c*. 4.5 mm long; filaments 1.5 mm long; anthers 3 mm long.
DISTRIBUTION. China, Sichuan (Szechwan) province.

13. EPIMEDIUM PAUCIFLORUM

Epimedium pauciflorum is a low-growing species with a few-flowered inflorescence, remarkable among Chinese species in having one leaf on the flower stem. It was first described as a variety of *E. platypetalum* which has yellow spurless flat petals whereas *E. pauciflorum* has white long-spurred petals. At present (2001) known only from the mountain Mao Xian in Sichuan province, it was introduced into cultivation in 1992 by Mikinori Ogisu. He found higher on the same mountain plants with two leaves on the stem and inflorescences with 5–11 flowers which probably belong to the same species.

Epimedium pauciflorum K.C.Yen in Guihaia 14: 124, fig.1 (1994); Stearn in Curtis's Bot. Mag. 12: 23 (1995). Type: China, Sichuan (Szechwan) province, Dagou, Mao Xian (31°42'N, 103°51'E), 1700 m, 16 April 1987, *K.C.Yen & S.L.Shao* 66535 (holotype GMXI).
E. platypetalum var. *tenuis* B.L.Guo & P.G.Hsiao in Acta Phytotax. Sin. 31: 195, fig.8–14 (1993). Type: China, Sichuan province, Mao Xian, 1990 m, 19 April 1989, *H.R.Xie* 89023 (holotype IMM).

ILLUSTRATIONS. Acta Phytotax. Sin. 31: 195, fig.8–14 (1993); Guihaia 14: 124 fig.1 (1995); Curtis's Bot. Mag. 12: 17 (1995).
DESCRIPTION. *Flowering stem* 12–16 cm long, bearing one leaf or possibly two leaves. *Rhizome* long creeping, slender (1–2 mm thick). *Basal and cauline leaves* with 3 leaflets,

13. EPIMEDIUM PAUCIFLORUM

Fig. 35. *Epimedium pauciflorum* in its natural habitat at Maoxian, western Sichuan, China, 1950 m. Photograph: Mikinori Ogisu.

the basal petioles 4–5 cm long; leaflets broadly ovate or almost orbicular, 1.5–4.5 cm long, 1–3.5 cm broad, the tip acute, the margin closely spinous with numerous spines 0.5–1 mm long, the base deeply cordate with the lobes rounded, almost touching or separated by a narrow sinus, almost coriaceous when mature, the underside glabrous when mature and with a close network of fine veinlets. *Inflorescence* simple, few-flowered (with 3–4 flowers but possibly to 11), *c.* 6 cm long; pedicels ascending, with very short glandular hairs, 1–2.5 cm long. *Flowers* large, *c.* 2–2.5 cm across. *Outer* sepals soon falling, oblong, obtuse, greenish, 2–4 mm long. Inner sepals lanceolate, faintly purple-tinged, acuminate, 8–10 mm long, *c.* 2.5 mm broad. *Petals* longer than inner sepals, white; spur elongated, subulate, slightly curved, *c.* 12 mm long, expanded at base into a lamina 6 mm high. *Stamens* included, 3 mm long; filaments 1 mm long; anthers 2 mm long, yellow, with yellow pollen.

DISTRIBUTION. China, Sichuan province, Dagou, Mao Xian (31°42'N, 103°51'E), 1850 m, 30 April 1992, *M.Ogisu* 92020 (K); *ibid.*, 1700 m, 16 April 1987, *K.C.Yen & S.L.Shao* 66535 (GXMI); Mao Xian, 1990 m, 19 April 1989, *H.R.Xie* 89023 (IMM).

THE GENUS EPIMEDIUM
13. EPIMEDIUM PAUCIFLORUM

Fig. 36. ***Epimedium pauciflorum.*** Drawn from *M.Ogisu* 92020 by Christabel King.

14. EPIMEDIUM MIKINORII

Epimedium mikinorii is distinctive among species of the series *Dolichocerae* and *Davidianae* for its narrow, almost straight to slightly curved purple petals, which have a yellow margin to the lamina.

The epithet *mikinorii* is given as an expression of gratitude to its collector, the Japanese botanist and horticulturist Mikinori Ogisu, for his major contribution to the knowledge of the genus *Epimedium* in China, by collecting, photographing and introducing as many as 15 species into cultivation. The name *E. ogisui* likewise commemorates his important work on the genus. The purple spurs of the petals of *E. mikinorii* markedly contrast with the almost white inner sepals.

Fig. 37. **Epimedium mikinorii.** Growing near Enshi, western Hubei, China, 670 m. Photograph: Mikinori Ogisu.

14. EPIMEDIUM MIKINORII

Fig. 38. *Epimedium mikinorii*. Photographed by John Fielding at Blackthorn Nursery, Hampshire.

Epimedium mikinorii Stearn in Kew Bull. 53: 214, fig.2 (1998). Type: China, Hubei (Hupeh) province, Ganxi, Enshi, 670 m, April 1995, *Ogisu* 95039, cultivated at Blackthorn Nursery, Kilmeston, Hampshire, 5 April 1997 (holotype K).

ILLUSTRATION. Kew Bull 53: 217, fig.1 (1998).
DESCRIPTION. *Flowering stem* 28–40 cm long, bearing two opposite leaves. *Rhizome* compact. *Leaves basal and cauline* with 3 leaflets; leaflets lanceolate, 8–11 cm long, 3–4 cm broad, the apex long-acuminate, the margin with numerous spines 1.5–2.0 mm long, the base cordate with an open sinus and acute lobes, coriaceous, persisting green all winter, the underside glaucous and glabrous. *Inflorescence* compound, loose, many-flowered (with about 30 flowers), about 18 cm long, glabrous; peduncles mostly 3-flowered, about 2 cm

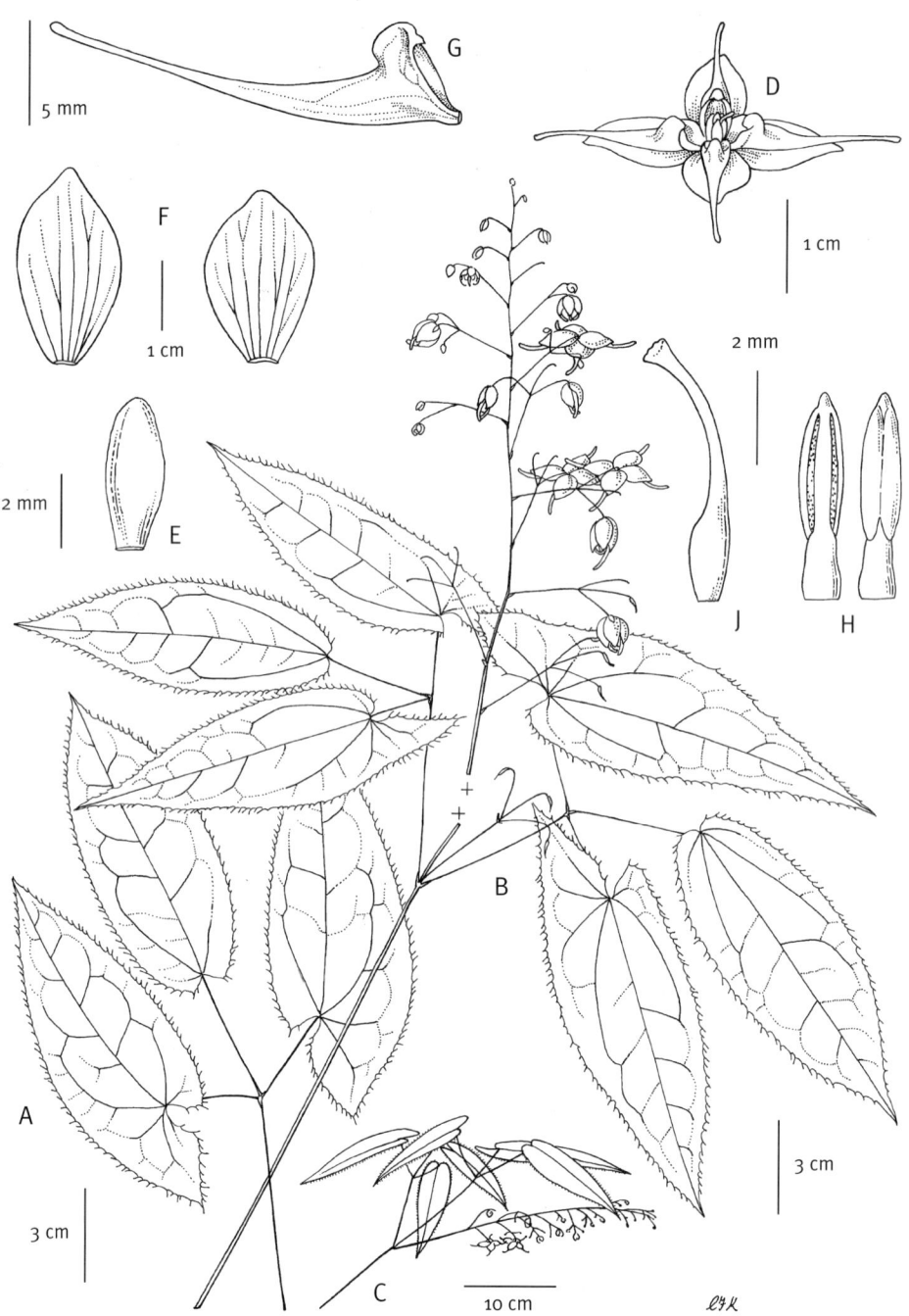

Fig. 39. ***Epimedium mikinorii.*** A, basal leaf; B, inflorescence; C, inflorescence showing natural poise; D, flower; E, outer sepal; F, inner sepal; G, petal; H, stamen, adaxial (left) and abaxial (right) views; J, gynoecium. Drawn from *M.Ogisu* 95039 by Christabel King.

long, pedicels about 1 cm long. *Flowers* large. *Outer* sepals soon falling, 3.5 mm long. *Inner sepals* elliptic, rose-tinged, 11–12 mm long, 4–5.5 mm broad. *Petals* much longer and narrower than inner sepals, with yellow-edged lamina 3.5 mm high; spur elongated, subulate, only slightly curved to almost straight, purple 1.7–2 cm long. *Stamens* enclosed, 3 mm long, whitish; filaments 1 mm long; pollen yellow.

DISTRIBUTION. China, Hubei (Hupeh) province, Ganxi, Enshui, 670 m.

Series C. DOLICHOCERAE

(Species number 15–30)

Series *Dolichocerae* Stearn in J. Linn. Soc., Bot. 51: 490 (1938). Type species: *E. membranaceum* K.I.Mey.

Series *Elongatae* Stearn, *op. cit.*: 509 (1938). Type species: *E. elongatum* Kom.

Petals without laminae and with long slender spurs.

15. EPIMEDIUM ELONGATUM

Epimedium elongatum is a variable species of western China notable for usually having three alternate stem leaves with thin leaflets which perish in late autumn or early winter, whereas most Chinese species of *Epimedium* have evergreen leaves. The flower stem usually carries three alternate leaves, less often two, and terminates in an elongated inflorescence of up to 30 pale yellow flowers. These have slender spurs without laminae and are characteristic of series *Dolichocerae* but the foliage differs from other members of this group in that the lower stem leaves often have nine or more ovate to almost orbicular leaflets. The specific epithet *elongatum* refers to the long inflorescence.

When the Russian botanist Vladimir Komarov, an authority on the plants of Eastern Asia, published the name in 1908 he provided a detailed description and excellent illustration, but nothing whatever about its provenance and collector. However, the type specimen in St. Petersburg bears the label 'G.N.Potanin, 19.vii.1843' and Bretschneider's account of the travels of the Cossack explorer Grigori Potanin (1835–1920) makes evident that he collected it in the mountains about 64 km (40 miles) north of Kangding (Tatsienlu, Tachenlu). Other specimens have been collected in this border area of eastern Sikang and western Sichuan. Komarov placed *E. elongatum* in his series *Polyphylla*, together with the Kashmir species *E. elatum* but they have little in common other than a slight resemblance in habit.

This species has twice been introduced into cultivation by Mikinori Ogisu.

Epimedium elongatum Kom. in Acta Horti Petrop. 29: 140, t.3 (1908); Stearn in J. Linn. Soc., Bot. 51: 509, t.30 (1938); T.S.Ying in Acta Phytotax. Sin. 13: 52 (1975); Stearn in Kew Bull. 51: 396 (1996). Type: China, Sichuan province, north of Kangding (Tachenlu), 19 July 1893, *Potanin* (holotype LE, isotype K).

15. EPIMEDIUM ELONGATUM

Fig. 40. *Epimedium elongatum* in the wild at Kangding, western Sichuan, China, 2720 m. Photograph: Mikinori Ogisu.

ILLUSTRATIONS. Acti Horti Petrop. 29: t.3 (1908); J. Linn. Soc., Bot. 51: t.30 (1938).
DESCRIPTION. *Flowering stem* 25–60 cm long. *Rhizome* moderately long-creeping, 2–4 mm thick. *Leaves basal and cauline*, tri- or biternate, rarely 3-foliolate; leaflets ovate to sub-orbicular, the tip rounded and apiculate to short-acuminate, the margin spinous-serrate, the base deeply cordate with the usually subequal lobes rounded or acute, thin and papery in texture, glabrous or with a few scattered hairs beneath, both sides finely reticulated with veins, 1.5–7.3 cm long, 1–6 cm broad. *Flowering stem* usually bearing three, less often two or one, irregularly spaced (5 mm to 9 cm apart) alternate or rarely opposite leaves, the lower tri- or bi-ternate, the upper bi-ternate or 3-foliolate. *Inflorescence* simple or compound with the lower peduncles 3–5-flowered, sparsely glandular, many-(17–30)flowered, elongated, 15–30 cm long, 4–8 cm across; pedicels appressed or spreading, 1–2 cm long (in flower) to 3 cm (in fruit). *Flowers* 2.5–3 cm across. *Outer sepals* ovate, blunt, whitish, 3–4 mm long, 2 mm broad. *Inner sepals* lanceolate, acute, purple, 4 mm long, 1.5 mm broad. *Petals* longer than the inner sepals, pale yellow, tapering from their swollen but lamina-less bases into slender horizontally spreading or curved spurs 1.5 cm long. *Stamens* slightly protruding, 3 mm long; *anthers* 2 mm long. *Capsule* elongated and slender, 2 cm long.
DISTRIBUTION. China, in the borderland of eastern Sikang and western Sichuan (Szechwan) provinces, around Kangding (Tatsienlu).

THE GENUS EPIMEDIUM
15. EPIMEDIUM ELONGATUM

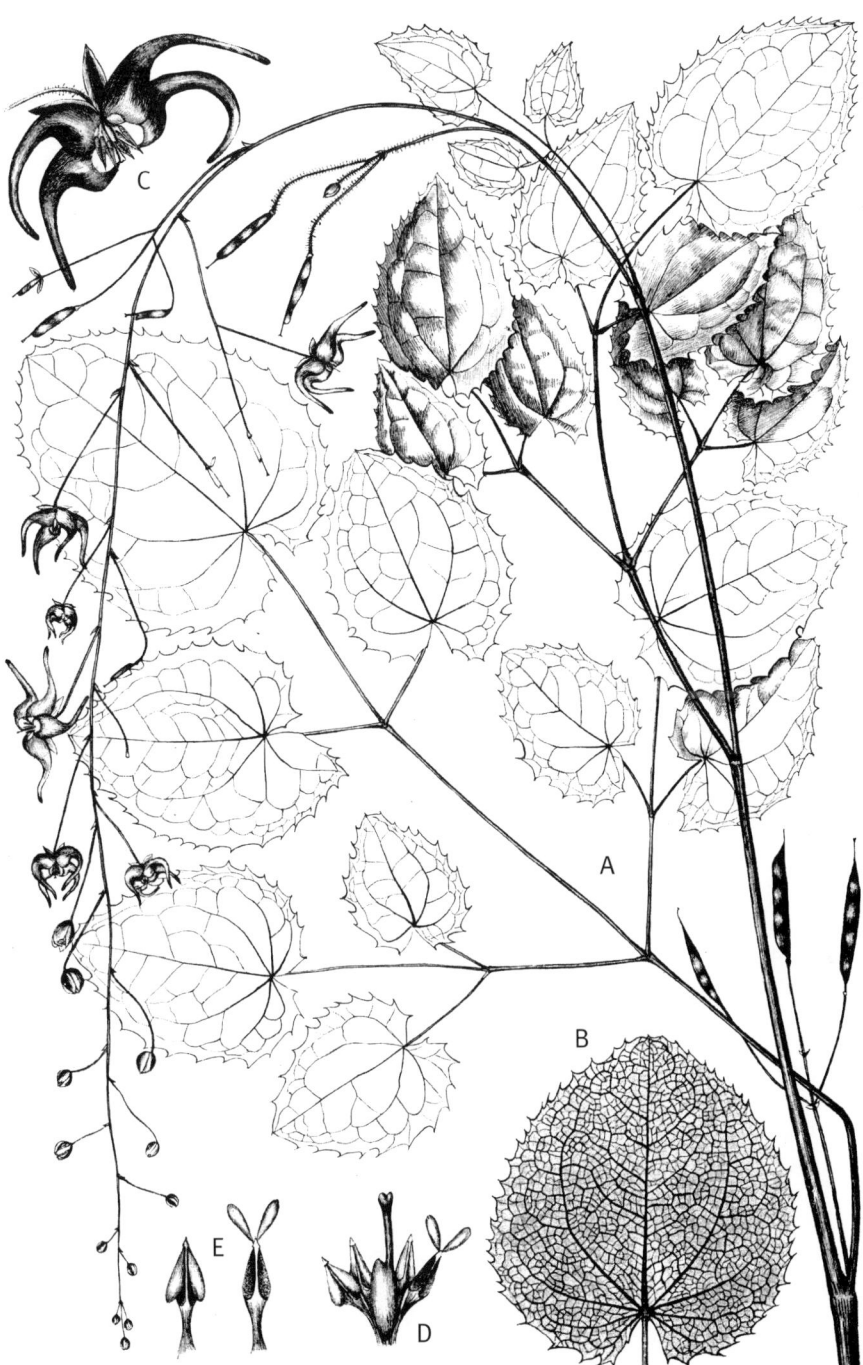

Fig. 41. ***Epimedium elongatum.*** A, flowering stem with cauline leaves, × 1; B, leaflet showing venation, × 1; C, flower, × 21; D, stamens and gynoecium, × 5; E, stamen, before and after dehiscence × 5. From: *Acta Horti Petropolitani* 29: t.3 (1908).

16. EPIMEDIUM MEMBRANACEUM

Cultivated at Kew, *Epimedium membranaceum* flowers longer than any other species of the genus, beginning in late April or in May and producing the last of its large, widely spaced, long-spurred pale yellow flowers in August, on ascending inflorescences 40–45 cm long. Its trifoliolate leaves remain green all winter.

First collected by Hans Wolfgang Limpricht in 1914 in north-central Sichuan (Szechwan), this species has also been collected by Handel-Mazzetti, C.K.Schneider, H.Smith and George Forrest and was introduced into cultivation by Mikinori Ogisu. Their collections demonstrate that it is widely distributed in the eastern half of Sichuan and adjoining northern Yunnan.

Epimedium membranaceum K.I.Mey. in Repert. Spec. Nov. Regni Veg. Beih. 12: 380 (1922); Handel-Mazzetti, Symb. Sinicae 7: 324 (1931); Stearn in J. Bot. 71: 346 (1933). Type. China, Sichuan province, Kwanhsien, *Limpricht* 1293 (isotypes K, WU). *E. membranaceum* subsp. *genuinum* Stearn in J. Linn. Soc., Bot. 51: 495 (1938).

Fig. 42. *Epimedium membranaceum* at Dujiangyan, western Sichuan, China, 1300 m. Photograph: Mikinori Ogisu.

16. EPIMEDIUM MEMBRANACEUM

Fig. 43. ***Epimedium membranaceum*** (*M.Ogisu* 93047). Photographed by John Fielding at Blackthorn Nursery, Hampshire.

ILLUSTRATION. J. Linn. Soc., Bot. 51: t.27 (1938).
DESCRIPTION. *Flowering stem* 20–65 cm long, ascending bearing two opposite or sometimes alternate leaves. *Rhizome* 2–4 mm thick. *Leaves basal and cauline*, 3-foliolate; leaflets broadly ovate or narrowly ovate or almost deltoid, the tip acute or short-acuminate, the margin very spinous-serrate, with spines 1–1.2 mm long, the base deeply or shallowly cordate with a narrow sinus and the lobes mostly rounded sometimes acute, those of the lateral leaflets rather unequal, thin when young and pubescent beneath, subcoriaceous when mature with the underside glaucescent or very glaucous, papillose and pubescent with spreading slender multicellular hairs, 3–10 cm long, 2–6 cm broad. *Inflorescence* compound with the lower peduncles usually 3-flowered, loose, glandular, few- or many-(5-35)flowered, 16–30 cm long; pedicels 1–1.5 cm or more long, 1.5–6 cm apart. *Flowers* 3–5 cm across, pale yellow (and white or pale rose according to collectors), pendulous. *Outer sepals* obovate, blunt, 4 mm long, 2 mm broad. *Inner sepals* ovate-elliptic, or narrowly ovate, acute, 6–7 mm long, 2.5–3 mm broad. *Petals* much longer than the inner sepals, horn-shaped, tapering from the swollen base, curving outwards, 1.5–2.5 cm long, with no lamina. *Stamens* 4 mm long; *anthers* 3 mm long, pale yellow. *Capsule* relatively long and slender, *c.* 2 cm long (incl. style).
DISTRIBUTION. China, in southwestern Sichuan (Szechwan) province and the adjoining region of northern Yunnan.

17. EPIMEDIUM RHIZOMATOSUM

Charles L.A.Royer (1831–1883) in his very original but little appreciated *Flore de la Côte d'Or avec Déterminations par les parties souterraines* (1881–1883) noted differences between allied species in their underground organs and provided keys using these. Such differences have always been more evident to gardeners propagating herbaceous perennials by division than to botanists studying their leaves and flowers. Experience with plants of *Epimedium* in cultivation has revealed that marked differences of rhizome character between plants which are otherwise much alike in flower and leaf are in fact correlated with subtle differences in above-ground organs. The species figured here and *E. membranaceum* provide an example. When *E. rhizomatosum* was introduced into cultivation by Mikinori Ogisu it was at first considered to be a 'running variety' of *E. membranaceum*. The major difference between the two lies underground in their rhizomes. The former has a compact rhizome with annual growth shoots 1–2 cm long, while the latter has an elongated rhizome and the base of each aerial shoot puts forth annual growth shoots about 3–5 cm long. The leaves of the two are much alike, as also are the pale yellow flowers, but the leaflets of *E. membranaceum* have about 9–11 spines to 3 cm of leaflet margin whereas leaflets of the same size in *E. rhizomatosum* have about 15–17 spines to the same length of margin; *E. membranaceum* has a 30–40 cm long inflorescence with numerous well-separated flowers (*cf.* Stearn, 1938, pl. 27 for illustration of type-specimen) whereas *E. rhizomatosum* has a much shorter inflorescence with fewer and more crowded flowers. The description of *E. membranaceum* subsp. *genuinum* (Stearn, 1938: 405), based entirely on herbarium specimens, covered both *E. membranaceum* (*sensu stricto*) and *E. rhizomatosum*; *E. mambranaceum* subsp. *orientale*, now also known from living material, has been given specific rank as *E. lishihchenii* (Stearn 1997: 664).

A Latin adjectival ending *-osus* indicates abundant or marked development, thus the epithet *rhizomatosum* refers to the numerous elongated rhizomes of this species.

Epimedium rhizomatosum Stearn in Kew Bull. 53: 226 (1998). Type: China, Sichuan province, Selenggong, Leibo Xian, 2040 m, *Ogisu* 92114, cult. July 1997 at Blackthorn Nursery, Kilmeston, Hampshire (K, holotype; PE, isotype).

ILLUSTRATION. Kew Bull. 53: 222, fig.5 (1998).

DESCRIPTION. *Flower stem* c. 38–42 cm long, erect, bearing two opposite or alternate leaves. *Rhizome* creeping, elongated, slender, 8 cm or more long, 1.5–2 mm thick. *Leaves basal and cauline* with 3 leaflets; leaflets narrowly ovate, c. 4–6 cm × 2–3 cm, the apex acuminate, the margin with numerous spines 1–1.5 mm long, the base shallowly cordate, with narrow sinus and the basal lobes rounded and diverging at an angle of 30°–60°, evergreen, the underside glaucous and with scattered minute erect hairs. *Inflorescence* compound, few- or many-flowered (with 5–30 flowers), 9–30 cm long with peduncles 2–5-flowered; pedicels glandular-hairy, 1.5–2 cm long. *Flowers* large, c. 4–6 cm across. *Outer sepals* soon falling, white with purple tinge, 4 mm long. *Inner sepals* appressed, narrowly ovate, white or reddish, 6 × 2.5 mm, acute, almost as wide as base of petal. *Petals*

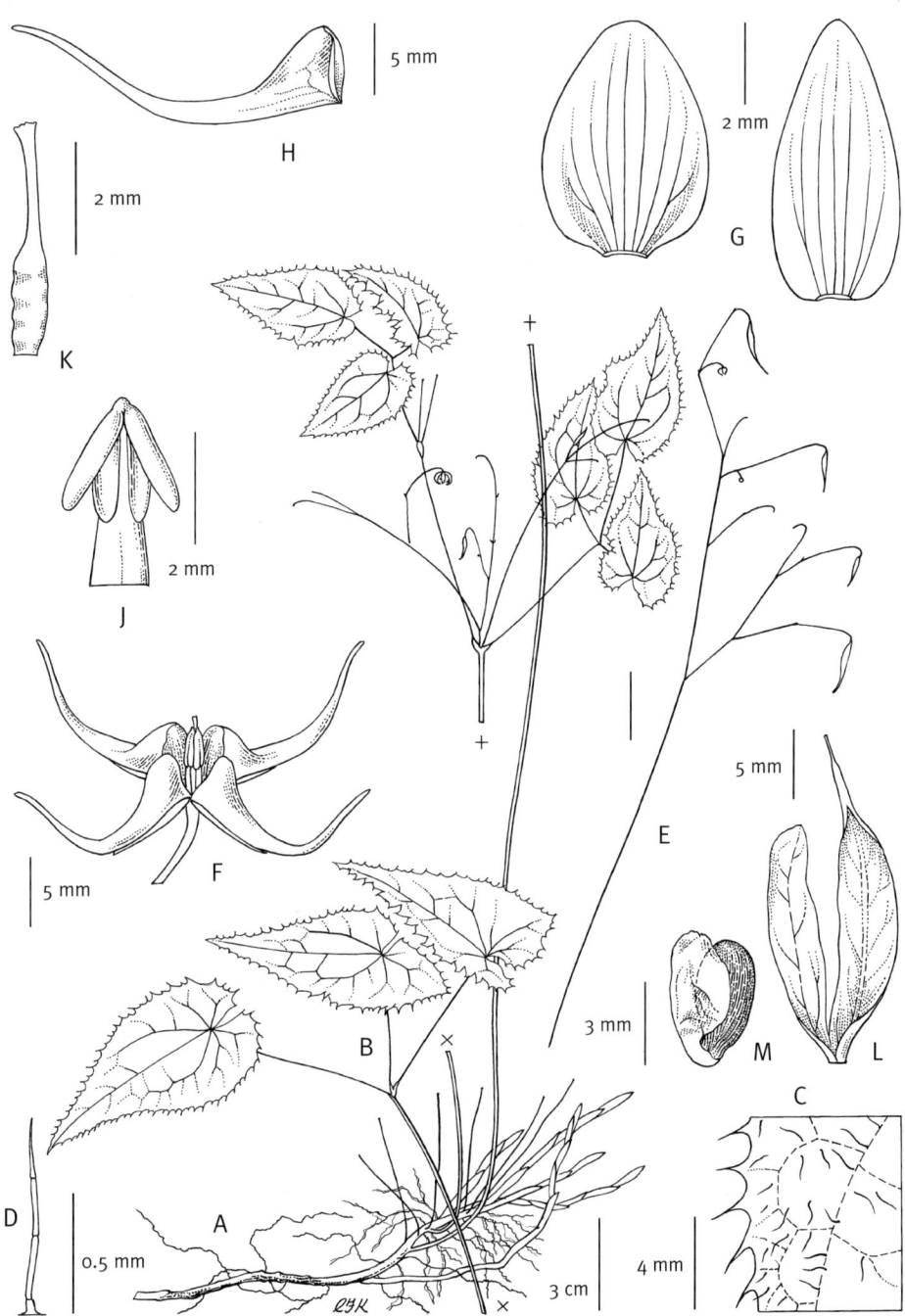

Fig. 44. **Epimedium rhizomatosum.** A, habit; B, basal leaf; C, detail of underside of basal leaf; D, detail of hair; E, inflorescence; F, flower; G, inner sepals; H, petal; J, stamen; K, gynoecium; L, fruit; M, seed with elaiosome. Drawn from *M.Ogisu* 92114 by Christabel King.

17. EPIMEDIUM RHIZOMATOSUM

Fig. 45. ***Epimedium rhizomatosum*** (*M.Ogisu* 92114) with *Eomecon chionantha* and *Arisaema formosanum*. Photographed by John Fielding in a London garden.

much longer than inner sepals, pale yellow; spur elongate, subulate, curved, 2–3.5 cm long, with no lamina. *Stamens* almost enclosed, 3.5 mm long, pale yellow; filaments 1.5 mm; anthers 3 mm long, with pale yellow pollen.

DISTRIBUTION. China, Sichuan (Szechwan) province, 2040–2200 m.

SPECIMENS SEEN. Sichuan, Leibo Xian, *Ogisu* 92114 (K); Mao Xian, *Ogisu* 93045 (K); Zhaoju, Xian, *Ogisu* 92111 (K).

18. EPIMEDIUM LISHIHCHENII

The name *Epimedium lishihchenii* commemorates a very remarkable man, the Chinese encyclopaedic Li Shih-chen (1518–1593), his name also rendered as Li Ši-čen, described by Joseph Needham in 1986 (*Science and Civilisation in China* 6, part 1: 308–321) as 'probably the greatest naturalist in Chinese history and worthy of comparison with the best of the scientific men contemporary with him in Renaissance Europe. His works are an unparalleled source of information on the development of biological and chemical knowledge in East Asia'. These include the huge *Pên Tschao Kang Mu* [Great Pharmacopeia] greatly influential in Japan as well as China and described by Berthold Laufer (*Sino-Iranica* 204: 1919) as 'a monumental work of great erudition and much solid information' which gives the first recognisable description of a Chinese *Epimedium*, the ying yang herb (*E. sagittatum*). Li Shih-chen was born in Hubei (Hupeh) province. Lu Gwei-Djen published a brief biography of 'China's greatest naturalist' in *Physis* (Florence) 8: 383–392 (1966).

18. EPIMEDIUM LISHIHCHENII

The existence of a species of *Epimedium* in the eastern Chinese province Jiangsi (Kiangsi) has been known since 1878 when Shearer collected mature leaves, but no flowers. Maries in 1878 and Miss E.M.Reid about 1899 collected flowering material in the same area. All these specimens must have come not from lowland Kiukiang (now Jiujiang) but from a mountainous area south of it, including the mountain Lushan visited by Maries. These specimens in the Kew Herbarium were variously identified as *E. sagittatum*, *E. macranthum* and *E. pubescens*; despite being poor both in quality and quantity, they represented none of these species but a new taxon which I tentatively identified as *E. membranaceum* subsp. *orientale* in 1938, suspecting it might prove a new species. To elucidate this matter Mikinori Ogisu obligingly visited Lushan in April and May 1996 and

Fig. 46. ***Epimedium lishihchenii*** in the wild in China at Lushan, Jiangxi Province, China, 1220 m. Photograph: Mikinori Ogisu.

18. EPIMEDIUM LISHIHCHENII

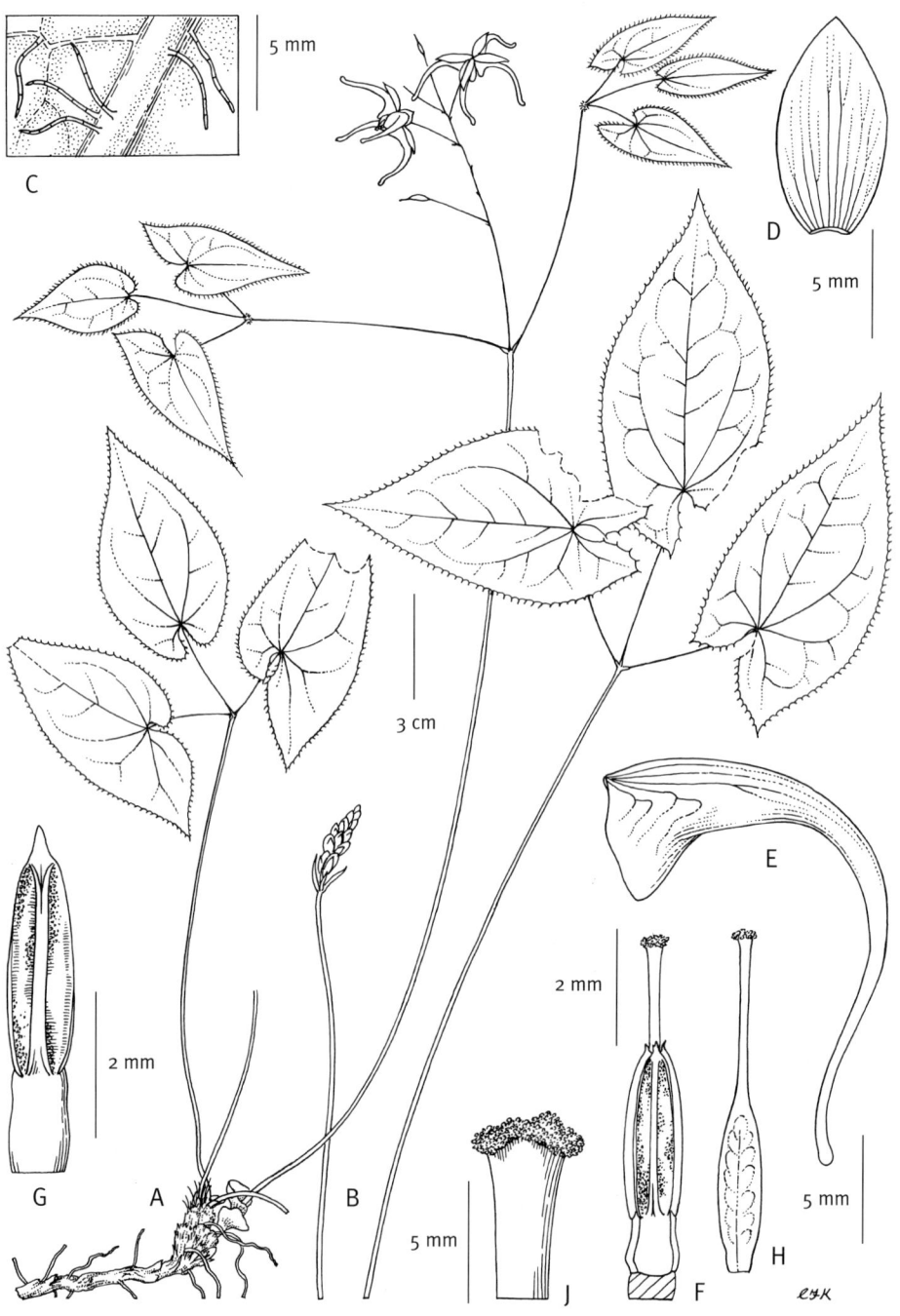

Fig. 47. **Epimedium lishihchenii.** A & B, habit; C, detail of underside of leaf; D, inner sepal; E, petal; F, stamens & gynoecium; G, stamen; H, gynoecium; J, apex of style. A, C from *M.Ogisu* 96305; B from *M.Ogisu* 96302; D–J from *M.Ogisu* 96304. Drawn by Christabel King.

collected good specimens with rhizome, mature leaves and in flower, which confirmed its status as a species and its distinctness from E. membraceum.

This species differs from *E. membranaceum* (*E. membranaceum* subsp. *membranaceum*) of Yunnan and Sichuan in its short, few-flowered inflorescence; that of *E. membranaceum* may reach 45 cm with numerous flowers. It differs form *E. franchetii* in having smaller leaflets with multicellular long hairs; *E. franchetii* has leaflets 9–4 cm long with short bristle-like hairs (*cf.* Stearn, 1996: 394, fig.1A, B).

Epimedium lishihchenii Stearn in Kew Bull. 52: 664, fig.4 (1997). Type: China, Jiangxi (Kiangsi) province, Jiujiang (Kiukiang), Lushan (29°32'N, 115°58'E), 5 May 1996, *Ogisu* 96305 (holotype K).
Epimedium membranaceum K.I.Mey. subsp. *orientale* Stearn in J. Linn. Soc., Bot. 51: 497
  (1938). Type: 'Kiangsi, Kiukiang, mountains', *Reid* 11 (holotype K).

ILLUSTRATION. Kew Bull. 52: 667, fig.3 (1997).
DESCRIPTION. *Flowering stem* 30–40 cm long, bearing 2 opposite leaves. *Rhizome* long-creeping, slender, 2–3 mm thick. *Basal and cauline leaves* with 3 leaflets; leaflets narrowly ovate, 5–11 × 3.5–5 cm, the tip acuminate or sometimes acute, the margin closely spinous with crowded spines 1.5 mm long, the base cordate with a narrow or open sinus, the lobes of the terminal leaflet equal and obtuse, those of the lateral leaflets very unequal with the inner one small rounded or obtuse, the outer one longer and acute, coriaceous when mature, persisting green all winter, the upper side dark green, the lower side glaucous and with long multicellular hairs, or almost glabrous. *Inflorescence* simple, few-flowered (with 5–11 flowers), 7–12 cm long; pedicels glandular-hairy, 1–2 cm long, 1–2 cm apart. *Flowers* large. *Outer sepals* soon falling, 4–5 mm long, with white margins. *Inner sepals* appressed to petals, ovate, narrowly oblong, yellowish, 10–11 mm long, 6–7 mm broad, acute. *Petals* much longer than inner sepals, without lamina, pale sulphur yellow; spur elongated, subulate, strongly curved, 20–25 mm long. *Stamens* exposed, 5 mm long, pale yellow; filaments 1 mm long.
DISTRIBUTION. China, Jiangxi (Kiangsi) province. The species may also occur in Sichuan (Szechwan), Zhejiang (Chekiang) and Hubei (Hupeh) provinces.
SPECIMENS SEEN. Jiujiang (Kiukiang) Mts, March–May, *Reid* 11 (K), *Shearer* (K); Lushan, *Maries* (K), 12 April 1996, *Ogisu* 96302 (K), 5 May 1996 *Ogisu* 96304 (K), *Ogisu* 96305 (holotype, K); Chinkiang and Kiukiang, *Maries* (K).

19. EPIMEDIUM ACUMINATUM

Herbarium specimens diligently collected in the Chinese province of Guizhou (Kweichow, Kouy-tchéou) between 1850 and 1860 by the French missionary Paul Hubert Perny (1818–1907) lay unappreciated, neglected and perishing, indeed the major part destroyed, in the Paris Muséum d'Histoire Naturelle until about 1880 when that assiduous botanist Adrien Franchet undertook their study. In the remnant of Perny's once

Epimedium acuminatum

Fig. 48. Maling, south-west Guizhou Province, China, where *Epimedium acuminatum* grows. Photograph: Mikinori Ogisu.

19. EPIMEDIUM ACUMINATUM

Fig. 49. *Epimedium acuminatum.* Photographed by John Fielding at Blackthorn Nursery, Hampshire.

extensive collection he found a new species of *Epimedium* collected in 1858 unlike any known from Japan, although it possessed leaves like those of *E. sagittatum* (*E. sinense*) and large long-spurred flowers, rather like those of *E. grandiflorum* (*E. macranthum*) but without petal laminae; there were two leaves on the flowering stem. On account of its long-pointed leaflets he named it *E. acuminatum* in 1886.

Later collectors have gathered this species not only in Guizhou but also in Yunnan and Sichuan (Szechwan), most frequently on Emei Shan (Mount Omei), whence Hiroshi Hara, Mikinori Ogisu and Roy Lancaster have independently introduced the purple form into cultivation. It flowered for the first time in Japan and in England in 1982. This has pale purplish inner sepals contrasting with the deep purple spurs. It has proved an elegant garden plant of moderately easy cultivation. In the wild on Emei Shan it has hybridised

Fig. 50. ***Epimedium acuminatum*** growing near Anlong, Guizhou Province, China. Photograph: Phillip Cribb.

with *E. fangii* producing a hybrid swarm (named *E.* × *omeiense*) which was discovered by Ogisu who has introduced distinct variants into cultivation; *E. acuminatum* has also hybridised in cultivation with *E. dolichostemon* producing *E.* 'Amanogawa' and *E.* 'Kaguyahime'.

Epimedium acuminatum Franch. in Bull. Soc. Bot. France 33: 109 (1886); & Pl. Delavay. 1: 40 (1889); Kom. in Acta Horti Petrop 29: 138 p.p. (1908); Stearn in J. Bot. 71: 246 (1933), J. Linn. Soc., Bot. 51: 493 (1938), Europ. Gard. Fl. 3: 393 (1989) & Kew Bull. 51: 396 (1996); T.S.Ying in Acta Phytotax. Sin. 13: 54 (1975); Hara in J. Jap. Bot. 57: 316 (1982), Lancaster, Travels in China: 110, 124 (1980). Type: China, Guizhou province, 1858, *Perny s.n.* (holotype P).

E. komarovii H.Lév. in Repert. Spec. Nov. Regni Veg. 7: 259 (1909). Type: China, Guizhou, Pin-Fa, 1908, *Cavalerie* 954 (holotype E).

ILLUSTRATION. J. Jap. Bot. 57: t.15 (1982); Lancaster, Travels in China: 110 (1989); Philipps & Rix, Early Perennials: 37 (1991).

DESCRIPTION. *Plant* in flower 25–50 cm high. *Rhizome* sometimes long-creeping, 2–5 mm thick. *Flowering stem* 25–50 cm high normally bearing two opposite trifoliolate leaves, abnormally three. *Leaves* basal and cauline, 3-foliolate; leaflets narrowly ovate to lanceolate, the tip long acuminate, the margin very spinous-serrate, the spines 1–2 mm long, the base deeply or shallowly cordate with the lobes rounded or acute, those of the lateral leaflets very unequal, thin and glabrous when young, coriaceous when mature, the underside glaucous, papillose and thickly or sparsely furnished with short appressed fairly stout bristle-like hairs, sometimes almost glabrous, 3–18 cm long, 1.5–7 cm broad. *Inflorescence* compound with the lower peduncles 2–5-flowered, loose, glabrous or rarely sparsely glandular, few or many (10–55)-flowered; pedicels 1–4 cm long. *Flowers* 3–5 cm across, yellow, white, rose-

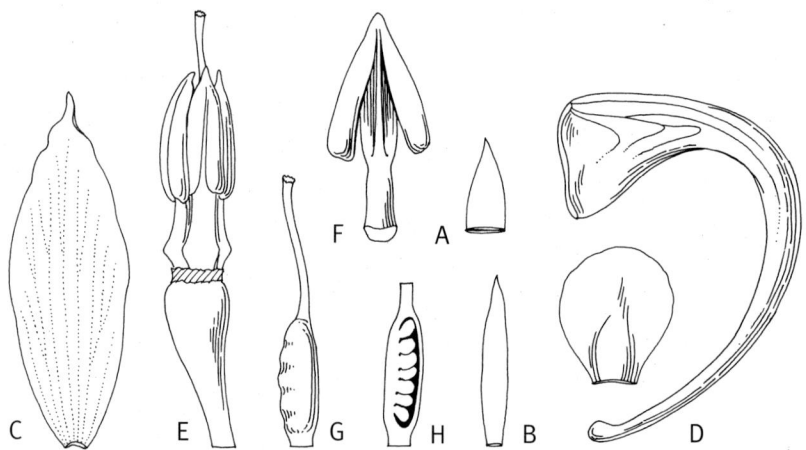

Fig. 51. **Epimedium acuminatum.** A, bracteole, × 6; B, sepal, × 6; C, petal, × 3; D, petal spur, × 3; E, flower with petals removed, × 6; F, stamen, × 6; G, gynoecium, × 6; H, gynoecium, l.s., × 6. Drawn by Christabel King.

purple, or pale violet (according to collectors). *Outer sepals* blunt, the outer pair ovate-oblong, 3 mm long, 2 mm broad, the inner pair broadly obovate, 4.5 mm long, 4 mm broad. *Inner sepals* ovate-elliptic, acute, 8–12 mm long, 3–7 mm broad. *Petals* much longer than the inner sepals, horn-shaped, tapering from the swollen but lamina-less base, curving outwards, 1.5–2.5 cm long. *Stamens* 3–4 mm long; anthers 2.5 mm long, pollen yellow. *Capsule* cylindric, comparatively long and slender, about 2 cm long (style included).

DISTRIBUTION. China, Guizhou (Kweichow), Yunnan and Sichuan (Szechwan) provinces, mountain woods, 1400–4000 m.

20. EPIMEDIUM FRANCHETII

Although closely resembling *E. acuminatum* Franch. in form of flowers and minute bristles on the underside of leaflets, *E. franchetii* differs primarily in having differently shaped leaflets, usually much larger than those of *E. acuminatum* grown under the same conditions, slightly narrower inner sepals and slightly longer stamens. Robust plants often have three whorled leaves on the flower stem.

The specific epithet *franchetii* commemorates the French botanist Adrien Franchet (1834–1900), whose acquaintance with the flora of Eastern Asia began with the identification and listing of nearly 1800 Japanese plants gathered by his friend Louis Savatier (1830–1891) when a medical officer in Japan. It resulted in Franchet and Savatier's *Enumeratio Plantarum in Japonica sponte crescentium* (2 vols., 1874–1879). This prepared Franchet for the study of the numerous Chinese specimens gathered by French missionaries, notably Armand David and Jean Marie Delavay, sent to the Muséum d'Histoire Naturelle, Paris (*cf*. Drake del Castillo 1900). Among these were new species of *Epimedium*, which led him to revise the genus as a whole in 1886.

This species with pleasing yellow flowers was introduced into cultivation by Mr Mikinori Ogisu in 1987 and at first identified and grown as *E. hunanense*, and then as a yellow variety of *E. acuminatum*.

Epimedium franchetii Stearn in Kew. Bull. 51: 396, fig.2 (1996). Type: China, Hubei province, Muayuping, Shonnongia Forest District, 1987, *Ogisu* 87001, cult. Kew 1995 (holotype K).

ILLUSTRATION. Kew Bull. 51: 397, fig.2 (1996).

DESCRIPTION. *Flowering stem* 20–60 cm long, with 2 opposite leaves or 3 whorled leaves. *Rhizome* compact, *c*. 7 mm thick. *Basal and cauline leaves* with 3 leaflets, the basal petioles 4–10 cm long; leaflets narrowly ovate, 9–14 cm long, the apex acute or acuminate, the margins spinous with spines 1.5–2.5 mm long, the base deeply cordate with a narrow sinus, the lobes of the terminal leaflet equal and obtuse or acute, those of the lateral leaflets very unequal with the inner lobe small and acute or obtuse, the outer much longer and acuminate, coriaceous, persisting all winter, glossy above, the underside somewhat glaucous,

20. EPIMEDIUM FRANCHETII

occasionally reddish tinged, with extremely minute appressed hairs. *Inflorescence* simple, many-flowered (with 14–25 flowers), 15–30 cm long; pedicels glandular-hairy, 1–3 cm long, the space between them 5–10 mm long. *Flowers* large, *c.* 4.5 cm across. *Outer sepals* soon falling, to 5 mm long, green. *Inner sepals* narrowly ovate, very pale yellow, acuminate, *c.* 10 mm long, 4–5 mm broad. *Petals* much longer than inner sepals, pale sulphur yellow; spur without lamina, elongate, subulate, much curved, *c.* 2 cm long. *Stamens* exposed, *c.* 4.5 mm long; filaments 2 mm long, pale yellow; anthers and pollen pale yellow.

DISTRIBUTION. China, Hubei and Guizhou provinces.

SPECIMENS SEEN. Hubei province, *Ogisu* 87001 (K); Guizhou province, Guiyang, (26°33'N, 106°45'E), 1250 m, *Ogisu* 87002 (K).

Fig. 52. ***Epimedium franchetii*** (*M.Ogisu* 87001). Photographed by John Fielding at Blackthorn Nursery, Hampshire.

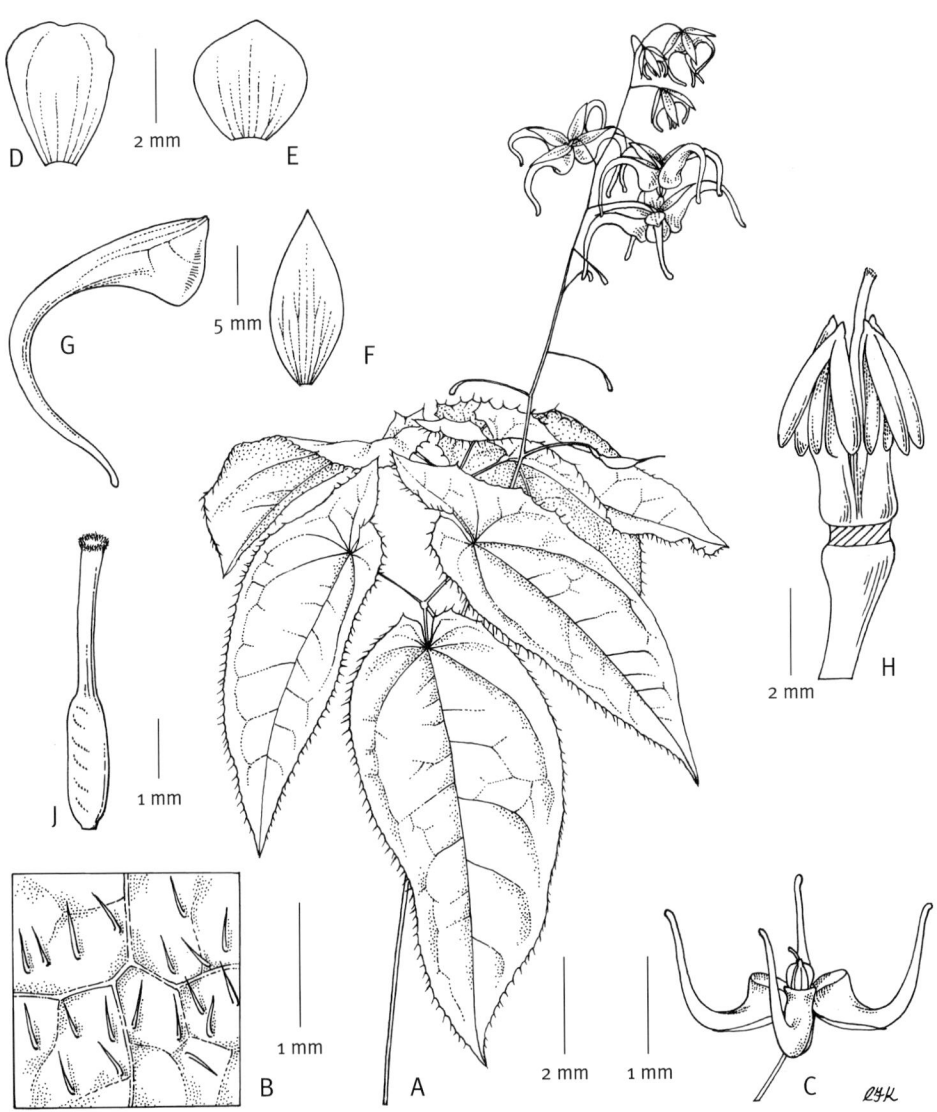

Fig. 53. *Epimedium franchetii*. A, inflorescence; B, detail of underside of leaf showing appressed hairs; C, flower; D & E, outer sepals; F, inner sepal; G, petal; H, stamen and gynoecium; J, gynoecium. Drawn by Christabel King.

21. EPIMEDIUM ENSHIENSE

Epimedium enshiense is a member of series *Dolichocerae* with broadly ovate leaflets almost glabrous beneath. Enshi, also spelled En-shih is in the southwest corner of Hubei (Hupeh) province bordered to the south by Hunan, and to the west and north by Sichuan. The description below is based on the protologue, no material being available.

Epimedium enshiense B.L.Guo & P.G.Xiao in Acta Phytotax. Sin. 31: 194, fig. (1993); Stearn in Kew Bull. 51: 396 (1996). Type: China, prov. Hubei at Enshi, 100 m, 1989, *Guo & Luo* 89013 (IMD).

ILLUSTRATION. Acta Phytotax. Sin. 31: 195 (1993).
DESCRIPTION. *Flowering stem* 25–70 cm long, bearing two opposite leaves. *Rhizome* creeping. *Basal and cauline leaves* with 3 leaflets; leaflets elliptic or broadly ovate, the basal 5–27 cm long, the cauline 3.2–7 cm long, 2.5–6 cm broad, apex abruptly acuminate, margin spinulose, base cordate, underside glaucous, almost glabrous but pilose near the base. *Inflorescence* many-flowered (with 11–20 flowers), glandular, 4–32 cm long. *Flowers* 1.3–2.5 cm across. *Outer sepals* obovate, 2.3 mm long, 1.5 mm broad. *Inner sepals* ovate, yellowish, 6–7 mm long, 3.2–3.7 mm broad. *Petals* longer than inner sepals, yellowish, without lamina; spur elongated 7–12 mm long. *Stamens* 3 mm long; anthers 2 mm long. *Capsule* 1–1.4 cm long, with 10 seeds.
DISTRIBUTION. China, Hubei province.

22. EPIMEDIUM SUTCHUENENSE

Epimedium sutchuenense takes its specific epithet from Sutchuen, a French rendering of the name of the province now known in Pinying as Sichuan. The French missionary Paul Farges collected it in 1893 at 1600 m near Tchenkéoutin (Chengkou) in the north-west. It belongs to series *Dolichocerae* and is notable for its elongated rhizomes and the scattered hairs on the underside of the leaflets. It was introduced into cultivation by Mr Darrell Probst.

Epimedium sutchuenense Franch. in J. Bot. (Morot) 8: 282 (1894); Stearn in J. Linn. Soc., Bot. 51: 498, t.28 (1938) & Kew Bull. 51: 399 (1996). Type: China, Sichuan province 'dans les bois de Kéoupin, près de Tchen-kéou-tin, 1893, *Farges* 1272 (holotype P).

ILLUSTRATION. J. Linn. Soc., Bot.: 51: t.28 (1938).
DESCRIPTION. *Flowering stem* 13–30 cm long, bearing two trifoliolate, opposite, equally developed leaves. *Rhizome* long-creeping, 1–3 mm thick with internodes up to 13 cm long. *Leaves basal and cauline*, 3-foliolate; leaflets ovate or narrowly ovate, the tip abruptly long-acuminate, the margin very spinous-serrate, the base deeply cordate with the lobes rounded or acute and diverging at 30°–70°, coriaceous when mature, the underside glaucous,

papillose and almost glabrous with only a few scattered grey hairs, 3–11 cm long, 2–5 cm broad. *Inflorescence* simple, glandular, few (4–8)-flowered; pedicels 1.5–2.5 cm long. *Flowers* 3–4 cm across, rose or mauve-purple. *Outer sepals* blunt, the outer pair ovate, 3 mm long, the inner pair broadly obovate, 4 mm long. *Inner sepals* narrowly lanceolate, long-acuminate, reflexing, 1.5–1.7 cm long, 3 mm broad near the base. *Petals* about as long as the inner sepals or slightly longer, horn-shaped, tapering from the swollen but lamina-less base, reflexing, 1.5–2 cm long. *Stamens* protruding, 4–5 mm long; anthers 3 mm long.
DISTRIBUTION. China, Sichuan province.

23. EPIMEDIUM CHLORANDRUM

Epimedium chlorandrum is notable among species of series *Dolichocerae* for the inner sepals being not appressed to the petals, as is usual, but slightly ascending. The anthers are green, hence the epithet *chlorandrum* from the Greek *chloros* (green) and *aner, andros* (man, in botanical Latin male, hence stamen and anther). The flowers are very pale primrose yellow. Most species of *Epimedium* have yellow anthers and pollen, green anthers and pollen being found only in *E. pubescens*, *E. fargesii*, this species and some of the garden hybrids. According to its collector Mikinori Ogisu, it grows in the same locality as *E. latisepalum*.

Fig. 54. ***Epimedium chlorandrum*** growing wild in China at Baoxing, western Sichuan, 930 m. Photograph: Mikinori Ogisu.

Fig. 55. *Epimedium chlorandrum* (*M.Ogisu* 94003). Photographed by John Fielding at Blackthorn Nursery, Hampshire.

Epimedium chlorandrum Stearn in Kew Bull. 52: 660, fig.3 (1997). Type: China, Sichuan province, Xiaoguanzi (30°18'N, 102°48'E), *Ogisu* 94003; cult. April 1996, Blackthorn Nursery, Kilmeston, Hampshire (holotype K).

ILLUSTRATION. Kew Bull. 52: 663, fig.2 (1997).

DESCRIPTION. *Flowering stem* 35 – 65 cm long, leaning, bearing two leaves. *Basal and cauline leaves* with 3 leaflets; leaflets narrowly ovate or almost lanceolate, 5 – 11 cm long, 2 – 4.5 cm broad, the tip shortly acuminate, the margin slightly undulate and spinous with spines 1.0 – 1.5 mm long, the base moderately cordate with an open sinus and the rounded lobes diverging at 60° – 80°, coriaceous, persisting green all winter, mottled with brown when young, the underside glaucous and with numerous minute appressed hairs. *Inflorescence* compound, loose, many-flowered (with 12 – 30 flowers), the lower peduncles 3-flowered, 25 – 36 cm long; pedicels glabrous, about 2.5 cm long. *Flowers* large, about 4 cm across. *Outer sepals* soon falling, green, 2 – 3 mm long. *Inner sepals* unequal, ascending, not appressed to petals, narrowly ovate, convex, greenish, the outer pair 8 mm long, 4.5 mm broad, the inner pair 10 mm long, 4.5 mm broad. *Petals* much longer than inner sepals, pale yellow; spur elongated, subulate, strongly curved with no lamina. *Stamens* protruding, 4.5 mm long; filaments white with slight pink tinge; anthers green, 3 mm long; pollen green.

DISTRIBUTION. China, Sichuan province.

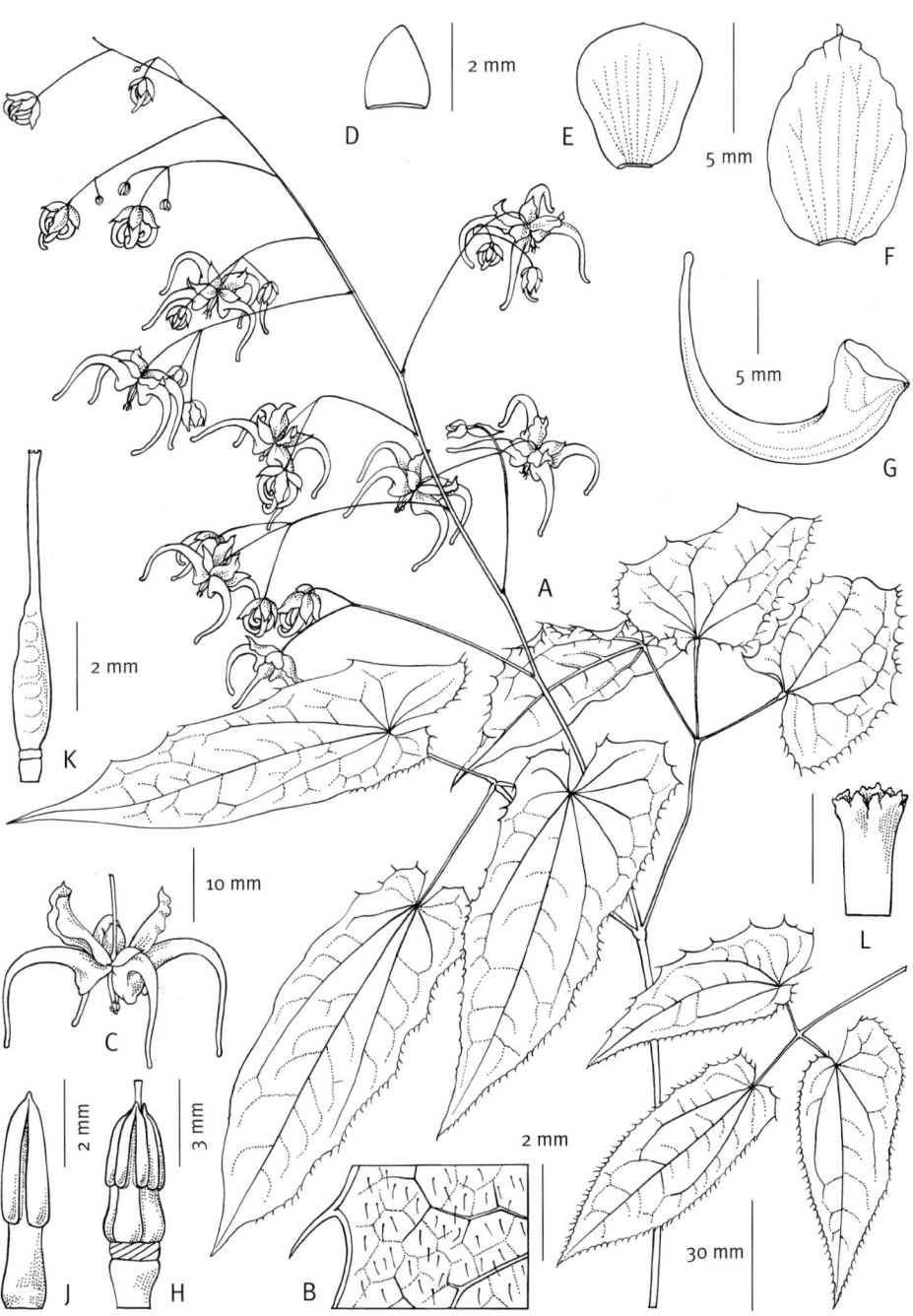

Fig. 56. ***Epimedium chlorandrum.*** A, habit; B, detail of underside of leaf; C, flower; D, outer (early caducous) sepal; E, F, inner sepal; G, petal with spur; H, stamens & ovary; J, stamen; K, gynoecium; L, apex of style. Drawn from *M.Ogisu* 94003 by Christabel King.

24. EPIMEDIUM WUSHANENSE

In his *The Explorer's Garden* (1999), David J.Hinkley has described himself as 'awe struck' on first seeing a well-grown *Epimedium wushanense* in full bloom in Robin White's Blackthorn Nursery. He was impressed by its unusual and elegant foliage with long narrow leaflets and its airy panicles that can carry up to 100 large, light pink to yellow or rusty orange flowers, rising to 4 ft. It is indeed a very distinctive species, named by the Chinese botanist Ying Tsün-shen who published a survey of the Chinese species of *Epimedium* in 1975. The elongate petals vary both in colour and poise, the spurs being slightly curved to almost straight. Mikinori Ogisu collected plants on Jinghang shan and on Changshi. The species takes its name from Wushan County, Sichuan.

Epimedium wushanense T.S.Ying in Acta Phytotax. Sin. 13 (2): 55 (1973); Stearn in Kew Bull. 45: 690, 691 (1990); Hinkley, The Explorer's Garden: 105 (1999). Type: Wushan, Sichuan province, *T.P.Wang* 10757 (holotype PE).

ILLUSTRATION. Hinkley, The Explorer's Garden: 105 (1999).
DESCRIPTION. *Flowering stem* 40–100 cm long, bearing two opposite leaves. *Rhizome* compact. *Basal and cauline leaves* with 3 leaflets; leaflets lanceolate or narrowly lanceolate,

Fig. 57. **Epimedium wushanense** growing at Wanyuan, north of Chonqing, Sichuan, 1070 m. Photograph: Mikinori Ogisu.

Fig. 58. ***Epimedium wushanense*** in the wild at Wanyuan, north of Chonqing, Sichuan, 990 m. Photograph: Mikinori Ogisu.

7–13 cm long, 2–7 cm broad, the tip long acuminate, the margin undulate and very spinous with spines 1.0–1.5 mm long, the base shallowly cordate, the lobes of the lateral leaflets very unequal, the larger lateral lobe acuminate, coriaceous and rigid, the underside glabrous or with profuse erect hairs. *Inflorescence* compound, paniculate, loose, many-flowered (with 20–100 flowers), 18–40 cm long, the rachis with sparse short erect hairs, many peduncles 3–5-flowered, glandular-hairy, almost horizontal; pedicels 1.5–2 cm, likewise glandular-hairy. *Flowers* large, 3.5–4 cm across. *Outer sepals* soon falling, *c.* 3.5–4 mm long. *Inner sepals* slightly ascending, lanceolate, acute to obtuse, white, *c.* 6 mm long, 2.5 mm broad. *Petals* longer than inner sepals, pale sulphur yellow or pink-tinged or orange, spur elongated, tapering, curved or almost straight *c.* 1.5–2.5 cm long, the mouth 6 mm high, with no lamina. *Stamens* protruding, 3.5–4 mm long, pale yellow; filament 0.5 mm long; pollen pale yellow.

DISTRIBUTION. China, Sichuan (Szechwan) province in the north-east near the border with Hubei province, in Wushan County, the principal city of which is Wushan (31°N, 109°8'E).

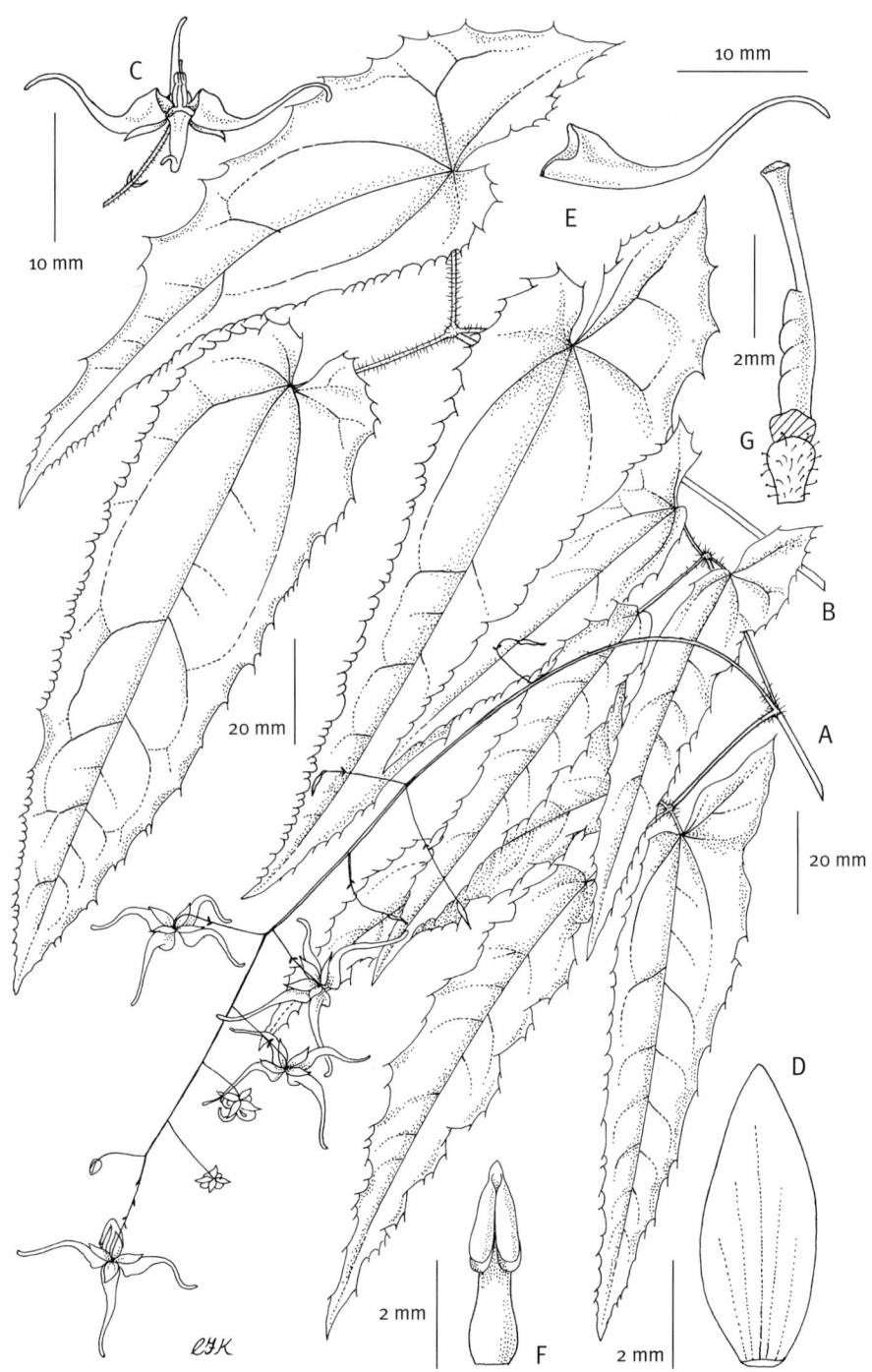

Fig. 59. **Epimedium wushanense.** A, flowering stem; B, mature leaf; C, flower; D, sepal; E, petal; F, stamen; G, gynoecium. Drawn from *M.Ogisu* 93019 by Christabel King.

25. EPIMEDIUM LEPTORRHIZUM

A French missionary, Emile Bodinier, collected specimens of *Epimedium leptorrhizum* in the Chinese province of Guizhou (Kweichow, Kouy-Chéou) in 1898 but Hector Léveillé misidentified his material and it was not until 1933 that it first received a botanical description and the epithet *leptorrhizum*. This is from the Greek *lĕptos*, 'slender, thin, weak' and *rhiza* 'root, rhizome', referring to its elongated, thin rhizome. No-one had any concept of the beauty of its exquisite pink, drooping flowers until Mikinori Ogisu introduced it from Sichuan (nos. 81012 and 93009). It requires light shade and moist, but well-drained, humus-rich soil where its rhizome never becomes parched.

Epimedium leptorrhizum Stearn in J. Bot. 71: 343 (1933); J. Linn. Soc., Bot. 51: 499, t.28 (1938). Type: China, Guizhou (Kweichow) province, 'environs de Kouy-yang'), *Bodinier* 2184 (holotype P).

ILLUSTRATION. J. Linn. Soc., Bot. 51: t.28 (1938).

DESCRIPTION. *Plant in flower* 12–30 cm high. *Rhizome* long-creeping, 1–2 mm thick, with internodes sometimes as much as 20 cm long. *Leaves basal and cauline*, 3-foliolate or abnormally 1-foliolate, the petioles furnished with spreading reddish hairs; leaflets

Fig. 60. ***Epimedium leptorrhizum*** growing at Shizhu, east of Chongqing, Sichuan, 1420 m. Photograph by Mikinori Ogisu.

Epimedium leptorrhizum

PAULINE DEAN

narrowly ovate, the tip long-acuminate, the margin very spinous with spines 1.0–1.5 mm long, the base deeply cordate with the sinus narrow and the usually rounded lobes almost touching, those of the lateral leaflets very unequal, coriaceous when mature, evergreen, the veins above much impressed with the underside glaucous, papillose, and pubescent with scattered spreading or curled reddish hairs which are densest at the insertion of the petiolules and along the primary veins, 3–11 cm long, 2–6 cm broad. *Flowering stem* usually bearing one trifoliolate leaf with occasionally a rudimentary leaf densely clothed in reddish hairs arising opposite; inflorescence simple, glandular or glabrous, few-(4–8)flowered; pedicels 1–2.5 cm long. *Flowers* 4 cm across, white tinged with rose or deep rose. *Outer sepals* blunt, ovate-oblong, 3–4 mm long. *Inner sepals* narrowly elliptic or lanceolate, acuminate, 1.1–2 cm long, 4–5 mm broad. *Petals* subequal to or longer than the inner sepals, horn-shaped, strongly curved, tapering from the swollen but lamina-less base, up to 2 cm long. *Stamens* 4 mm long; anthers 3 mm long, yellow.

DISTRIBUTION. China, Guizhou (Kweichow) and Sichuan provinces.

26. EPIMEDIUM BRACHYRRHIZUM

Experience with related taxa of *Epimedium* differing in rhizome has shown that associated with this difference are other characters, sometimes very evident, sometimes more subtle, indicative of their specific distinctness. Between *E. leptorrhizum* and *E. brachyrrhizum* the major difference is in the elongated, very slender rhizome of *E. leptorrhizum* and the more compact clump-forming rhizome of *E. brachyrrhizum*. Associated with that are minor differences: *E. leptorrhizum* with about 10–13 spines to each 5 cm of leaflet margin, narrowly elliptic inner sepals, stamen filaments 0.5 mm broad, *E. brachyrrhizum* with 17–20 spines to each 5 cm of margin of comparable leaflet, lanceolate inner sepals, stamen filaments 0.3 mm broad.

The epithet *brachyrrhizus* is from the Greek *brachys* (short) and *rhiza* (root, rhizome), to contrast with the epithet *leptorrhizus* (The initial *r* of a Greek word, transliterated as *rh*, is doubled in a compound when preceded by a vowel, i.e. has an additional *r* inserted, as in *Glycorrhiza*).

Epimedium brachyrrhizum Stearn in Kew Bull. 52: 659, fig.1 (1997). Type: China, Guizhou province, Fanjingshan Mts, *Beijing Botanical Garden Expedition*, cult. 1995 at Hubbardston, Massachusetts, by *Darrell Probst* CPC 94.0495 (holotype K).

ILLUSTRATION. Kew Bull. 52: 661, fig.1 (1997).

DESCRIPTION. *Flower stem c.* 23 cm long, bearing one leaf. *Rhizome* compact, clump-forming. *Leaves basal and cauline* with 3 leaflets; leaflets narrowly ovate, *c.* 6–10 × 3.5–5 cm, the apex acuminate, the margin with numerous spines 1–1.5 mm long, the base deeply cordate with an open sinus and the basal lobes acute or almost rounded and

26. EPIMEDIUM BRACHYRRHIZUM

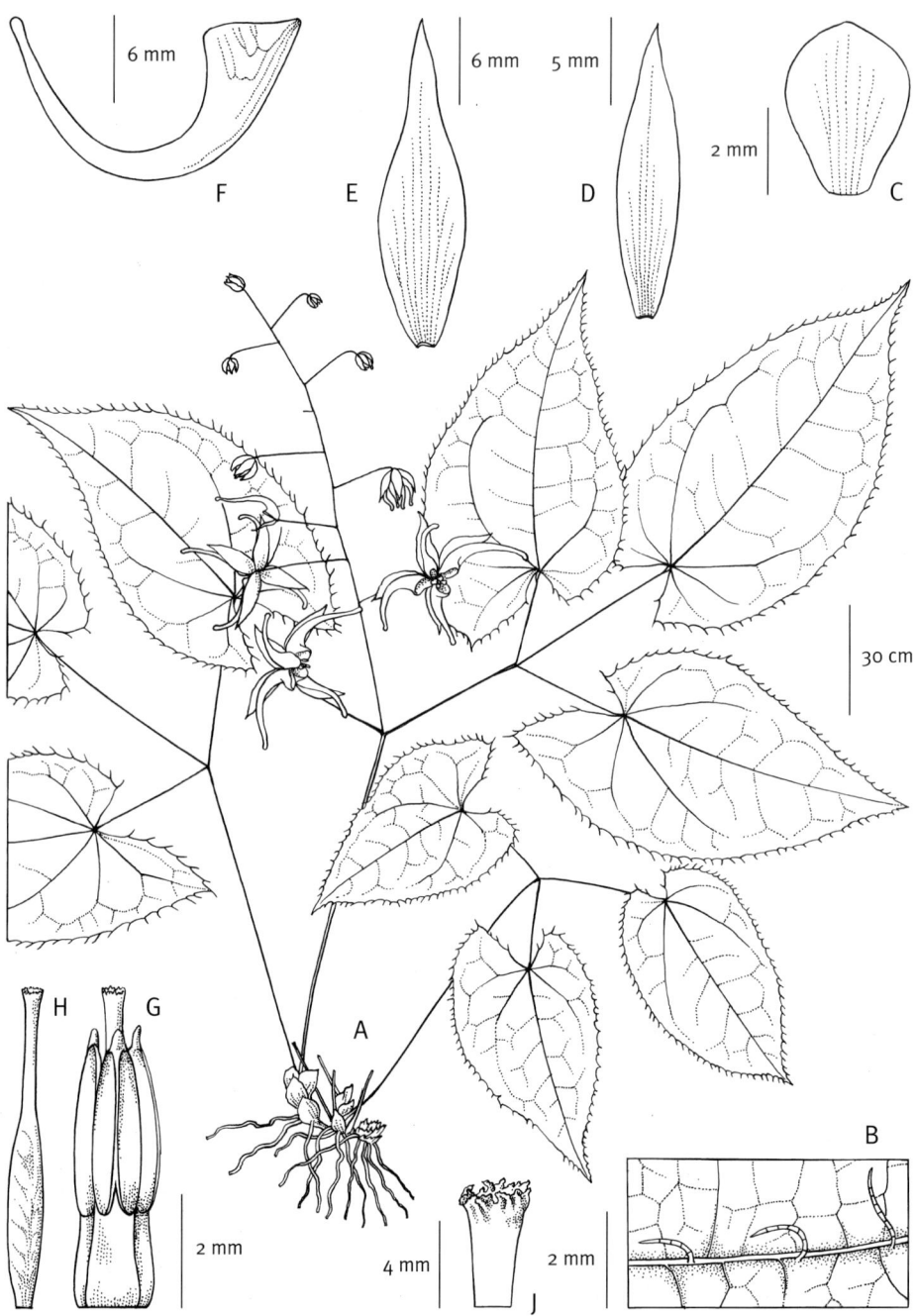

Fig. 61. ***Epimedium brachyrrhizum.*** A, habit; B, detail of underside of leaf; C, outer sepal; D & E, inner sepals; F, petal; G, stamens & gynoecium; H, gynoecium; J, apex of style. Drawn from *D.Probst* CPC 94.0495 by Christabel King.

26. EPIMEDIUM BRACHYRRHIZUM

Fig. 62. *Epimedium brachyrrhizum* (*D.Probst* CPC 94.0412). Photographed by John Fielding at Blackthorn Nursery, Hampshire.

diverging at an angle of 45°–60°, the underside glabrous except for the hairy thickened insertion of the secondary petiole, evergreen. *Inflorescence* simple, loose, few-flowered (with 6–12 flowers) *c*. 14 cm long; pedicels sparsely glandular-hairy, mostly 2 cm long. *Flowers* large, *c*. 4 cm across. *Outer sepals* 3.5–4 mm long, obtuse. *Inner sepals* lanceolate, pale rose, 20–22 × 6 mm, slightly wider than base of petal. *Petals* a little longer than the inner sepals; spur elongate, *c*. 22–26 mm long, strongly curved, the base rose, the rest white, with almost no lamina. *Stamens* protruding, *c*. 4 mm long; filaments whitish, *c*. 1 mm long, very slender; anthers yellow.

DISTRIBUTION. China, Guizhou province, in Fanjingshan mountains, 600–1200 m.

27. EPIMEDIUM SIMPLICIFOLIUM

Most species of *Epimedium* have leaves with nine, five or three leaflets, so T.S.Ying's description and illustration in 1975 of a species, appropriately named *E. simplicifolium*, with leaves each consisting of only one leaflet, revealed unanticipated extreme reduction from a compound leaf hitherto unknown in the genus. Since then three other Chinese species with leaves having a single leaflet have been found: *E. baojinense*, *E. glandulosopilosum* and *E. zhushanense*. Parallel variation exists within *Corydalis*, a genus with leaves more elaborately compound than in *Epimedium*, one species of which, *C. ludlowii*, has the leaf reduced to a single leaflet.

Epimedium simplicifolium T.S.Ying in Acta Phytotax. Sin. 13, 2: 51, fig.1 (1975); Stearn in Kew Bull. 45: 690 (1990). Type: Guizhou, (Kweichow) province, Wachuan, *P.X.Tsoung* 606 (holotype PE).

ILLUSTRATION. Acta Phytotax. Sin. 13, 2: 51, fig.1 (1975).
DESCRIPTION. *Flowering stem c.* 50 cm high, bearing 2 opposite leaves. *Basal and cauline leaves* all with only one leaflet; leaflets ovate to broadly elliptic-ovate, 17–19 cm long, 8–10 cm wide, the base cordate with lobes about equal, the underside white-sericeous. *Inflorescence* densely flowered; peduncles glabrous; pedicels sparsely pubescent. *Flower* large, yellow. *Outer sepals* obtuse, 4 mm long, 2 mm broad. *Inner sepals* ovate, acute, *c.* 6 mm long, 4 mm broad. *Petals* with spur longer than inner sepals, spur slender. *Stamens* 3 mm long, pale yellow.
DISTRIBUTION. China, Guizhou (Kweichow) province.

28. EPIMEDIUM ZHUSHANENSE

Epimedium zhushanense, like *E. simplicifolium*, is a Chinese species in which the usual three leaflets to a leaf have been reduced to a single large leaflet. It is stated to differ from *E. simplicifolium* by its simple inflorescence with glandular hairs, petaloid inner sepals, purple petals and leaflet shape. While most anthers dehisce in the normal way by two up-rolling valves, the authors have observed longitudinal and lateral dehiscing.

Epimedium zhushanense K.F.Wu & S.X.Qian in Acta Phytotax. Sin. 23: 71, fig.1 (1985). Type: China, Hubei (Hupeh) province, Zhushan, 1220 m, 15 June 1979, *S.X.Qian* 2021 (holotype East China Normal University, Shanghai).

ILLUSTRATION. Acta Phytotax. Sin. 23: 71, fig.1 (1983).
DESCRIPTION. *Flowering stem* with two opposite leaves. *Leaf* consisting of one short-petioled leaflet; leaflets almost orbicular or very broadly ovate 10–14 cm long, 7–11 cm broad, the tip obtuse, the margin sparsely spinose, the base deeply cordate with the two lobes almost equal, rounded and sometimes overlapping, the underside densely tomentose; petioles 5.5–6 cm long. *Inflorescence* simple, loosely many-flowered, to 30 cm long; peduncle and

pedicels with glandular hairs; pedicels 1.5–5 mm long. *Outer sepals* ovate, 2 mm long. *Inner sepals* ovate-lanceolate, 9–11 mm long, 3.5 mm broad, obtuse. *Petals* saccate, purse-like, purple, without basal hairs. *Stamens* purple, 17 mm long, with very short filaments.
DISTRIBUTION. China, Hubei (Hupeh) province, Zhushan at 1220 m.

29. EPIMEDIUM BAOJINGENSE

The leaf in most Chinese species of *Epimedium* consists of three stalked leaflets at the end of a common petiole; the terminal leaflet is symmetrical with equal lobes at the base but the two lateral leaflets are asymmetrical with the two basal lobes markedly unequal. Species of *Epimedium* that normally have leaves consisting of three leaflets may occasionally produce a basal leaf with only one leaflet, but *E. baojingense* in Hunan, together with *E. simplicifolium* in Guizhou, and *E. glandulosopilosum* and *E. zhushanense* in Hubei, are exceptional in having stem leaves with only one leaflet. It is known only from the type specimen collected without flowers at Baojing, Hunan province, eastern China. According to the authors, it differs from *E. simplicifolium* in its simple inflorescence with dark yellow (*atroflavus*) hair covering, as also the underside of the leaflets.

Epimedium baojingense Q.L.Chen & B.M.Yang in Acta Phytotax. Sin. 20: 482 (1982). Type: China, Hunan province, Baojing, Lüdong Shan, April 1956, *B.M.Yang 34* (holotype Hunan Teachers' College)

ILLUSTRATION. Acta Phytotax. Sin. 20: 483, fig.1 (1982).
DESCRIPTION. *Flower stem* 30–50 long, bearing two opposite unifoliolate leaves. *Rhizome* compact, 5 mm thick. *Basal and cauline leaves* each with a single leaflet; leaflets ovate or ovate-elliptic, 7–13 cm long, 4.5–8 cm broad, the apex acute or shortly acuminate, the margin spinous, the base deeply cordate with rounded lobes separated by a narrow sinus, the underside sparsely puberulous with dark yellow hairs. *Inflorescence* simple, few- (5–9) flowered, 9–18 cm long, dark yellow-puberulous; pedicels 0.5–2 cm long, dark yellow-puberulous. *Flowers* unknown.
DISTRIBUTION. China, Hunan province.

30. EPIMEDIUM GLANDULOSOPILOSUM

As its specific epithet partly implies, *Epimedium glandulosopilosum* has abundant long multicellular glandular hairs on the inflorescence and the stem below, as indeed have some other species of *Epimedium*. However, it is one of the few Chinese species in which the usual three leaflets have been reduced to a single leaflet, as in *E. simplicifolium*. From the latter it is distinguished by the hairs on the underside of leaves being golden-yellow, not silky white.

The following description is based on the protologue by Hai-Rui Liang.

Epimedium glandulosopilosum H.R.Liang in Acta Phytotax. Sin. 28: 323 (1990). Type: China, Sichuan, Wushan, March 1987, *H.R.Liang* 144 (holotype Beijing College of Traditional Chinese Medicine).

ILLUSTRATION. Acta Phytotax. Sin. 28: 324, fig.3 (1990).

DESCRIPTION. *Flower stem* 20–50 cm long, covered with multicellular glandular golden hairs, bearing two opposite unifoliolate leaves. *Rhizome* compact. *Basal and cauline leaves* simple; leaflets ovate or lanceolate, 5–8.5 cm long, 2.3–5 cm broad, the apex acuminate, the margin spinous, the base deeply cordate with a wide sinus and the lobes equal and acute, the underside densely covered with very long golden yellow hairs. *Inflorescence* simple, to 23 cm long, with 8–24 flowers, glandular-hairy; pedicels 1–3 cm long. *Flowers* large, about 2.5 cm across. *Inner sepals* ovate, acute, 8–9 mm long, 4–6 mm broad. *Petals* longer than inner sepals, yellow, *c.* 13 mm long, elongated into a slender spur without lamina. *Stamens* protruding, *c.* 5 mm long; anther *c.* 4 mm long.

DISTRIBUTION. China, Sichuan (Szechwan) province.

Series D. **BRACHYCERAE**

(Species number 31–43)

Series *Brachycerae* Stearn in J. Linn. Soc., Bot. 51: 500 (1938). Type species: *E. sagittatum* (Sieb. & Zucc.) Maxim.

Petals very short, much shorter than the inner sepals, usually only 1–4 mm long.

31. EPIMEDIUM PUBESCENS

Although *Epimedium sagittatum* was the first Chinese species of the genus to be named by European authors, it was described from plants cultivated in Japan whereas *E. pubescens*, its close ally, was the first to be described from native Chinese material. It had been collected by a Russian military surgeon, David J. Piaretski and concisely described by Maximowicz in 1877. The usually abundant long, slender, mostly spreading hairs on the underside of the leaflets distinguish this species from *E. sagittatum* in which the leaflet underside is either glabrous or beset with appressed short bristle-like hairs. In addition, the relative proportions of the flower parts differ since its longer and more acuminate inner sepals distinctly exceed the outer sepals, petals and stamens. It is distinct from *E. fargesii* in its shorter stamens, etc., and from *E. brevicornu* in its trifoliolate leaves, etc.

Epimedium pubescens has been divided into two subspecies: subsp. *pubescens*, with the inflorescence 5–6 cm across towards the base and the leaflets often densely hairy beneath, and subsp. *cavaleriei*, with the inflorescence larger, up to 12 cm across towards the base, and the leaflets more sparsely hairy (nearly glabrous) beneath. The last-named commemorates the French missionary Pierre Julien Cavalerie (b. 1869).

Fig. 63. ***Epimedium pubescens.*** Growing at Dujiangyan, western Sichuan, 850 m. Photograph: Mikinori Ogisu.

Epimedium pubescens Maxim. in Bull. Acad. Imp. Sci. Saint-Pétersbourg 23: 309 (1877), 29: 222, t.1 (1883); reimpr. in Mélanges Biol. Bull. Phys.-Math. Acad. Imp. Sci. Saint-Pétersbourg 9: 712 & 11: 868 (1883); Franch. in Bull. Soc. Bot. France 33: 111 (1886); Maxim. in Acta Horti Petrop. 11: 43 (1890); Kom. in Acta Horti Petrop. 29: 135 p.p. (1908); Stearn in J. Linn. Soc., Bot. 51: 501 (1938). Type: China, Shaanxi (Shensi), near the river Han Kiang, 1875, *Piaretski* (holotype LE).

subsp. **pubescens**
E. pubescens subsp. *primarium* Stearn in J. Linn. Soc., Bot. 51: 503 (1938).

ILLUSTRATION. Bull. Acad. Imp. Sci. Saint-Pétersbourg 29: t.1 (1883).
DESCRIPTION. *Plant in flower* 20–60 cm high. *Rhizome* sometimes elongated, 3–4 mm thick. *Leaves* basal and cauline, 3-foliolate (or the basal leaves sometimes 1-foliolate); leaflets ovate, narrowly ovate or lanceolate, the tip acuminate (abnormally rounded), the margin very spinous-serrate, the base deeply or shallowly cordate with usually rounded lobes, those of the lateral leaflets very unequal, coriaceous when mature, the underside persistently pubescent to tomentose with numerous, fine, multicellular, spreading or curled grey hairs densest along the three primary veins and at the insertion of the petioles, 3–15 cm long, 2–8 cm broad. *Flowering stem* bearing two (abnormally three), opposite, trifoliolate leaves. *Inflorescence* compound, loose, usually glandular, many-(to 30)flowered, 10–20 cm long, 5–6 cm across towards base, with the lower peduncles 3–5-flowered: pedicels 1–2 cm long. *Flowers* 1 cm across. *Outer sepals* broadly ovate, purplish, 2–3 mm long. *Inner sepals* lanceolate or narrowly lanceolate, acute or acuminate, white, several-nerved, 5–7 mm long, 1.5–3.5 mm broad. *Petals* minute, much shorter than the inner sepals, saccate, blunt, brownish, 2 mm long, with no basal laminae. *Stamens* protruding, 4 mm long: anthers 2 mm long.
DISTRIBUTION. China, Anhui (Anhwei), Shaanxi (Shensi) and Sichuan (Szechuan) provinces.

subsp. **cavaleriei** (Stearn) Stearn in J. Linn. Soc., Bot. 51: 504, t.3 (1938). Type: Guizhou (Kweichow), Anshun (Ganchouen), *Cavalerie* 1849 (holotype K).
E. pubescens var. *cavaleriei* Stearn in J. Bot. 71: 345 (1933).

ILLUSTRATION. J. Linn. Soc., Bot. 51: t.30 (1938).
DESCRIPTION. *Plant in flower* 40–60 cm high. *Rhizome* apparently short, 5 mm thick. *Leaves* 3-foliolate; leaflets narrowly ovate or lanceolate, the tip long-acuminate, the base shallowly cordate with usually rounded lobes, those of the lateral leaflets very oblique, coriaceous when mature, finely reticulated with veins above and below, the lower surface sparsely pubescent with scattered slender hairs or subglabrous, 5–13 cm long, 2–4.4 cm long (possibly up to 25 cm long, 6 cm broad). *Flowering stem* bearing two (abnormally three) trifoliolate, opposite, leaves. *Inflorescence* compound, panicled and diffuse, glabrous, many-(about 60-) flowered, about 10–30 cm long, towards the base up to 12 cm across, with the spreading lower peduncles often 5–7-flowered; pedicels 1–2.5 cm long. *Flowers* 1 cm across,

Fig. 64. *Epimedium pubescens.* Photographed by John Fielding at Blackthorn Nursery, Hampshire.

with almost black outer sepals, white inner sepals 5 mm long, minute petals 2 mm long. *Stamens* 3–4 mm long, more or less as in *E. pubescens* subsp. *pubescens*. *Capsule* 1 cm long.

DISTRIBUTION. China, Guizhou (Kweichow) province.

NOTE. In addition to the type specimen the Kew Herbarium contains two specimens, without flowers, collected by Cavalerie at Ganchouen; these apparently belong to subsp. *cavaleriei* and represent its mature leaf-state. On one the leaflets (probably basal) are about 25 cm long, 6 cm broad, with the common petiole about 10 cm or more long, the petiolules 5 to 8 cm long. On the other, the leaflets (certainly cauline) are 16 to 19 cm long, 3 to 4 cm broad. These huge leaflets are sparsely pubescent and distinctly papillose beneath.

32. EPIMEDIUM BREVICORNU

Epimedium brevicornu compensates for the smallness of its starry white flowers by producing them in profusion. It is indeed a charming plant when in full bloom, the loose compound inflorescences laden with up to 50 flowers rising above small elegant leaves. Reginald Farrer, who found it in the hills of southern Gansu, described it as a 'beautiful white *Epimedium* with flowers like snowy butterflies aflutter loosely up the spike' (*Eaves of the World* 1: 98, 1917). It differs from the other small-flowered Chinese species in having usually biternate leaves, with nine leaflets, but sometimes also on the same plant leaves with three or five broadly ovate leaflets, instead of leaves consistently with only three leaflets. In its distribution, *E. brevicornu* is the most northern species extending across the provinces of Gansu (Kansu), Shaanxi (Shensi), Shanxi (Shansi) and into north-western Sichuan (Szechwan), among bushes at 800–2100 m. In cultivation, into which it was introduced from Shanxi and north-western Sichuan by Mikinori Ogisu, it has proved a very hardy species.

The epithet *brevicornu* is from *brevis* (short) and *cornu* (horn), with reference to the short petals, and is a noun in apposition; an adjective of the same meaning would be *brevicornualis* or *brevicornuatus* not *brevicornus*.

Epimedium brevicornu Maxim. in Acta Horti Petrop. 11: 42 (1890); Kom. in Acta Horti Petrop. 29: 137 (1908); Stearn in J. Linn. Soc., Bot. 51: 500 (1938); Ying in Acta Phytotax. Sin. 13: 53 (1975); Stearn in Europ. Gard. Fl. 3: 393 (1989). Type: China, Gansu province, 'circa monasterium Dshoni' *Potanin* (holotype LE).

E. *rotundatum* Hao in Repert. Spec. Nov. Regni Veg. 36: 233 (1934); reimpr. in Contr. Bot. Nat. Acad. Peiping 1: 3 (1935). Type: China, Gansu province, near Wu'-tu Hsien, *K.S.Hao* 468 (holotype PE).

ILLUSTRATION. J. Linn. Soc., Bot. 51: t.29 (1938).

DESCRIPTION. *Plant in flower* 20–60 cm high. *Rhizome* short, clumped, 3 mm thick. *Leaves* basal and cauline, usually biternate (9 leaflets), rarely 5- or 3-foliolate; leaflets ovate to broadly ovate, the tip acute or short-acuminate, the margin spinous-serrate, the base

Fig. 65. ***Epimedium brevicornu.*** In the wild at Jiuzhaigou, northern Sichuan, China, 1410 m. Photograph: Mikinori Ogisu.

deeply cordate with the acute lobes of the lateral leaflets only slightly unequal, small and thin at flowering time, firm and parchment-like when mature with the underside almost glabrous, ± 2 cm long, 1.5 cm broad at flowering time but later up to 8 cm long, 6.5 cm broad. *Flowering stem* bearing two opposite, usually biternate, leaves. *Inflorescence* compound, loose, glandular, many-(20–50)flowered; pedicels 5–20 mm long. *Flowers* 1.5 cm across. *Outer sepals* narrowly ovate, dark, 1–3 mm long. *Inner sepals* lanceolate, acute, white or yellowish, 10 mm long, 4 mm broad. *Petals* much shorter than the inner sepals, orange with a very slight laminae and narrow, conical, blunt, white spurs 2–3 mm long. *Stamens* exserted, 3–4 mm long; anthers 2 mm long. *Capsule* 1 cm long.

DISTRIBUTION. China, Gansu (Kansu), Shaanxi (Shensi), Shanxi (Shansi) and north-western Sichuan (Szechwan) provinces.

Fig. 66. **Epimedium brevicornu** (*M.Ogisu* 88010). Photographed by John Fielding at Blackthorn Nursery, Hampshire.

33. EPIMEDIUM RETICULATUM

Although described as akin to *Epimedium davidii* by the author C.Y.Wu, *E. reticulatum* has very little in common with that species, its inflorescence being paniculate and its flowers minute and lacking long spurs. The very spinose leaflets have a network of very fine secondary veins, hence the epithet *reticulatus*, 'made like a net, net-like' from *reticulum*, 'a little net, a fishing net'.

Epimedium reticulatum C.Y.Yu in Acta Phytotax. Sin. 25: 156 (1987). Type: China, Sichuan province, Jinyin, 1100 m, June 1959, *Exped. Sichuan Economic Pl.* 3357 (holotype KUN).

ILLUSTRATION. Acta Phytotax. Sin. 25: 157, fig.6 (1987).
DESCRIPTION. *Flower stem* 30–40 cm long, bearing 2 opposite leaves. *Rhizome* compact. *Leaves basal and cauline* with 3 leaflets; leaflets ovate, 5–7 cm long, 3.5–5 cm broad, the apex acute or acuminate, the margin spinous with spines 2.5–3 mm long, the base deeply cordate with a narrow sinus and the somewhat rounded lobes not markedly unequal, coriaceous, the underside pilose, the fine reticulate veining evident on both sides. *Inflorescence* compound, loose, paniculate, many-flowered (at least 30 flowers), 30–35 cm long; pedicels to 1.5 cm long, glandular-hairy. *Flowers* small, yellow. *Outer sepals* c. 5 mm long. *Inner sepals* elliptic, obtuse, c. 6 mm long. *Petals* saccate, 1–2 mm long. *Stamens* 3 mm long, with very short filaments.
DISTRIBUTION. China, Sichuan (Szechwan) province, at about 1100 m.

34. EPIMEDIUM SAGITTATUM

Mentioned in the ancient Chinese classical pharmacopeia *Shen Nung Pên Tshao Ching* (Han Dynasty, 206 BC to 219 AD) and in later Chinese herbals, notably the renowned *Tshao Kang Mu* of Li Shih-Chen (1518–1593) under the name 'yin yang huo', *Epimedium sagittatum* has been known much longer than any other member of the genus. The real and reputed medicinal properties of its leaves, certainly not its inconspicuous flowers, led to its introduction long ago from China into Japan. The name 'yin yang huo' has been variously explained as 'herb causing goats to copulate excessively' or as referring to a mythical beast, the yinyang, copulating many times daily. Experiment has shown that it increases the seminal secretion of dogs. Works on Chinese traditional medicine agree on its aphrodisiacal effect and its use against impotence by promoting the secretion of sperm, together with anti-rheumatic properties. *Epimedium sagittatum* first came to European attention in 1845 when Siebold and Zuccarini described it as a Japanese species and named it *Aceranthus sagittatus*.

As a garden plant *E. sagittatum* has little to recommend it, the flowers being not more than 8 mm across, with white inner sepals about 4 mm long and the yellow petals no bigger. The leaves, however, are distinctive. The underside of leaflets may be glabrous, as described by Siebold and Zuccarini, or develop, as it matures, very short appressed sparse

Fig. 67. **Epimedium sagittatum.** In the wild at Xiuning, Anhui Province, China, 980 m. Photograph: Mikinori Ogisu.

Fig. 68. *Epimedium* leaves, probably including those of **E. sagittatum**, being prepared for use in Chinese traditional medicine in a market in Mao Lan, Guizhou Province, China. Photograph: Phillip Cribb.

bristle-like hairs, as described by Regel, with an almost invisible base of minute condensed cells and a stout apical cell. The only other species with like hairs are *E. acuminatum* and *E. franchetii*. The leaflets vary greatly in size, being sometimes up to 19 cm long, and with a profusion of hairs. It is a widespread species but nevertheless rather scarce in nature, as noted by Handel-Mazzetti.

It is surprising that Siebold and Zuccarini placed this species in the genus *Aceranthus*, as it bears no resemblance to the type species *A. diphyllus* (*Epimedium diphyllum*) in leaf and flower. The name *Aceranthus macrophyllus* was based on a specimen with leaflets '7 poll. longa et supra basin 3 poll. lata et latiora', i.e. 18 cm long and 8 cm or more broad above the base. The type sheet of *Aceranthus triphyllus* has leaves of *E. grandiflorum* and an inflorescence of *E. sagittatum* mounted together.

The epithet *sagittatus* (from *sagitta*, arrow), which in classical Latin means 'wounded by an arrow', means in botanical Latin 'shaped like an arrow-head'. It refers in this species to the pointed terminal leaflet with equal basal lobes, which may be pointed but are more frequently rounded.

Epimedium sagittatum (Siebold & Zucc.) Maxim. in Bull. Acad. Imp. Sci. Saint-Pétersbourg 23: 310 (1877); Baker in Gard. Chron. 13: 683 (1880); Handel-Mazz., Symb. Sinicae 7(2): 324 (1931); Stearn in J. Linn. Soc., Bot. 51: 505 (1938). Type: Japan, cult. St. Petersburg (holotype LE).

Fig. 69. The habitat of **Epimedium sagittatum**, in forest at Fanjingshan, Guangxi Province, China. Photograph: Phillip Cribb.

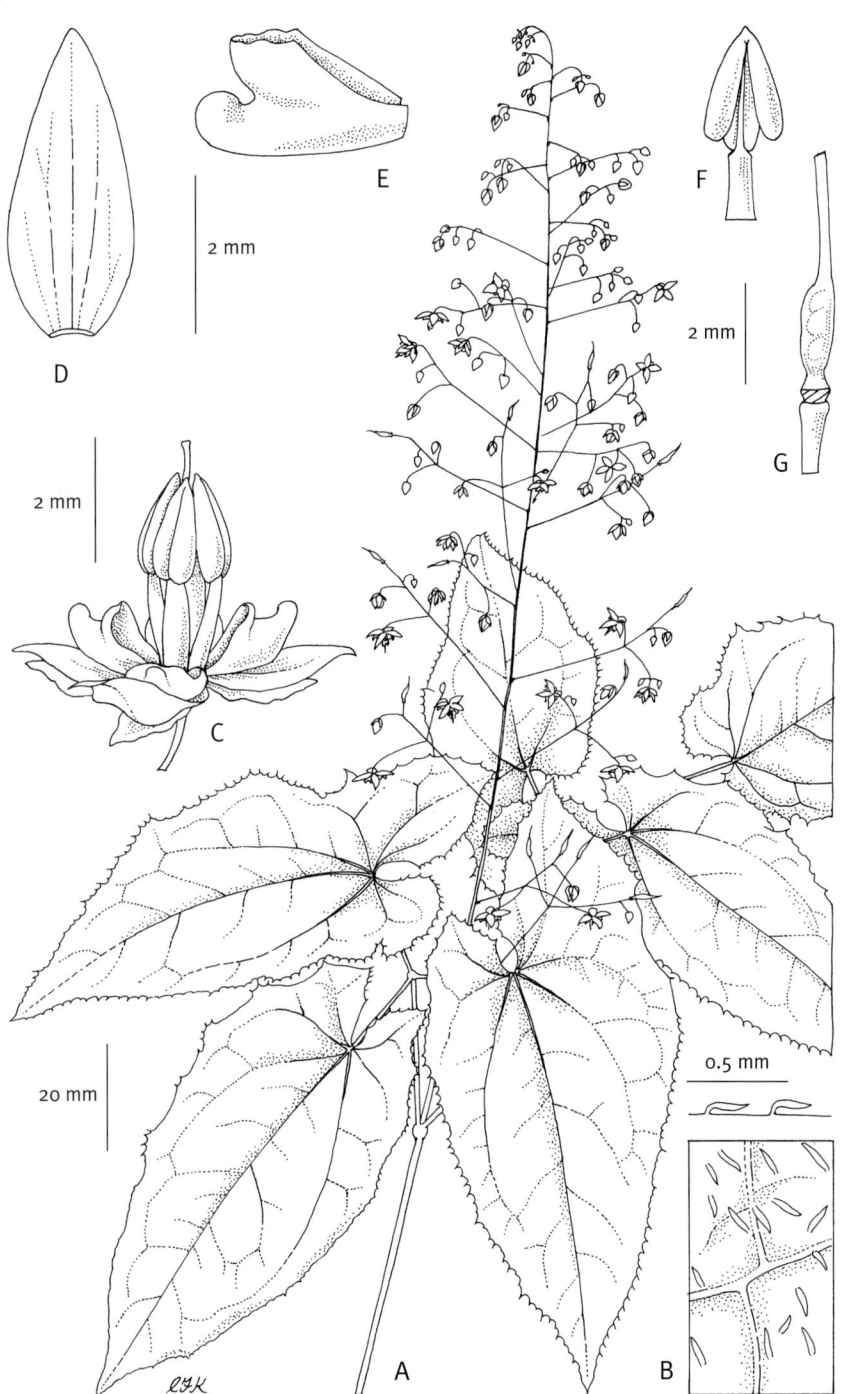

Fig. 70. ***Epimedium sagittatum.*** A, flowering stem; B, underside of leaf showing appressed hairs; C, flower; D, sepal; E, petal; F, stamen; G, gynoecium. Drawn by Christabel King.

34. EPIMEDIUM SAGITTATUM

Aceranthus sagittatus Siebold & Zucc. in Abh. Math.-Phys. Cl. Königl. Bayer. Akad. Wiss. 4(2): 175 (1846) Miquel, Ann. Mus. Bot. Lugduno-Batavum 2: 71 (1865).

A. triphyllus C.Koch in Miquel, Ann. Mus. Bot. Lugduno-Batavum 1: 253 (1864). Type: 'Japonia', *Bürger'* (holotype L).

A. macrophyllus Blume ex C.Koch in Miquel, Ann. Mus. Bot. Lugduno-Batavum 1: 253 (1864). Type: 'Japonia', *Siebold* (holotype L).

Epimedium ikariso Siebold ex Regel, Index Sem. Hort. Bot. Petrop. 1868: 89 (1868).

E. sinense Siebold ex Hance in J. Bot. 20: 2 (1882); Franch. in Bull. Soc. Bot. France 33: 110 (1886); Ito in J. Linn. Soc., Bot. 22: 432 (1887); Kom. in Acta Horti Petrop. 29: 134 (1908). Type: as for *Aceranthus sagittatus*.

E. sagittifolius Farrer, Engl. Rock Garden 1: 8 (1919).

E. sagittatum subsp. *typicum* Stearn in J. Linn. Soc., Bot. 51: 505 (1938).

ILLUSTRATIONS. Iinumai, Somoku Dzus. ed. 3 (Makino) 2: 123, t.44 (1907); Iwasaki, Honzo Dzufu 6: 9 verso, 10 recto (1828); Terasaki, Nippon Shokob. Zuhu: 153 (1933); World Health Organization, Med. Pl. China: 122 (1989).

DESCRIPTION. *Flower stem* 25–70 cm long, bearing 2 opposite or 3 whorled leaves. *Rhizome* short, nodose, 3–5 mm thick. *Basal and cauline leaves*, 3-foliolate (the basal rarely biternate); leaflets narrowly ovate to lanceolate, very variable in size, often 5 cm long, 3 cm broad, but sometimes up to 19 cm long, 8 cm broad; the apex acute or acuminate, the margin undulate and very spinous with spines 1–1.5 mm long, the base deeply or shallowly cordate, the terminal leaflet with equal rounded or (rarely) acute lobes, the lateral leaflets very oblique with the outer lobe large, deltoid and acute, the inner lobe smaller and rounded, coriaceous when mature, with the underside at first glabrous and sometimes remaining so but usually becoming (as the leaflet expands and matures) sparsely or densely pilose with stout short unicellular appressed hairs. *Inflorescence* compound, usually glabrous, many- (10–30) flowered, oblong and comparatively narrow in outline, 10–37 cm long, with the lower peduncles 3–5-flowered; pedicels 5–15 mm long. *Flowers* 8 mm or less across. *Outer sepals* blunt, purple-spotted, the outer pair narrowly ovate, 3.5 mm long, 1.5 mm broad, the inner pair oblong-ovate, 4.5 mm long, 2 mm broad. *Inner sepals* ovate-deltoid, acute, white, 4 mm long, 2 mm broad. *Petals* minute, almost as long as the inner sepals, saccate, blunt, brownish-yellow or yellow, with slight lateral flanges at base, 2–4 mm long. *Stamens* protruding, 5 mm long; *anthers* 3 mm long. *Capsule* 1 cm long.

DISTRIBUTION. China, Guangxi (Kwangsi), ?Guizhou (Kweichow), Gwangdong (Kwangtung), Fujian (Fukien), Zhejiang (Chekiang), Jiangsu (Kiangsi), Anhui (Anhwei) and Hubei (Hupeh) provinces.

35. EPIMEDIUM MYRIANTHUM

In flower *E. myrianthum* closely resembles *E. sagittatum* (Siebold & Zucc.) Maxim. with only a slight difference in the form of the petal, and could perhaps be regarded as a subspecies of this, with a many-flowered inflorescence. In a living state *E. sagittatum* has leaves with markedly undulate margins, green when young, whereas *E. myrianthum* has leaves with flat margins, red-mottled when young, the underside having hairs which are much more slender than the short, stout hairs of *E. sagittatum*. The epithet *myrianthum* from Greek *myrias* 'numberless' and *anthes* 'flower' refers to the many-flowered inflorescence.

The species was collected in the Tianpinshan mountains of Hunan by an expedition of the Beijing Botanic Garden and distributed by Mr Darrell Probst. It first flowered in England in April 1997 at the Blackthorn Nursery, Hampshire.

Epimedium myrianthum Stearn in Kew Bull 53: 218, figs.3, 4 (1998). Type: China, Hunan province, Tianpingshan mountains; cultivated by *Darrel Probst*, Hubbardston, Mass., U.S.A., 1996 under the number CPC 94.0102 (holotype K).

E. sinense var. *pyramidale* Franch., Pl. Delavay, 1: 40 (1889). Type: 'Yunnan', 31 March 1882, *Delavay* [hence in eastern Sichuan near the Hubei border on the bank of the Yangtse-kiang (*cf.* Stearn, 1938: 509].

E. sagittatum var. *pyramidale* (Franch.) Stearn in J. Bot. 71: 346 (1933).

E. sagittatum subsp. *pyramidale* (Franch.) Stearn in J. Linn. Soc., Bot. 51: 508 (1938).

ILLUSTRATION. Kew Bull. 53: 219, fig.3; 220, fig.4 (1998).

DESCRIPTION. *Flowering stem* 30–60 cm long, with 2 opposite or sometimes 3 whorled long-petioled leaves. *Rhizome* compact, *c.* 5 mm thick. *Basal and cauline leaves* with 3

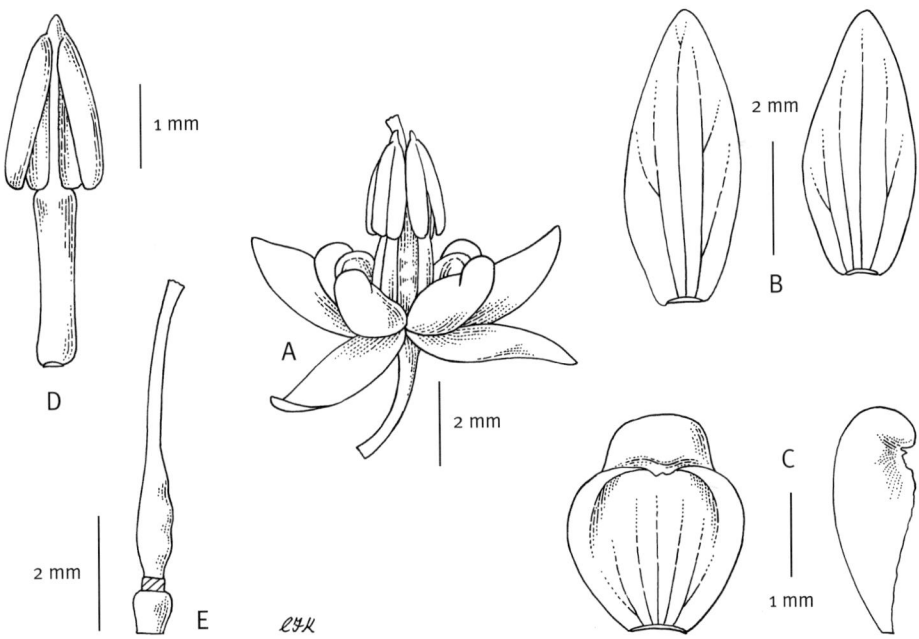

Fig. 71. **Epimedium myrianthum.** A, flower; B, sepals; C, petal, side and front views; D, stamen; E, gynoecium. Drawn by Christabel King.

leaflets, the cauline petioles 6–8 cm long; leaflets of basal leaves mostly ovate, 5–6 cm long, 3–4 cm broad, the apex acute, leaflets of cauline leaves mostly narrowly ovate, sometimes elliptic or lanceolate, 6–11 cm long, 2–6 cm broad, the apex long acuminate, the margin flat with spines 0.8–2 mm long, the base moderately cordate with narrow sinus, the lobes of the terminal leaflet rounded, those of the lateral leaflets very unequal with the lobes rounded or acute, coriaceous, persisting all winter, red-mottled when young, glossy above, the underside glaucous and with extremely minute appressed hairs. *Inflorescence* compound, loose, paniculate, many-flowered (with 70–210 flowers), 18–34 cm long, 7–9 cm broad, glabrous with peduncles 3- or 5-flowered, 2–4 cm long; pedicels 0.5–1.5 cm long. *Flowers* minute. *Outer sepals* soon falling, unequal, one pair 2 mm long, the other pair 3.5 mm long, obtuse, blackish. *Inner sepals* narrowly ovate, acute, white, 4 mm long, 1.5–2 mm broad. *Petals* much shorter than inner sepals, slipper-shaped, orange-yellow and red, *c.* 2–2.5 mm long, obtuse. *Stamens* exposed, 4 mm long, pale yellow; filaments 2 mm long; anthers 2 mm long, with yellow pollen.

DISTRIBUTION. China, Hunan and Sichuan provinces.

36. EPIMEDIUM STELLULATUM

Epimedium stellulatum was introduced into cultivation in England by Roy Lancaster who collected it on 17 May 1983 on a hillside beneath the Purple Clouds Temple at approximately 2800 feet [860 m] on the mountain Wudang Shan in the north-western corner of Hubei (Hupeh) province. It has become known in gardens as 'Temple Star' and 'Wudang Star', whence the epithet *stellulatum* (from *stellula* 'a little star' and *-atus* indicating likeness), the inflorescence suggesting a constellation of little white stars. In his book *Travels in China*: 410 (1989), Lancaster records that 'the crevices between the stone blocks of ruined walls and buildings supported a flora of their own in which ferns were prominent. Even more plentiful was an *Epimedium*, possibly *Epimedium sagittatum*, with arrow-shaped leaves and airy branched heads of star-shaped white flowers. It literally filled the cracks and was equally common in rocky places around'.

The same species has also been introduced by Mikinori Ogisu from Jiangyou, north-western Sichuan.

Epimedium stellulatum Stearn in Kew Bull. 48: 810, fig.1, 2 (1993). Type: specimen prepared by *W.T.Stearn*, 19 April 1992, of *Lancaster* 1193 from China, Hubei province, Wudang Shan, cultivated at the Royal Botanic Gardens, Kew (holotype K).

ILLUSTRATION. Kew Bull. 48: 811, fig.1, 812, fig.2 (1993).

DESCRIPTION. *Plant in flower c.* 20–35 cm high. *Rhizome* short-creeping. Leaves basal and cauline 3-foliolate; leaflets ovate, the tip acute or shortly acuminate, the margin spinous-serrate with spines *c.* 0.7–1 mm long, the base deeply cordate with the lobes acute, those of the lateral leaflets moderately unequal, coriaceous when mature, persisting green

Fig. 72. ***Epimedium stellulatum*** growing in northern Sichuan, China, at Pingwu, 961 m. Photograph: Mikinori Ogisu.

Fig. 73. ***Epimedium stellulatum* 'Wudang Star'** (*R.Lancaster* 1193). Photographed by John Fielding at Blackthorn Nursery, Hampshire.

36. EPIMEDIUM STELLULATUM

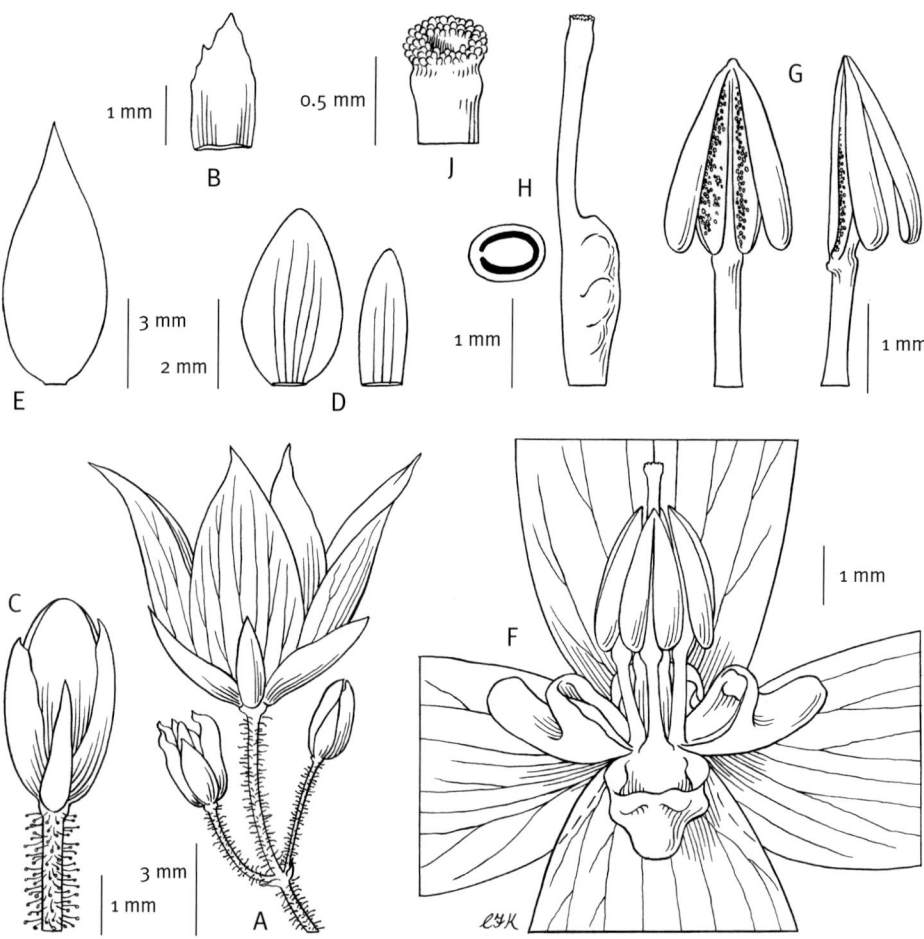

Fig. 74. **Epimedium stellulatum.** A, portion of inflorescence; B, bract; C, bud; D, outer sepals; E, inner sepal; F, centre of flower with petals and gynoecium; G, stamen, two views; H, gynoecium & t.s.; J, stigmatic surface. Drawn from *R.Lancaster* 1193 by Christabel King.

through the winter, glossy above, with the underside pubescent, its hairs profuse, very short, slightly ascending, *c.* 8–9 cm long, 4–7 cm broad. *Flowering stem* bearing two 3-foliolate leaves, rarely one. *Inflorescence* compound, loose, *c.* 15–20 cm long, with lower peduncles to 3 cm long, each bearing 3 flowers, many-flowered (with about 20–40 flowers altogether); pedicels with numerous glandular hairs 5–15 mm long. *Flowers* small, with spreading, not reflexed, inner sepals. *Outer sepals* soon falling, 2.5–3 mm long. *Inner sepals* lanceolate, acuminate, white, *c.* 12 mm long, 3 mm broad. *Petals* much shorter than inner sepals, almost straight, *c.* 2.5 mm long, brown-suffused but orange towards the base, with slight lamina and blunt nectariferous spur. *Stamens* conspicuously protruding, *c.* 3.5 mm long; filaments yellowish *c.* 1.5 mm long; anthers *c.* 2 mm long, yellow like the pollen.
DISTRIBUTION. China, Hubei (Hupeh) province, Wudang Shan; Sichuan province, Jiangyou.

37. EPIMEDIUM DOLICHOSTEMON

Epimedium dolichostemon has an historical interest as being the first of the many new or little-known Chinese species of *Epimedium* introduced into cultivation by Mikinori Ogisu, under the name *E. fargesii* Franch. This one was brought from Sichuan, central China, in 1988. It differs primarily from the latter species, which is from north-eastern Sichuan, in its shorter inner sepals (*c.* 8–9 mm instead of 15–18 mm long and reflexing) and its short, strongly incurved petals (*c.* 3 mm instead of 7 mm long and almost straight). The stamens, 8–10 mm long, of both species have the filaments longer than the anthers; this also occurs in *E. pinnatum*, *E. perralderianum* and *E. elatum*, with stamens 5–6 mm long. The epithet *dolichostemon* (from *dolichos*, long, *stemon*, thread, hence stamen) refers to its relatively long stamens.

Epimedium dolichostemon Stearn in Kew Bull. 45: 685, fig.2 (1990); 48: 808 (1993). Type: specimen prepared from plant collected China, Sichuan by *M.Ogisu*, cultivated by *E. Strangman*, April 1989 at Washfield Nursery, Hawkhurst, Kent, England (holotype K). [Note: The type locality was erroneously stated in the protologue as 'Emei Shan (Mount Omei)'. This should be corrected to Sichuan: Shizhu, 1400 m', according to its collector Mikinori Ogisu (in litt.). Shizhu (Shih-chu) is about 150 km north-east of Chongqing (Chungking), near the border with Hubei province.]

DESCRIPTION. *Plant in flower c.* 30 cm high. *Rhizome* short-creeping. *Leaves* basal and cauline, 3-foliolate; leaflets narrowly ovate, the tip acuminate, the margin spinous-serrate with spines 0.5–1.5 mm long, the base deeply cordate with the lobes acute, those of the lateral leaflets very unequal, coriaceous when mature with the underside glabrous, persisting through the winter, to 8 × 3 cm. *Flowering stem* bearing two 3-foliolate opposite leaves; inflorescence compound with peduncles 1–2 cm long each bearing 2 or 3 flowers, loose, up to 38-flowered altogether; pedicels glabrous, 1–1.5 cm long. *Flowers* small with spreading inner sepals. *Outer sepals* soon falling, *c.* 2.5–3 mm long. *Inner sepals* narrowly elliptic, white, *c.* 8–9 × 2.5 mm. *Petals* much shorter than the inner sepals, reddish purple, *c.* 3 mm long, cucullate, with short lamina and blunt incurved spur. *Stamens* conspicuously protruding, *c.* 8 mm long; filaments yellowish, c 4.5–5 mm long; anthers *c.* 2.5 mm long.
DISTRIBUTION. China, Sichuan (Szechwan) province.

38. EPIMEDIUM FARGESII

A French missionary in China, Paul Guillaume Farges (1844–1912), collected in north-eastern Sichuan the species which Franchet named *Epimedium fargesii* in 1894. This remained unique in the genus, on account of its spreading or reflexing, narrow lanceolate inner sepals, and stamens 8–10 mm long in which the filaments are longer than the anthers, until the introduction of *E. dolichostemon* in 1988. From that species, which has short, coiled petals, its longer, straight petals are an obvious distinction. The flowers suggest a miniature shuttlecock.

37. EPIMEDIUM DOLICHOSTEMON

Fig. 75. ***Epimedium dolichostemon.*** Habit, × $^2/_3$. Drawn by Ann Farrer.

Mr Mikinori Ogisu's introduction into cultivation of plants closely resembling Farges's type-material in having inner sepals 15 mm long, straight petals 8–9 mm long and projecting stamens 8–9 mm, but differing between themselves in having pale purple or white inner sepals and white or purple filaments, indicates that *E. fargesii* is unexpectedly variable; Farges's dried material appears to have white filaments. All the variants are attractive.

Epimedium fargesii Franch. in J. Bot. (Morot) 8: 281 (1894); Kom. in Acta Horti Petrop. 29: 137 (1908); Stearn in J. Linn. Soc., Bot. 51: 502 (1938). Type: China, Sichuan province, 'distr. Tchen-Keou-tin', *Farges* 506 (holotype P, isotype K).

ILLUSTRATIONS. J. Linn. Soc., Bot. 51: 457, fig.47 & t.29 (1938).

DESCRIPTION. *Flower stem* 20–50 cm long, bearing 2 opposite, or occasionally 3, whorled leaves, or no leaves. *Rhizome* short and clumped, 2–5 mm thick. *Leaves basal and cauline*, 3-foliolate; leaflets narrowly ovate, 4–10 cm long, 1.5–3 cm broad; the apex acuminate, the margin very spinous-serrate, spines *c*. 1–1.5 mm long, the base deeply cordate with usually acute lobes, those of the lateral leaflets unequal, coriaceous when mature with the underside white-glaucous, papillose, and glabrous or sparsely pilose. *Inflorescence* simple or compound with the lower peduncles 3-flowered, to 27 cm long, glandular, few- or many-flowered; pedicels 1.5–4 cm long. *Flowers* small, pendulous, with spreading or reflexing inner sepals and petals. *Outer sepals* narrowly ovate, obtuse, violet-tinged, 2–4 mm long, 1.5 mm broad. *Inner sepals* narrowly lanceolate, white or pale purple, acuminate, 15–18 mm long, 2.5–4 mm broad. *Petals* slightly curved (but almost straight),

Fig. 76. **Epimedium fargesii** at Chengkou, north of Chongqing, north-eastern Sichuan. Photograph: Mikinori Ogisu.

much shorter than inner sepals, dark purple with whitish tip, 6.5–9.5 mm long, the lamina slightly bilobed. *Stamens* conspicuous, protruding, 7–9 mm long; filaments white or purple; anthers yellow or green 2–4 mm long.

DISTRIBUTION. China, Sichuan (Szechwan) province.

39. EPIMEDIUM ELACHYPHYLLUM

According to the original description, *Epimedium parvifolium* is a low-growing plant with small leaves broader than long, a few-flowered inflorescence and minute flowers, the inner sepals yellow, the petals spurless and purple.

Komarov, in his 1905 *Revisio Critica*, did not maintain *Vancouveria* as a genus distinct from *Epimedium*, as shown by his combination *E. parvifolium* (Greene) Kom. (1908), which is now regarded as a synonym of *V. planipetala* Calloni (1887). Thus, *Epimedium parvifolium* S.Z.He & T.L.Zhang (1994) is a later homonym requiring a new name, for which *Epimedium elachyphyllum* is here proposed, from the Greek *ĕlachys* ('elaxus') meaning small, short, and *phyllon*, leaf.

Epimedium elachyphyllum Stearn, **nom. nov.**

E. *parvifolium* S.Z.He & T.L.Zhang in Guihaia 14: 25 (1994) non *E. parvifolium* (Greene) Kom. in Acta Horti Petrop. 29: 141 (1908). Type: China, Guizhou (Kweichow) province, Songtao, Ganlong, 1300–1400 m, April 1922, *T.L.Zhang* 9202 (holotype Guizhou Institute of Chinese Materia Medica).

DESCRIPTION. *Flowering stem* 15–25 cm long. *Rhizome* elongate, about 15 cm or more long, 1.5–3 mm thick. *Leaves basal and cauline* with 3 leaflets; leaflets 2.5–3.5 cm long, 3–4 cm broad, the tip acuminate, the margin spinous, the underside glabrous, pruinose. *Inflorescence* simple, 8–10 cm long, few-flowered (with 8–12 flowers), glabrous; pedicels 1–2 cm long. *Outer sepals* purple 3–5 mm long. *Inner sepals* lanceolate, yellow, 3 mm long, 1.2 mm broad. *Petals* purple, 1.2 mm long, without spurs. *Stamens* yellow, 2.5 mm long; anthers 1.8 mm long. *Capsule* 2 cm long, seeds brown, *c.* 3.5 mm.

DISTRIBUTION. China, Guizhou (Kweichow) province, 1300–1400 m.

40. EPIMEDIUM TRUNCATUM

Epimedium truncatum is a species with minute flowers borne in a many-flowered panicle and has distinctive leaves. The leaflets are not cordate with a sinus at their base; the middle leaflet is rounded at its base, while the two lateral leaflets are very asymmetrical with the base truncate as if cut off with a pair of scissors, whence the epithet from Latin *trunco*, 'to maim, cut off'. Unlike related species, this one is completely glabrous.

The following description is based on the prologue by Hai-Rui Liang.

Epimedium truncatum H.R.Liang in Acta Phytotax. Sin. 28: 322 (1990). Type: China, Hunan province, Baojing, 600–1000 m, May 1986, *S.S.Yang* (syntypes Beijing College of Traditional Chinese Medicine).

ILLUSTRATION. Acta Phytotax. Sin. 28: 323, fig.2 (1990).

DESCRIPTION. *Flower stem* 50–80 cm long, bearing 2 opposite leaves or 3 whorled leaves. *Leaves basal and cauline* with 3 leaflets; leaflets ovate, 7.5–15 cm long, 3.5–9 cm broad, the apex acute or acuminate, the margin spinous-serrate, the base of the terminal leaflet rounded, the base of the lateral leaflets obliquely truncate, glabrous. *Inflorescence* compound, paniculate, many-flowered (with about 40 flowers), to 28 cm long, 7 cm broad, most peduncles 3-flowered, glabrous, 2.5–5.5 cm long. *Flowers* minute, *c.* 4 mm across. *Outer sepals* rounded, 2–2.5 mm long. *Inner sepals* broadly ovate (according to illustration) or lanceolate (according to description), *c.* 2 mm long. *Petals* very broadly ovate, rounded, slightly saccate, *c.* 1 mm long. *Stamens* protruding, *c.* 2 mm long; filament 0.7 mm long; anther 1.5 mm long. *Seeds* 5 mm long.

DISTRIBUTION. China, Hunan province, 600–1000 m.

41. EPIMEDIUM COACTUM

Epimedium coactum, according to its authors Liang Hai-hui and Yan Wen-mei, is a Chinese species native of Guizhou (Kweichow) province, closely allied to *E. pubescens* but with leaflets densely brown to lanate beneath, the glabrous, paniculate inflorescence, the inner sepals and stamens the same length and the petals flat, not saccate.

The epithet *coactus*, 'with a thick covering', or 'felted', presumably refers to the hair covering on the underside of the leaflets; the variety *longtouhum* H.R.Liang has fusiform appressed white hairs on the leaflets.

Epimedium coactum H.R.Liang & W.M.Yan in Acta Phytotax. Sin. 28: 321 (1990). var. **coactum**. Type: China, prov. Guizhou, Kaili, 600–880 m, April 1987, *D.Q.Xiu* (syntypes Beijing College of Traditional Chinese Medicine).

ILLUSTRATION. Acta Phytotax. Sin. 28: 322, fig.1 (1990).

DESCRIPTION. *Flowering stem* 30–90 cm long, with 2 opposite leaves. *Rhizome* compact. *Basal and cauline leaves* with 3 leaflets; leaflets ovate, 6–13 cm long, 3–7.5 cm broad, the tip acute or acuminate, the margin spinous, the base deeply cordate with the lateral lobes acute, the underside densely brown lanate. *Inflorescence* compound, loosely many-flowered (with up to 100 flowers), to 40 cm long, glabrous; lower peduncles 6-flowered. *Flowers* small, 6 cm across. *Outer sepals* purple. *Inner sepals* lanceolate, white, acute, 4 mm long, 1.5 mm broad. *Petals* elliptic, yellow, obtuse, 2 mm long, 1.5 mm broad, spurless. *Stamens* protruding, c. 6 mm long; anthers 4 mm long, filaments c. 2 mm long.

DISTRIBUTION. China, Guizhou (Kweichow) province.

var. **longtouhum** H.R.Liang in Acta Phytotax. Sin. 28: 322 (1990). Types: China, Guizhou province, Kaili, 1100 m, April 1986, *H.R.Liang* 86024, 86025, 86028 (syntypes Beijing College of Traditional Chinese Medicine).

DESCRIPTION. From var. *coactum* this is stated to differ in the leaflets being covered beneath with fusiform, appressed white hairs.
DISTRIBUTION. China, Guizhou (Kweichow) province.

42. EPIMEDIUM BOREALIGUIZHOUENSE

Although stated by its authors to be near *E. wushanense*, a large-flowered species in series *Dolichocerae*, there is no resemblance whatever between that species and *E. borealiguizhouense* as described and illustrated. In habit the latter more closely resembles *E. pubescens* subsp. *cavalierei*, but the petals differ. Both are natives of the province of Guizhou (Kweichow).

Epimedium borealiguizhouense S.Z.He & Y.K.Yang in J. Pl. Resources & Environm. 2 (4): 51–53 (1993) (as *E. baiealiguizhouense*). Type: China, Guizhou (Kweichow) province, Yannan, *S.Z.He* 93001 (holotype Guizhou Institute of Chinese Materia Medica).

ILLUSTRATION. J. Pl. Resources & Environm. 2 (4): 52, fig.1 (1993).
DESCRIPTION. *Flowering stem* 40–60 cm long, bearing 2 opposite or 3 whorled leaves. *Rhizome* compact. *Basal and cauline leaves* with 3 leaflets; leaflets lanceolate or narrowly lanceolate, 1.3–18 cm long, 2.5–4 cm broad, the apex long acuminate, the margin serrate-spinous, the base deeply cordate with a narrow sinus and the lobes of the lateral leaflets very unequal, the larger lobe acute, the underside lanate. *Inflorescence* compound, to 35 cm long, many-flowered (with about 150 flowers), the lower peduncles 5-flowered, the upper 3-flowered, glabrous. *Flowers* very small, *c.* 6 mm across. *Outer sepals* purple, *c.* 3.5 mm long. *Inner sepals* ovate, white, 2.5 mm long, 1 mm broad. *Petals* flat, broadly ovate, rounded, spurless, yellow, 2 mm long. *Stamens* protruding, 4 mm long; anthers 2.5 mm long. Capsule *c.* 1 cm long.
DISTRIBUTION. China, Guizhou (Kweichow) province, 300–500 m.

43. EPIMEDIUM LOBOPHYLLUM

Named for the division of its terminal leaflet into three (or less often two or four) acuminate apical lobes, *Epimedium lobophyllum* belongs to series *Brachycerae*, but according to its authors differs in leaf and floral details. It is close to *E. myrianthum*.

Epimedium lobophyllum L.H.Liu & B.G.Li in Acta Phytotax. Sin. 37: 288 (1999). Type: Hunan province, Sangzhi, Tianpingshan, 1450 m, May 1978, *L.H.Lui* 17296 (holotype Hunan Normal University).

ILLUSTRATION. Acta Phytotax. Sin. 37: 289, fig.7 (1999).
DESCRIPTION. *Flowering stem* 30–70 cm long, bearing two opposite leaves. *Rhizome* compact. *Leaves basal and cauline* 3-foliolate; lateral leaflets narrow, ovate, entire, the tip acuminate, the margin spinous, the base shallowly cordate with wide open sinus, the lobes very unequal, the outer lobe acute, the inner rounded, the terminal leaflet obovate or ovate, the tip divided into 3 (or 2, 4 or 5) acuminate lobes, the margin spinous, the base shallowly cordate with an open sinus, the lobes rounded, 9–14 cm long, 4–6.4 cm broad, the underside pruinose and glabrous except for the pilose main veins. *Inflorescence* compound, paniculate to many-flowered, 25–34 cm long, glabrous; peduncles mostly 5-flowered, 7–17 cm long. *Flowers* minute, 3–4.5 mm across, white or yellowish. *Outer sepals* ovate or elliptic, 2–3.5 mm long. *Inner sepals* obovate or ovate-lanceolate, 2–3.5 mm long, *c.* 1 mm broad. *Petals* almost orbicular, 1.5–3 mm long, 1.3 mm broad, slightly saccate. *Stamens* 3–4 mm long, anthers longer than filaments.
DISTRIBUTION. China, Hunan province.

Section ii. MACROCERAS

(Species number 44–49)

Section *Macroceras* C.Morren & Decne. in Ann. Sci. Nat. Bot. 51(2): 349 (1834). Type species: *E. grandiflorum* C.Morren.

Stem bearing one leaf; petals with a large spur, or spurless (*E. diphyllum*).

44. EPIMEDIUM GRANDIFLORUM

The only species of *Epimedium* known to European gardeners before 1830 was the small-flowered inconspicuous *E. alpinum*, so that when two variants of a Japanese species with large, long-spurred flowers were introduced they created great interest. *Epimedium grandiflorum* remained the only long-spurred species in cultivation until the introduction of *E. acuminatum* and *E. membranceum* in the 1980s.

When Philipp Franz von Siebold returned to the Netherlands from Japan in 1830, only 260 of the 1200 living plants taken aboard in Nagasaki had survived the long voyage to Antwerp, which took six months. The survivors included these two variants of *E. grandiflorum* which were planted in the Ghent botanic garden in July 1830. The revolution which broke out in August and resulted in the separation of Belgium from the Dutch Netherlands caused Siebold to move his herbarium and ethnographical material to Leiden but he left his living plants in Ghent. The surprisingly large, elegant flowers of Siebold's two epimediums caught the attention of Charles Morren (1807–1859) at Ghent and led him and his compatriot Joseph Decaisne (1804–1882), then working in the Paris Museum of Natural History, to prepare the first monographic account of *Epimedium*. This was published in December 1834. Here, they named one of Siebold's plants, with almost white flowers, *E.*

Epimedium grandiflorum (three variants) — CHRISTABEL KING

macranthum and the other, with pale violet flowers, *E. violaceum*. However, Morren had already, in September 1834, described and illustrated them as *E. grandiflorum* and *E. violaceum* respectively. The living plants evidently thrived and were rapidly propagated and distributed, for illustrations appeared in English botanical periodicals as early as 1836 and 1838.

The species has an extensive geographical range and varies greatly in flower colour: white, light purple, pale yellow and red-purple. The various colour forms are now in cultivation and accordingly merit naming. In Japan there are also numerous hybrids of *E. grandiflorum* with other species, while in Europe it has long been known as a parent of the hybrids described here as *E.* × *rubrum*, *E.* × *versicolor* and *E.* × *youngianum*. It is, however, a beautiful species in its own right. Although easy to cultivate, all variants thrive only on non-calcareous soil and should be watered with water from a source known not to be alkaline; usually rain-water is suitable.

The epithets *grandiflorum* (Latin) and *macranthum* (Greek) both mean 'large-flowered'. The Japanese name 'Ikariso' is from *ikari* (grapnel or anchor) and *so* (plant), the four long curved spurs of the flower suggesting the four-fluked grapnel (*Nottsuzume ikari*, four-claw anchor) which is used by Japanese fishermen.

Fig. 77. ***Epimedium grandiflorum* 'Lilafee'.** Photographed by John Fielding at Blackthorn Nursery, Hampshire.

44. EPIMEDIUM GRANDIFLORUM

Epimedium grandiflorum C.Morren, *sensu lato*.

E. grandiflorum C.Morren in L'Horticult. Belge 2: 141, t.35A (Sept. 1834); Stearn in J. Linn. Soc., Bot. 51: 479 (1938) & Europ. Gard. Fl. 3: 391 (1989); Suzuki, Nippon no Ikawso: 159 (1990). Type: cult. Belgium (iconotype, loc. cit. t.35A).

E. violaceum C.Morren in L'Horticult. Belge 2: 142, t.35B (Sept. 1834); C.Morren & Decne. in Ann. Soc. Nat. Bot. II. 2: 352, t.12 (Dec. 1834). Type: (iconotype, loc. cit. t.35B).

E. macranthum C.Morren & Decne. in Ann. Sci. Nat. Bot. II. 2: 352, t.13 (Dec. 1834); Baker in Gard. Chron. N.S. 13: 683 (1880); Franch. in Bull. Soc. Bot. France 33: 105 (1886); Kom. in Acta Horti Petrop. 22: 324 (1903) & 29: 131 (1908). Type: (iconotype, loc. cit., t.13.

DESCRIPTION. *Flowering stem* 12–35 cm long, bearing one biternate or triternate leaf. *Rhizome* compact, 3–5 mm thick. *Leaves basal and cauline*, biternate or triternate; leaflets narrowly ovate to broadly ovate, the tip acute or acuminate, the margin very spinous-serrate, the base deeply cordate with usually rounded lobes, membranous in texture, glaucescent beneath, bronzy or light green when young, at length glabrous or sparsely pubescent beneath, 3–13 cm long, 2–8 cm broad. *Inflorescence* simple or compound with the lower peduncles 3-flowered, loose, glabrous or rarely pilose, 4–16-flowered, usually overtopping the stem-leaf; pedicels 1–2 cm long. *Flowers* 2–4.5 cm across, white, pale yellow, deep rose, red-purple or violet. *Outer sepals* red-tinged, oblong, 4–5 mm long. *Inner sepals* narrowly ovate to

Fig. 78. *Epimedium grandiflorum* **'White Queen'**. Photographed by John Fielding at Blackthorn Nursery, Hampshire.

THE GENUS EPIMEDIUM
44. EPIMEDIUM GRANDIFLORUM

Fig. 79. *Epimedium grandiflorum* **'Yellow Princess'**. Photographed by John Fielding at Blackthorn Nursery, Hampshire.

lanceolate, acute, flat, 8–18 mm long, 3–6 mm broad. *Petals* usually much longer than the inner sepals, with distinct, petaloid, rounded laminae 5–8 mm deep and slender tapering subulate spurs 1–2 cm long. *Stamens* included, 5 mm long; anthers 4 mm long.

DISTRIBUTION. Japan (Kyushu, Shikoku, Honshu, Hokkaido) northern Korea and southern Manchuria, in deciduous woodland.

Infra-specific taxa are recognised here as follows:

forma **grandiflorum**
E. grandiflorum C.Morren in L'Horticult. Belge 2: 142, t.35A (Sept. 1834).
E. macranthum C.Morren & Decne. in Ann. Sci. Nat. Bot. II. 2: 352, t.13 (Dec. 1834).
E. macranthum var. *normale* H.R.Wehrh. in Bonstedt, Pareys Blumengärtn. 1: 621 (1931).
E. grandiflorum f. *normale* (H.R.Wehrh.) Stearn in J. Linn. Soc., Bot. 51: 482 (1938).

ILLUSTRATIONS. L'Horticult. Belge 2: 35A (1834); Bot. Reg. 22: t.1906 (1836); Maund & Henslow, Botanist, 2: t.90 (1838).
DESCRIPTION. *Flowers* almost white, the inner sepals tinged with pale violet, the spurred petals white.

forma **flavescens** Stearn in J. Linn. Soc., Bot. 51: 482 (1938). Type: Nishijima, Hishui Hakku Fu t.10 (1982).

44. EPIMEDIUM GRANDIFLORUM

Fig. 80. The yellow-flowered Japanese species often referred to as "***Epimedium koreanum***", but probably a form of **E. grandiflorum.** Photographed in the wild in Japan at Katsuyama-shi, Fukui-ken, by Kazuo Mori.

E. sulphurellum Nakai in J. Jap. Bot. 20: 75 (1944). Type: Japan, Honshu (no type cited).
E. cremeum Nakai & F.Maek. ex Nakai in J. Jap. Bot. 20: 76 (1944). Type: Japan Honshu, prov. Etchu, mount Tateyama, Asimoto-Hakma (holotype TI).

ILLUSTRATIONS. Nishijima, Hishui Hakku Fu, t.10 (1934) [as *E. macranthum*]; Satake *et al.*, Wild Flowers of Japan t.90, figs.3, 4 (1982).
DESCRIPTION. Flowers light yellow.
NOTE. This pale yellow Japanese form is in cultivation under the name *E. koreanum* but is seemingly not identical with the plant native in Korea. It is mentioned by the Japanese botanist Shimada, alias Yonan, (Savatier, 1875) as far back as 1763 when he described the 'Ikariso' (*E. grandiflorum*) as having 'clear violet or white or clear yellow' flowers (as translated by Savatier in 1875).

forma **violaceum** (C.Morren) Stearn in J. Linn. Soc., Bot. 51: 483 (1938). Type: (iconotype L'Horticult. Belge 2: 142, t.35B (1834).
E. violaceum C.Morren in L'Horticult. Belge 2: 142, t.35B (1834); C.Morren & Decne. in Ann. Sci. Nat. Bot. II. 2: 354, t.12 (1834).
E. macranthum var. *thunbergianum* Miq. in Ann. Mus. Bot. Lugduno-Batavum 2: 70 (1865).
E. violaceum var. *grandiflorum* hort. in Ann. Hort. Bot. 2: 8, t. (1869).

E. macranthum var. *violaceum* (C.Morren & Decne.) Franch. in Bull. Soc. Bot. France 33: 106 (1886).

E. macranthum f. *violaceum* Voss, Vilmorin's Blumengärtn. ed. 3, 1: 51 (1896).

E. grandiflorum var. *violaceum* (C.Morren) Stearn in Kew Hand-list Rock Garden Pl., 4th ed. 53 (1934).

E. grandiflorum var. *thunbergianum* (Miq.) Nakai in J. Jap. Bot. 20: 69 (1944); K.Suzuki, Nippon no Ikariso: 160 (1990).

ILLUSTRATIONS. L'Horticult. Belg. 2: t.35B (1834); Ann. Sci. Nat. Bot. II. 2: t.12 (1834); Paxton, Mag. Bot. 5: 123 (1838); Curtis's Bot. Mag. 66: t.3751 (1839); Bot. Reg. 24: t.43 (1840); Honzo Dzufu 6: fol. 11 verso (1828).

DESCRIPTION. *Flowers* with light violet inner sepals and petals in the type, but now including variants with red-purple flowers. An especially notable variant is *E. grandiflorum* 'Rose Queen', a very beautiful form with intense red-purple flowers, alternatively known in English gardens as *E.* 'Rose Queen' and *E. macranthum* 'Rose Queen' (see Stearn in *Gardening Illustr.* 34: 31 (1932) & *J. Linn. Soc., Bot.* 51: 483 1938). After the Second World War, when it became rare or possibly extinct, it was known as 'Crimson Beauty'. In the RHS Colour Chart (1966) the outer sepals match 72B. The colour seems nearest to the 'carmine cramoisi' or 'crimson-carmine' of Dauthenay's *Répertoire des Couleurs*, p. 159, no. 1, the petals being slightly paler than the inner sepals, with the spurs becoming white at their tips. It is the most conspicuously distinct of the *E. grandiflorum* colour forms.

Fig. 81. **Epimedium grandiflorum** forma **violaceum** at Otari-mura, Nagano, Japan. Photograph: Kazuo Mori.

45. EPIMEDIUM SEMPERVIRENS

Although like *Epimedium grandiflorum* in having biternate leaves, with acuminate leaflets, and large long-spurred flowers, *E. sempervirens* has been distinguished from this as a species on account of its more elongated rhizome and evergreen leaves which, if not glabrous, have fine appressed (not erect) hairs below. These hairs have an erect base of about two very short cells, then a very long terminal cell which turns abruptly from this and lies parallel with the leaflet surface, not arising erect from it. Kazuo Suzuki (1990) distinguishes two varieties: var. *sempervirens* and var. *rugosum* (see below).

Epimedium sempervirens is a Japanese species endemic to the western (or Sea of Japan) side of southern Honshu, where the depth and duration of winter snow cover is very much greater and better protects the leaves than on the eastern (or Pacific Ocean) side, where non-evergreen *E. grandiflorum* has its main area. The name *sempervirens* (evergreen) refers to the over-wintering leaves, from *semper* (always, at all times) and *vireo* (to be green).

Epimedium sempervirens requires the same conditions in the garden as *E. grandiflorum*.

Epimedium sempervirens Nakai ex F.Maek. in Bot. Mag. (Tokyo) 46: 582 (1932); Nemoto, Fl. Jap. Suppl.: 236 (1936); Stearn in J. Linn. Soc., Bot. 51: 483 (1938); Ohwi, Fl. Jap.: 464 (1965); Ohwi, Elick & Booth, Japonica Magnifica: 58 (1982); K.Suzuki, Nippon no Ikariso, 161 (1990). Type: Japan, Honshu (Honda), prov. Noto, *Hara*, cult. Tokyo, 1928 (holotype TI).

Fig. 82. ***Epimedium sempervirens*** growing wild in Japan at Kinomoto-chō, Shiga-ken. Photograph: Kazuo Mori.

Fig. 83. *Epimedium sempervirens* in its natural habitat at Monzen-chō, Ishikawa-ken, Japan. Photograph: Kazuo Mori.

E. macranthum var. *hypoglaucum* Makino in J. Jap. Bot. 7: 13 (1931). Type: Japan, Honshu, prov. Wakasa, Mt. Aoba, *J.Takada* (holotype TI).

E. sempervirens var. *hypoglauceum* (Makino) Ohwi in Bull. Nat. Sci. Mus. Tokyo 33: 73 (1953).

E. grandiflorum subsp. *sempervirens* (Nakai ex F.Maek.) Kitam. in Acta Phytotax. Geobot. 20: 202 (1962).

ILLUSTRATIONS. Hara & others, Makino's Illustr. Fl. Japan Suppl.: t.207 (1955); Bot. Mag. (Tokyo) 46: 582 (1932); K.Suzuki, Nippon no Ikariso: 166 (1990).

DESCRIPTION. *Flowering stem* 30–60 cm long, bearing one leaf. *Rhizome* shortly-creeping (but more elongate than that of *E. grandiflorum*). *Leaves basal and cauline* biternate; leaflets narrowly ovate to broadly ovate, 6–10 cm long, 3–5 cm broad, the tip acuminate, sometimes trilobed, the margin very spinous, the base deeply cordate with narrow sinus, the lateral leaflets with very unequal lobes, the outer one larger and often acuminate, the inner rounded, coriaceous, evergreen, the underside glaucous and glabrous or with fine appressed hairs below. *Inflorescence* simple or compound, 5.5–7 cm long, with 8 to 12 or more flowers, glabrous; pedicels 10–15 mm long. *Flowers* large, to 4 cm across, white (var. *sempervirens*) or reddish purple (var. *rugosum*). *Outer sepals* purplish, 2–5 mm long. *Inner sepals* acute, flat, *c.* 11 mm long. *Petals* longer than inner sepals, with distinct petaloid rounded laminae 7 mm deep and downwards curved slender tapering spurs 15–20 mm long. *Stamens* included, *c.* 4.5 mm long, the filament very short.

DISTRIBUTION. Japan, Honshu, western side.

The two varieties recognised by Suzuki are:

var. **sempervirens**

ILLUSTRATION. Y.Satake *et al.*, Wild Flowers of Japan: 2: t.91 (1981).
DESCRIPTION. *Flowers* white, with the spur less than 18 mm long

var. **rugosum** (Nakai) K.Suzuki, Nippon no Ikariso: 162 (1990).
E. rugosum Nakai in J. Jap. Bot. 20: 81 (1944). Type: Japan, Honshu.

ILLUSTRATION. Elick & Booth, Japonica Magnifica: 58 (1992).
DESCRIPTION. *Flowers* reddish purple, with the spur more than 18 mm long.

46. EPIMEDIUM KOREANUM

Epimedium grandiflorum and *E. koreanum* are very much alike in flower structure and foliage, but there is an as yet unresolved problem with the yellow-flowered *E. koreanum*. The pale yellow Japanese epimedium that is in cultivation under the name '*E. koreanum*' appears to be a yellow-flowered form of *E. grandiflorum* and is seemingly not identical with the true *E. koreanum* native to Korea. Provisionally the Korean plant is here kept apart from both *E. grandiflorum* and *E. sempervirens*.

Epimedium koreanum Nakai, Fl. Sylv. Koreana 21: 55, 60 & 63 (1936); J. Jap. Bot. 20: 78 (1944). Type: North Korea, Heihoku province, Mount Hakuhekizan, *Ishidoya* (holotype TI).
E. grandiflorum subsp. *koreanum* (Nakai) Kitamura in Acta. Phytotax. Geobot. 20: 202 (1962).

DESCRIPTION. Similar to that of *E. sempervirens* (above) but flowers yellow.
DISTRIBUTION. Korea.

47. EPIMEDIUM MACROSEPALUM

Although allied to *Epimedium grandiflorum*, *E. macrosepalum* is remarkably distinct in having leaves with only three to seven leaflets instead of nine or more, a slender elongated rhizome, and oblong-ovate inner sepals 8–11 mm broad, instead of narrowly ovate inner sepals 3–6 mm broad as in *E. grandiflorum*.

Epimedium macrosepalum Stearn in J. Linn. Soc., Bot. 51: 485 (1938). Type: Russia, Maritime Province, Tigrovaya Station, Suchansk Railway, *Desoulavy* 3583 (holotype LE).

ILLUSTRATION. J. Linn. Soc., Bot. 51: pl.26 (1938).

47. EPIMEDIUM MACROSEPALUM

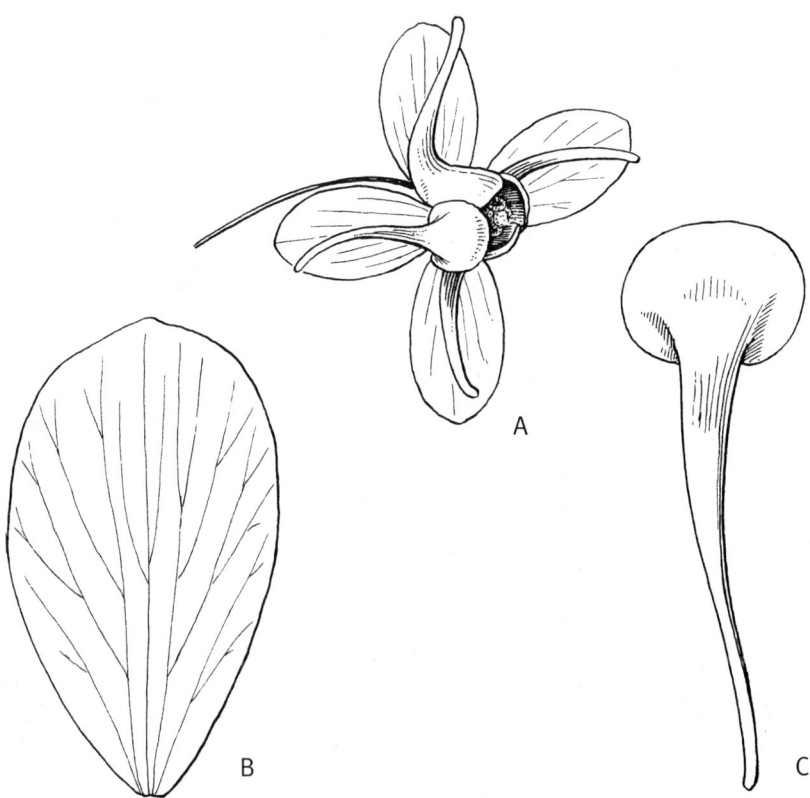

Fig. 84. *Epimedium macrosepalum.* A, flower, × 1; B, sepal, × 3; C, petal, × 3. Drawn by Stella Ross-Craig from the type herbarium specimen, *N.Desoulavy* 3583 (LE).

DESCRIPTION. *Flowering stem c.* 25 cm high, bearing a 3-foliolate leaf. *Rhizome* elongated, 1–2 mm thick. *Leaves* basal and cauline, normally 3- (less often 7-) foliolate; leaflets broadly ovate to almost orbicular, the tip blunt, the margin sparsely spinous-serrate or almost entire, the base deeply cordate with rounded or acute lobes, membranous but firm in texture, above dark green, beneath glaucous and pubescent with short reddish hairs, 4–7 cm long, 3.3–5.5 cm broad. *Inflorescence* simple, glabrous, few-(1–3)flowered, overtopped by the stem-leaf; pedicels 2–3 cm long, or the peduncle, if one-flowered, 6 cm long. *Flowers* large, yellowish, 4 cm across. *Inner sepals* oblong-obovate, blunt, 20 mm long, 8–11 mm broad. *Petals* long-spurred, with distinct petaloid rounded laminae *c.* 6 mm high, and slender tapering spurs *c.* 20 mm long. *Stamens* included, 5 mm long; anthers 4 mm long. *Capsule* (including the 7 mm long style) 35 mm long, 3 mm wide.

DISTRIBUTION. Russia (Far Eastern Area), Maritime Province, Southern district, approx. 43°–44°N, 133°–135°E, east of the Sikhota-Alin main range.

48. EPIMEDIUM TRIFOLIOLATOBINATUM

The small white flowers of *Epimedium trifoliolatobinatum* resemble in general appearance those of *E. diphyllum* except for their slender spurs 10–15 mm long. Morphologically intermediate between *E. diphyllum* and *E. grandiflorum*, and undoubtedly derived from their hybridization, *E. trifoliolatobinatum* is accepted as a species by Kazuo Suzuki, the monographer of Japanese *Epimedium*. It consists of intermediate populations kept stable by small, short-tongued pollen-collecting bees which fly only short distances and pollinate only plants within a single population. One stable population is found only on serpentine (*E. trifoliolatobinatum* subsp. *trifoliolatobinatum*) and another (subsp. *maritimum*) in maritime regions; these, Suzuki describes as 'unusual habitats that may have allowed the stabilization of its distinctive features and allowed its persistence'.

Subsp. *trifoliolatobinatum*, according to Suzuki, has summer-green leaves, usually with two sets of three papery leaflets; subsp. *maritimum* has evergreen leaves, usually with two sets of two coriaceous leaflets. Both come within the range of variation of the garden hybrids grouped under *E.* × *youngianum* and a specimen of unknown provenance would be identified as one of these.

Epimedium trifoliolatobinatum Koidz. in Acta Phytotax. Geobot. 8: 55 (1939); Ohwi, Fl. Japan: 464 (1965); K.Suzuki in Iwatsuki, Raven & Bock, Mod. Aspects of Species: 195 (1986) & Nippon no Ikariso: 159 (1990).

E. musschianum var. *trifoliolato-binatum* Koidz. in Acta Phytotax. Geobot. 5: 126 (1936); Stearn in J. Linn. Soc., Bot. 51: 488 (1938). Type: Japan; Shikoku island, 'prov. Tosa prope oppidum Kochi' *Yoshinaga* (KYO).

ILLUSTRATION. J. Jap. Bot. 57: 66 (1982); Revised Makino's New Illustr. Fl. Japan: 162 (1989); Suzuki, Nippon no Ikariso: 6 fig.f, 15 fig.d (1990).

DESCRIPTION. *Flowering stem* 20–50 cm long, bearing 1 leaf. *Rhizome* compact, short-creeping. *Leaves* basal and cauline, consisting of 2 divisions each with 3 or 2 leaflets, thus with 6 or 4 leaflets; leaflets narrowly ovate, 3–7 cm long, 2–4 cm broad, the apex acute, the margin sparsely spinous with spines 0.2–0.5 mm long, the base shallowly cordate, the lobes rounded or acute; the lateral ones very unequal, membranous or coriaceous, the underside with very minute hairs beneath. *Inflorescence* simple or with lower peduncles 2-flowered, loose, 13–18 cm long, few-flowered (with 6–12 flowers), glabrous; pedicels 2–3 cm long. *Flowers* small, campanulate, pendulous, white. *Outer sepals* to 4 mm long, obtuse. *Inner sepals* lanceolate, acute, white *c.* 8 mm long, 2 mm broad. *Petals* white, with lamina 4.5 mm high and slender curved spurs 10–15 mm long. *Stamens* enclosed, 3.5 mm long; anthers yellow.

DISTRIBUTION. Japan, on the west side of Shikoku island.

The two subspecies of *E. trifoliolatobinatum* distinguished by Suzuki are as follows:

Fig. 85. *Epimedium trifoliolatobinatum* subsp. *maritimum*. Photographed by Kazuo Mori in Japan at Sata-misaki, Kōchi-ken.

subsp. **trifoliolatobinatum**

E. trifoliolatobinatum Koidz. in Acta Phytotax. Geobot. 8: 55 (1939).

DESCRIPTION. *Leaves* summer green, perishing in autumn usually with 2 sets of 3 papery leaflets.

subsp. **maritimum** K.Suzuki in J. Jap. Bot.: 57 & 65 fig. (1982) & Nippon no Ikariso: 159 (1990).

DESCRIPTION. *Leaves* evergreen, with 2 sets of 2 coriaceous leaflets or sometimes 3 sets.

49. EPIMEDIUM DIPHYLLUM

Epimedium diphyllum (*sensu stricto*) was the first Japanese member of the genus to reach Europe, being included in Philipp Franz von Siebold's first consignment of plant species from Nagasaki to Leiden. Thanks to the cordial relations between the nursery firm of George Loddiges and Sons of Hackney, London, and continental botanic gardens and nurseries, it came into the hands of George Loddiges and was illustrated in their periodical the *Botanical Cabinet* [19: t.1, 858 (1832)], but without a description. Its pair of leaflets, with no median leaflet, and its flat white petals, led Morren and Decaisne to create in 1834, a new genus named *Aceranthus*, formed from the Greek *a* ('without'), *keras* ('horn') and *anthus* ('flower'). Meanwhile the plant had been acquired by the Edinburgh Botanic

Epimedium diphyllum

MARK FOTHERGILL

Garden and was described as *Epimedium diphyllum* by Robert Graham (1786–1845), without reference to Morren and Decaisne's *Aceranthus diphyllum*.

Epimedium diphyllum is a small, elegant and distinctive plant of easy cultivation but is liable to be overwhelmed by more vigorous species. In Japan it has hybridised with *E. grandiflorum* producing the garden hybrids known as *E.* × *youngianum* and, in the wild, the stable species *E. trifoliolatobinatum*.

Epimedium diphyllum Graham in Curtis's Bot. Mag. 62, t.3448 (1835). Type: Described from a cultivated plant at Edinburgh Botanic Garden in 1834 (iconotype, Curtis's Bot. Mag. t.3448).

DESCRIPTION. *Flowering stem* 10–20 cm high, rarely more, bearing one, normally 2-foliolate, leaf. *Rhizome* short-creeping, caespitose, 1–2 mm thick. *Leaves basal and cauline*, normally 2-foliolate (? sometimes 3-foliolate with a small median leaflet or the two secondary petioles each 3-foliolate); leaflets ovate-deltoid or narrowly ovate, obtuse or acute, the margin ± entire or slightly spinous, the base shallowly cordate and usually very unequal-sided, thin and membranous in texture, light green, glabrous or with minute spreading hairs on upper surface, sparingly pilose beneath, 2–5 cm long, 1–2 cm broad. *Inflorescence* simple or the lower peduncle 2-flowered, almost glabrous, few-(4–9) flowered; pedicels 1–2 cm long. *Flowers* campanulate, pendulous, white, but pink forms (perhaps hybrids referable to *E.* × *youngianum* 'Roseum') are said to occur. *Outer sepals* oblong, blunt, 3 mm long. *Inner sepals* narrowly ovate, bluntish, horizontally spreading, 6 mm long, 2.5 mm broad. *Petals* slightly longer and broader than the inner sepals, obovate, rounded, flat, spurless but with a slight median furrow, 7 mm long, 3–4 mm broad. *Stamens* included, 3 mm long: anthers nearly 2.5 mm long.

DISTRIBUTION. Japan: Kyushu.

Kazuo Suzuki, the monographer of the Japanese taxa of *Epimedium*, considers that *E. diphyllum* and *E. kitamuranum* T.Yamanaka are best treated as conspecific; the latter, which commemorates Siro Kitamura, is maintained as a subspecies of *E. diphyllum* and is distinguished as follows:

Leaves with 2 leaflets or with 2 sets of 2 leaflets, glabrous on upper
    surface, obtuse, entire . . . . subsp. **diphyllum**
Leaves with 2 or 3 leaflets, with minute spreading hairs on upper surface,
    acute, slightly spinous-margined . . . . subsp. **kitamuranum**

subsp. **diphyllum**

E. diphyllum Graham in Bot. Mag. 62, t.3448 (1835) & in Edinburgh New Philos. J. 20: 191 (1835); Baker in Gard. Chron. II. 12: 683 (1880); Franch. in Bull. Soc. Bot. France 33: 108 (1886); Ito in J. Linn. Soc., Bot. 22: 432 (1887); Makino in Iinuma Somoku Dzus., ed. 3, 2: t.46 (1907); Kom. in Acta Horti Petrop. 29: 133 (1908); Matsumura,

THE GENUS EPIMEDIUM
49. EPIMEDIUM DIPHYLLUM

Fig. 86. ***Epimedium diphyllum.*** Photographed by John Fielding in a garden near London.

Index Pl. Jap. 2 (Dicot.): 130 (1912); Stearn in J. Linn. Soc., Bot. 51: 487, t.31, fig.6 (1938) & Europ. Gard. Fl. 3: 392 (1989); Ohwi in Fl. Japan: 463 (1965); Satake *et al.*, Wild Flowers of Japan 2: 391 (1984); Suzuki, Ikariso no Nippon: 158, fig.4 (1990).

E. diphyllum Lodd., Bot. Cabinet, 19: t.1858 *sine descr.* (1832).

Aceranthus diphyllus C.Morren & Decne. in Ann. Sci. Nat. Bot. II, 2: 350, t.14 (1834); Spach, Hist. Veg. Phan. 8: 61 (1839); Wehrhahn, Gartenstauden 1: 452 (1930).

Vindicta begonifolia Raf., Fl. Tellur. 2: 52 (1837).

E. grandiflorum Marnock, Floricult. Mag. 4: 41: t.40 (1839), *non* C.Morren.

E. japonicum Siebold ex Miq. Prolug. Fl. Jap.: 3 (1865) & Miq. in Ann. Mus. Lugdano-Batavum 2: 71 *pro syn. Aceranthi diphylli* (1865), *non* Makino.

ILLUSTRATIONS. Lodd., Bot. Cabinet, 29: t.1858 (1832); Bot. Mag. 62: t.3448 (1835); Ann. Sci. Nat. Bot. II, 2: t.14 (1834); Iinuma, Somoku Dzus. ed. 3, 2: t.46 (1907); Marnock Floricult. Mag. 4: t.40 (1839); S.Satake *et al.*, Wild Flowers of Japan, 2: t.90 (1981).

DESCRIPTION. *Leaves* with 2 leaflets, or with 2 sets of 2 leaflets, obtuse, with non-spinous margins, glabrous on upper surface.

DISTRIBUTION. Japan: Kyushu, Shikoku and south-eastern Honshu.

subsp. **kitamuranum** (T.Yamanaka) K.Suzuki in J. Jap. Bot. 57: 67 (1982) & Nippon no Ikariso: 159 (1990). Type: Japan, Shikoku, Sanuki province, Sanuki-fuji, 5 May 1951, *Yamanaka s.n.* (holotype TI).

Epimedium elatum

PAULINE DEAN

E. kitamuranum T.Yamanaka in Acta Phytotax. Geobot. 15: 26 (1953).

ILLUSTRATION. Makino, New Illustr. Fl. Japan 162 (1984).
DESCRIPTION. *Leaves* with 2 or 3 acute, slightly spinous-margined leaflets, with minute spreading hairs on upper surface.
DISTRIBUTION. Japan, north-eastern Shikoku prefectures of Kayaw and Tokushima.

Section iii. POLYPHYLLON

(Species number 50)
Section *Polyphyllon* (Kom.) Stearn **stat. nov.** Type species: *E. elatum* C.Morren & Decne.
Series *Polyphylla* Kom. in Acta Horti Petrop. 29: 139 (1908).
Subsection *Polyphyllon*, series *Elatae* Stearn in J. Linn. Soc., Bot. 51: 509 (1938).

Although in its flowers akin to section *Epimedium*, this monotypic Himalayan section is unique in the genus by having up to eight ternately compound leaves.

50. EPIMEDIUM ELATUM

Epimedium elatum is the tallest species of the genus, reaching a height of 140 cm ($4^1/_2$ ft) and has the general habit of a *Thalictrum*, the stem bearing up to eight ternately compound leaves with 50 or more leaflets. It terminates in a paniculate inflorescence of up to 100 flowers. The species is widespread in the whole of Kashmir and also extends into Swat. The French traveller Victor Jacquemont (1801–1832), invited into Kashmir by its ruler Rangit Singh and the first botanist to visit the area, discovered this remarkable species near Hirpur in June 1831. Although formerly associated with the Chinese species *E. elongatum* it is isolated geographically and morphologically and is evidently a relict of an ancient western centre of evolution long ago eliminated by climatic change.

The epithet *elatum* (tall) is entirely appropriate. Although in flower approaching *E. alpinum* and *E. pubigerum* it is so different in habit as to merit being placed in a monotypic section, *Polyphyllum*. Falconer also described the species, using the epithet *hydaspidis*; this refers to the river Hydasper of ancient geographers, now the Jhelum.

Epimedium elatum C.Morren et Decne. in Ann. Sci. Nat. Bot. II. 2: 356 (1834); Decne. in Jacquemont, Voyage dans l'Inde (Bot.) 9: t.8 (1841); Hook.f. & Thomson, Fl. Ind.: 231 (1855); Hook.f., Fl. Brit. India 1: 112 (1872); Baker in Gard. Chron. II. 13: 620 (1880); Franch. in Bull. Soc. Bot. France 33: 111 (1886); Kom. in Acta Horti Petrop. 29: 139 (1908); Stearn in J. Linn. Soc., Bot. 51: 511, fig.1 (1938). Type: Kashmir, Hirpur, 1831 *Jacquemont* 575 (holotype P).
E. hydaspidis Falconer in Proc. Linn. Soc. London 1: 18 (1839), *nomen subnudum*.

ILLUSTRATION. Jacquemont, Voyage dans l'Inde (Bot.) t.8 (1841); J. Linn. Soc., Bot. 51: 416 (1951).

DESCRIPTION. *Flowering stem* 50–140 cm high. *Rhizome* moderately short, nodose, 3–5 mm thick. *Leaves basal and cauline*, ternately compound, the larger leaves with 50 or more leaflets; leaflets ovate or broadly, deeply to shallowly, cordate, with the lobes usually subequal and rounded, thin and papery in texture, sparingly pilose beneath, with a few hairs near the veins or ultimately glabrous, very variable in size, 2–6 cm long, 1.5–4.5 cm broad. *Stem* ribbed towards the base, bearing 3–8 irregularly-spaced alternate leaves, the lower leaves large, the others smaller, those immediately below the inflorescence being reduced to 5 or 3 (or even 1) leaflets. *Inflorescence* compound, glabrous or glandular, many-(50–100)flowered, usually very diffuse, 20–30 cm long, 15–25 cm across, the lower branches subtended by small 5-, 3- or 1-foliolate leaves or large, cuneate-based, lanceolate, spinous-margined green bracts; pedicels filiform, 5–20 mm long. *Flowers* scarcely 1 cm across. *Outer sepals* ovate, blunt, whitish, 1–2 mm long. *Inner sepals* narrowly ovate, acute, hyaline or yellowish, soon reflexing, 5 mm long, 2.5 mm broad. *Petals* shorter than the inner sepals, slipper-like, cylindric, blunt, 2–3 mm long, with no basal laminae. *Stamens* protruding, 5 mm long; anthers purplish, 1.5 mm long. *Capsules* comparatively short and stout, 1 cm long.

DISTRIBUTION. Western Himalaya: Kashmir, in mountain woods. Usually at around 2500 m (*c.* 8000 ft) but ranging between 1370 m (4500 ft) and 2750 m (9000 ft) in the upper valleys of the Chenab and the Jhelum rivers and their tributaries; also in Swat.

Section iv. EPIMEDIUM

(Species number 51–52)

Section *Epimedium*. Type species: *E. alpinum* L.

Section *Microceras* C.Morren & Decne. in Ann. Sci. Nat. Bot. II. 2: 355 (1834), *E. elato* excluso. Lectotype: *E. alpinum* L.

Series *Microcerae* (C.Morren & Decne.) Stearn in J. Linn. Soc., Bot. 51: 460 (1938).

51. EPIMEDIUM ALPINUM

The Italian herbalist Luigi Anguillara (d. 1570) was the first to find and to record, in 1561, a plant later named *Epimedium alpinum* by Linnaeus (in 1739), growing in Italy and 'Schiavonia' (? Carniolia) which he associated with the *Epimedium* of Dioscorides. In this he was certainly mistaken. The plant was soon afterwards introduced into cultivation, presumably from the Colli Euganei east of Padua, and, under the name *Epimedium* easily found its way into the gardens of Central and Western Europe. Pena and L'Obel received it from the garden of a Venetian apothecary when in Italy between 1558 and 1568. Camerarius the Younger grew it at Nüremberg in 1588, Gerard in London had received

50. EPIMEDIUM ELATUM

Fig. 87. *Epimedium elatum.* Photographed in the wild at 2460 m in the district of Hazara, Pakistan, by Mikinori Ogisu.

it from Robin in Paris by 1597 and gave it the name 'Barren Woort' in English "bicause, as some authors affirme, being drunke it is an enimie to conception." There is no evidence that this or any *Epimedium* has contraceptive properties.

The continued use of the name *Epimedium* for this species led Tournefort in 1694 and 1700, and Linnaeus from 1737 to 1754, to base their generic descriptions upon it; *Epimedium alpinum* L. is accordingly the type species (generitype) of the genus *Epimedium*. Thus a Dioscoridean name became permanently linked with a plant unknown to Dioscorides. Such a misapplication would not have troubled Linnaeus who deliberately gave classical names such as *Ceanothus* and *Ptelea* and *Silphium* to American genera.

THE GENUS EPIMEDIUM
PLATE 8

Epimedium alpinum

FERDINAND BAUER

51. EPIMEDIUM ALPINUM

Epimedium alpinum is an easily grown plant, spreading by long rhizomes, but its small flowers tend to be obscured by overtopping foliage. It has given rise to two hybrids, *E.* × *cantabrigiense* (*E. alpinum* × *E. pubigerum*) and *E.* × *rubrum*, syn. *E. alpinum* var. *rubrum* (*E. alpinum* × *E. grandiflorum*).

Epimedium alpinum L., Sp. Pl. 1: 117 (1753), *et auct. plur.*: Kniphof, Herb. Viv., cent. 10: t. sine no. (1758): Miller, Gard. Dict. ed. 8 art. *Epimedium* (1768); Sowerby & Smith, Engl. Bot. 7, t.438 (1798); ed. 2, 2, p. 13, t.226 (1835); ed. 3, of Syme, 1, t.52 (1863); Sibthorp & Smith, Fl. Graeca, 2: 39, t.150 *quoad icone, descript. et syn., sed loc. excl.* (1813); DC., Veg. Syst. 2: 28, excl. var. ß (1821); Bertoloni, Fl. Ital. 2: 192 (1835); Reichb., Icon. Fl. Germ. 3, t.18, no. 4485 (1838–9); Seringe, Fl. Jard. 3: 272 (1849); Schlosser & Vukotinovic, Fl. Croat.: 190 (1869); Baker in Gard. Chron. II. 13: 620, excl. var. (1880); Schlechtendal, Langethal & Schenk, Fl. Deutschl., ed. 5, of Hallier, 12, 17 (1883); Franch. in Bull. Soc. Bot. France 33: 106, excl. var. (1886); Fiori & Paoletti, Icon. Fl. Ital.: 188 (1895); Fiori, Nuov. Fl. Analit. Ital. 1: 691 (1924); Pampanini, Essai Géogr. Bot. Alpes: 95, map vi., no. 85 (1903); Pampanini in Mém. Soc. Fribourg. Sci. Nat. ser. Géol. III, fasc. 1; Pampanini in Nuovo Giorn. Bot. Ital. N.S. 14: 105 (1907); Hayek, Fl. Steiermark 1: 358 (1908); Kom. in Acta Horti Petrop. 29: 130, excl. syn. '*E. rubrum*' (1908); Bonnier, Fl. Illustr. France, Suisse 1: 43, t.23, fig.101 (1911); Hegi, Illust. Fl. Mitt.-Europa 4: 9, fig.721 (1913); Moss, Cambridge Brit. Fl. 3: 166, t.165 (1920); Stearn in J. Linn. Soc., Bot. 51: 470 (1938), in Plantsman: 188 (1979) & Europ. Gard. Fl. 3: 391 (1989). Lectotype: Herb. Clifford.: 37, Epimedium 1 (BM) — see Chamberlain (1993).

E. alpinum f. *normale* Voss in Vilm. Blumengärtn. ed. 3, 1: 90 (1896).
Epimedium dodonaei [Dalechamps] Hist. Gen. Pl.: 1095, *cum icone* (1587); Valentini, Viridarium Reform.: t.66 (1719).
Epimedium quorundam J.Bauhin, Hist. Pl. Univ. 2, 391, *cum icone* (1651).
Epimedium dioscoridis [Marchant in] Recueil Pl. Louis 14: t.127 (*c.* 1788).

ILLUSTRATIONS. Kniphof; Sowerby and Smith; Sibthorp and Smith; Reichenbach; Schlechtendal, Langethal and Schenk; Fiori and Paoletti; Bonnier; Hegi; Moss. See above references; others are enumerated in Stapf, Index Londin. 3: 54 (1930).

DESCRIPTION. *Flowering stem* 6 cm (i.e. dwarfed alpine specimens) to 30 cm high, bearing one biternate leaf. *Rhizome* long-creeping, 2–4 mm thick. *Leaves basal and cauline*, usually biternate, rarely trifoliolate; leaflets ovate, the tip acute or acuminate, the margin spinous, the base deeply or shallowly cordate, membranous in texture, bright green above, often marginally red-tinged when young, at first pubescent beneath but usually becoming subglabrous, variable in size but capable of attaining 13 cm long, 8.5 cm broad, under cultivation. *Inflorescence* compound, loose, glandular, usually many-(8–26)flowered, nearly always slightly or much shorter than the stem-leaf; lower peduncles 2–3-flowered; pedicels usually 5–15 (sometimes up to 25) mm long. *Flowers* 9–13 mm across. *Outer sepals* grey, speckled red, oblong to obovate, 2.5–4 mm long, 2 mm broad. *Inner sepals* narrowly ovate, blunt or subacute, concave, dull garnet-red, 5–7 mm long, 3 mm broad. *Petals* slightly shorter than the inner sepals, slipper-like, cylindric, 4 mm long, canary-yellow, with no basal laminae. *Stamens* protruding, 3 mm long; anthers 2 mm long. *Capsule* about 15 mm long.

DISTRIBUTION. South Europe‡. From isolated stations in Albania and Hercegovina, *E. alpinum* spreads northward over the highlands of Serbia and Bosnia, having here its area of greatest abundance. Apparently very local and rare in south-west Croatia, it reappears in northern Croatia and north-eastern Istria, becomes fairly widespread in Carniolia northwards to Celje, and then spreads westward across northern Italy along the southern fringe (prealpi) of the Julian Alps and Friuli into the southern Dolomites, here becoming fairly widespread, and ascending, from around Bassano and Verona through the Val Sugana and Val Lagarina, into the Tirol along both sides of the valley of the Adige (Etsch) up to Salorno, but halting to the west at Monte Baldo by the Lago di Garda; it also occurs in two groups of wooded hills, the Colli Euganei near Padua and the Monte Berici near Vicenza, which rise like islands from the valley of the Po; in Lombarida it has only a few isolated stations by the Lago d'Iseo and the Lago di Como, but in the lower regions of the western Alps at the north-west corner of Piemonte between Lago Maggiore and Ivrea it again becomes fairly abundant and occurs in the hills of central Piemonte at Casale. It is also recorded to the south at Garessio and in the Etruscan Apennines near Monte Cimone, but these need confirmation.

Evidently *E. alpinum* is a plant which has spread from the Balkan peninsula — an area of refuge during the Quaternary Ice-Age — into northern Italy up the valley of the Po and thence penetrated into the southern valleys of the Alps; its range is paralleled in whole or in part by various other species with Balkan affinities, e.g. *Lilium carniolicum*, *Festuca spectabilis*, *Genista sericea*, *Omphalodes verna*, *Saxifraga elatior* and *Valeriana saxatilis* (*cf.* Pampanini, 1903). Its altitudinal range is mostly between 100 and 1000 m, and its customary habitat the shade of mixed deciduous woods dominated by *Quercus* and *Castanea*; in Bosnia and Croatia, however, it occurs in woods of beech (*Fagus*) and of conifers (*Abies*, *Picea*, *Pinus*) and it ascends to 1250–1300 m in south-western Serbia and to 1500 m on the Veliki Stolak in eastern Bosnia. In neither ecology nor general range is it an alpine species.

Under cultivation *E. alpinum* thrives in almost any soil and is hardy as far north as Sweden and Scotland; it has also become naturalised here and there in Belgium, Czechoslovakia, France, Germany and Switzerland.

For a more detailed account of distribution, with sources of information, see J. Linn. Soc., Bot. 51: 471–475 (1938), with map (p. 474) used in Hegi, Illust. Fl. Mitt.-Europa 4 (1913).

‡**NOTE**. In preparing this account I used, in addition to herbarium records, information kindly supplied by correspondents, notably A. Béguinot (Genova), A. Froti (Verona), L. Glia i (Beograd), Ivo Horvat (Zagreb), K. Maly (Sarajevo), O. Mattirolo (Turin), and V. Val Nero (Verona); the absence of any other Epimedium or similar plant from this region makes it possible to accept such records. Many specimens were seen at Florence, Prague, Vienna, and elsewhere.

Epimedium pubigerum — ANN FARRER

52. EPIMEDIUM PUBIGERUM

The obvious characters which distinguish *Epimedium alpinum* and *E. pubigerum* in a living state seem to have escaped, until 1938, the attention of botanists working only in the herbarium. Despite the assertion of Franchet and other authors that *E. pubigerum* differs from *E. alpinum* only in the denser and more persistent pubescence of its leaflets, the two are really very distinct. The rhizome of *E. pubigerum* forms a compact woody mass which grows slowly, extending 2–3 cm a year, as the new shoots are short and thus arise near one another, while *E. alpinum* has more elongated slender rhizomes, sometimes growing 6–7 cm a year; under favourable conditions the latter runs into wide patches, so for a small garden, *E. pubigerum* is a much better subject. The leaflets of *E. pubigerum* are firmer in texture and remain green all winter, whereas those of *E. alpinum* have usually perished in late autumn or early winter. *Epimedium pubigerum* has an elongated inflorescence rising well above the stem leaf and thus prominently displays its small flowers whereas the stem leaf of *E. alpinum* usually overtakes its flowers.

The two species also differ greatly in distribution, their nearest stations being some 480 km (300 miles) apart. *Epimedium alpinum* is widespread in the north-west of the Balkan Peninsula whereas *E. pubigerum* gets no further into the Balkan Peninsula than the Strandja mountains of Bulgaria, and has its main area in northern Asia Minor eastward to the Caucasus.

The type material of *E. alpinum* was collected by Olivier near Istanbul (Constantinople), presumably in the Belgrad Forest, where many later botanists have

Fig. 88. ***Epimedium pubigerum.*** Photographed by John Fielding at Blackthorn Nursery, Hampshire.

collected it. It was introduced into English gardens in 1881 by Ellen Ann Willmott from Edmund Boissier's garden in Geneva. This original form had almost white inner sepals but the species varies and other forms have been introduced; for example, one variant from near Trabzon in northern Turkey has redder inner sepals.

The epithet *pubigerum* is from *pubes* (hair, particularly of adolescence) and *gero* (to bear or carry), referring to the persistent hairs covering the lower side of the leaflets.

Epimedium pubigerum (DC.) C.Morren & Decne. in Ann. Sci. Nat. Bot. II. 2: 355 (1834); Seringe, Fl. Jard. 3: 273 (1849); Boiss. Fl. Orient. 1: 101 (1867); Rouy, Ill. Pl. Europ., fasc. 6: t.127 (1896); N.Busch in Fl. Cauc. Crit. 3: 3: 209 (1903); Kom. in Act Horti Petrop. 29: 131 (1908); Stef. & Stoj., Fl. Bulg. 1: 468, fig.565 (1925); Turrill in Hooker's Icon. Pl. 32: (5th ser. 2), t.3116 (1927) & Pl. Life Balkan Penins. 139, 269, 420, 463 (1929); Grossheim, Fl. Kavkaza 2: 124 (1930), ed. 2, 4: 80, map 91 (1950); Stearn in J. Linn. Soc., Bot. 51: 476 (1938), Plantsman 1: 188 (1979) & Europ. Gard. Fl. 3: 393 (1989); Coode in Davis, Fl. Turkey 1: 211 (1965). Type: Turkey, 'circa Constantinopolem', *Olivier* (holotype G).

? *E. orientale, flore albo* Tourn., Corollarium, p. 17 (1703).

? *E. orientale, flore albo flavescente* Tourn., loc. cit. (1703).

E. alpinum var. *pubigerum* DC., Syst. Nat.: 28 (1821); Baker in Gard. Chron. N.S. 13: 620 (1880); Franch. in Bull. Soc. Bot. France 33: 107 (1886).

ILLUSTRATIONS. Hooker's Icon. Pl. 32: t.3116 (1927); Stef. & Stoj., Fl. Bulg. 1: fig.565 (1925); Phillips & Rix, Perennials 1: 32, 33 (1993).

DESCRIPTION. *Flowering stem* 20–75 cm high, bearing one biternate leaf. *Rhizome* comparatively short and stout, nodes *c.* 5 mm or more thick. *Leaves basal and cauline*, usually biternate, occasionally trifoliolate or triternate; leaflets ovate to broadly ovate or almost rotund, the tip acute, the margin sparsely spinous, the base deeply cordate, firm in texture, dark green above, blue-glaucescent and persistently pubescent beneath with crowded soft white hair, usually very dense at the insertion of the petiole, about 4.5 cm long, 2.5 cm broad, but attaining 9.5 cm long, 6 cm broad. *Inflorescence* compound, loose, very glandular, usually many-(12–30)flowered, overtopping the stem-leaf; lower peduncles often 3–5-flowered; pedicels usually 3–10 (sometimes up to 20) mm long. *Flowers* 8–12 mm across. *Outer sepals* red-tinged, broadly ovate, blunt, about 3 mm long, sometimes with short glandular red-tipped hairs. *Inner sepals* narrowly ovate, blunt, very concave, pale rose or nearly white, 5–7 mm long, 2 mm broad. *Petals* slightly shorter than the inner sepals, slipper-like, cylindric, blunt, 3.5–4 mm long, canary yellow, with no laminae. *Stamens* protruding, about 3 mm long; anthers about 2 mm long. *Capsule* short and fairly plump, 5–15 mm long, including the persistent style.

DISTRIBUTION. South-eastern Bulgaria, European Turkey near Istanbul, northern Asiatic Turkey along the Black Sea coastal region to western Georgia (Adzharya), in damp deciduous woods of *Fagus orientalis* and other Pontic trees and shrubs which clothe the sides of the deep and narrow valleys.

Fig. 89. The moist forest of the Black Sea region of Turkey is the home of **Epimedium pubigerum** and **E. pinnatum** subsp. **colchicum.** Photograph: Brian Mathew.

Subgenus II. RHIZOPHYLLUM

(Species number 53–54)

Subgenus *Rhizophyllum* (Fisch. & C.A.Mey.) Stearn **stat. nov.** Type species: *E. pinnatum* Fisch.

Section *Rhizophyllum* Fisch. & C.A.Mey., Sert. Petrop. 1: sub. t.1 (1846).

Section *Dimorphophyllum* Baill. in Adansonia, 2: 270 (1862); Hist. Pl. 3: 56 *in adnot.* (1871) & Nat. Hist. Pl. 3: 56, *in adnot.* (1874).

[Subgen.] *Euepimedium* [sect.] *Gymnocaulon* Franch. in Bull. Soc. Bot. France 33: 40 (1886); N.Busch in Fl. Cauc. Crit. 3, 3: 208 (1903); Kom. in Acta Hort. Petrop. 29: 128 (1908).

This well-defined subgenus is easily recognised by its leafless flower-stalk, conspicuous yellow inner sepals, very small petals, and protruding stamens.

DESCRIPTION. *Rhizome* elongated. *Leaves* normally all basal, with subcoriaceous leaflets. *Inflorescence* simple. *Inner sepals* yellow, rounded at tip, obovate, elliptic or ovate, up to 1 cm long, 8 mm broad. *Petals* reduced to small nectaries with blunt nectariferous sacs (spurs) up to 3 mm long and yellow usually dentate laminae up to 3 mm deep, 3 mm across. *Stamens* protruding, about 5 mm long, with filaments equalling or exceeding anthers.

DISTRIBUTION. Western Asia and North Africa.

53. EPIMEDIUM PINNATUM

The collective species *Epimedium pinnatum* is easily recognisable by its combination of leaves with usually five, occasionally nine, broad evergreen leaflets, leafless flowering stems, conspicuous yellow petal-like inner sepals and minute brownish petals. The species inhabits two widely separate areas: the western Caucasus near the Black Sea and the eastern Caucasus and northern Iran near the Caspian Sea; it is absent from the central Caucasus. These Caspian and Black Sea populations of *E. pinnatum* are some 650 km apart and no *Epimedium* occurs in the intervening area. Since they differ slightly in the shape of the inner sepals and in the length of petals, they are best treated as subspecies: subsp. *pinnatum* from the Caspian region and subsp. *colchicum* from the Black Sea region. Most of the literature on *E. pinnatum* refers to subsp. *colchicum*. This distribution pattern corresponds to that of other Tertiary plants in the Caucasus which, from being spread over the whole area, have retreated into the widely separated maritime districts of Lenkoran and Pontus-Colchis; *cf.* Kusnezow (1909) or Palibin in Bull. Soc. Bot. Genève II. 2: 22–24 (1910). Three subspecies are recognised here: subsp. *pinnatum*, subsp. *colchicum* and subsp. *circinatum*.

Epimedium pinnatum subsp. *colchicum*

ANN FARRER

Epimedium pinnatum subsp. *pinnatum*

This is the eastern and 'typical' subspecies of *E. pinnatum*, *sensu lato*, distinguished by its often biternate leaves, slightly obovate inner sepals, and minute-spurred petals. Grossheim (1927: 19) includes it in his group of 'autochthonous Hyrcanian species, relics of an eastern Old-Mediterranean centre which extended throughout the Pontic region, the Caucasus and the southern Caspian provinces, the Hyrcania of the Ancients'.

Although long known, this plant is poorly represented in herbaria; there are few specimens and these mostly lack flowers. De Candolle described it as *E. pinnatum* from a tracing of a specimen in Pallas' herbarium at St. Petersburg which was sent to him by F.E.L.Fischer; this drawing, now in the Prodromus herbarium of the Conservatoire de Botanique at Geneva, bears Fischer's note: '*Epimedium pinnatum* m. in prov. Persiae Gilan lecta a cl. Hablitzl. Hb. Pallasii'. The discovery of the species thus stands to the credit of the German botanical explorer Carl Ludwig von Hablitzl (1752–1820) for whom Marshal von Bieberstein named the genus *Hablitzia* (Chenopodiaceae).

C.A.Meyer introduced it into cultivation from the Talysh area probably in 1830, and, from a living plant grown at St. Petersburg, he and Fischer published a detailed description and coloured plate (Sert. Petrop.: t.1). Since then this original Caspian plant has been almost lost sight of, the '*E. pinnatum*' of later authors being mostly the western subspecies, subsp. *colchicum*. However, in about 1955, Rear Admiral Paul Furse (1904–1978) and his wife Polly introduced it anew from northern Iran.

Koidzumi's *E. sieboldianum* was described as a new Japanese species from a specimen (Rijks. Herb., Leiden, sheet no. 898, 196–202!) bearing the printed label 'Herb. Lugd. Batav. Japonia' and determined by Miquel as '*Aceranthus sagittatus* Sieb. & Zucc. var.'; a later hand added the name 'Siebold' as collector. This is the only evidence of its supposed Japanese origin; there is no label in Siebold's own hand-writing. I cannot distinguish it from the Caspian *E. pinnatum* and think that it is a cultivated specimen of this species, probably grown in Siebold's nursery near Leiden and by accident labelled as coming from Japan. No eastern Asiatic species resembles it.

E. pinnatum Fisch. ex DC., Syst. Nat. 2: 29 (1821) subsp. **pinnatum** C.A.Mey., Verz. Pflanz. Cauc. Casp. 175 (1831); Fisch. & Mey., Sert. Petrop. 1: t.1 (1846); Boiss., Fl. Orient. 1: 102 (1867), excl. ß; N.Busch, Fl. Caucas. Crit. 3: 209 (1903), excl. ß; Kom. in Acta Horti Petrop. 29: 129 (1908); Grossheim in Beih. Bot. Centralbl. 43, 2: 19 (1927); Grossheim, Fl. Kavkaza 2: 124 (1930) & ed. 2, 4: 79, t.9, fig. 1 (1950). Type: drawing of specimen collected in northern Iran by *Hablitzl* (G).

E. pteroceras C.Morren *secundum* Baker in Gard. Chron. II. 13: 683 (1880), *vix vel non* C.Morren.

E. sieboldianum Koidz. in Bot. Mag. (Tokyo) 44: 95 (1930); Nemoto, Fl. Jap. Suppl.: 236 (1936).

E. pinnatum subsp. *originarum* Stearn in J. Linn. Soc., Bot. 51: 462 (1938).

ILLUSTRATIONS. Fisch. & Mey., Sert. Petrop.: t.1; Gardening Illustr. 54: 31 (Jan. 1932); Grossheim, Fl. Kavkaza ed. II. 4: 89, fig.1 (1950).

DESCRIPTION. *Flowering stem* leafless, 20–30 cm high. *Rhizome* long-creeping, somewhat nodose, 3–4 mm thick. *Leaves* all basal, biternate (9 leaflets), or often pinnate (5 or 11 leaflets) rarely trifoliolate, covered when young with long white or reddish hairs, later nearly glabrous; leaflets narrowly ovate to broadly ovate, the tip acute, the margin spinous-serrate, the base deeply cordate with usually rounded lobes, at length subcoriaceous, dark green above, glaucescent beneath, small at flowering time but later up to 8 cm long, 7 cm broad. *Flowering stem* equalling the leaves, ending in a simple loose glandular or glabrous raceme of 12–30 flowers; pedicels 1–1.5 cm long. *Flowers* about 1.6 cm across. *Outer sepals* ('flower-bracts') oblong-ovate, blunt, brownish with hyaline margin, about 3–5 mm long, 1–2 mm broad. *Inner sepals* broadly obovate, rounded, yellow, sometimes streaked with red near the short basal claw, about 8 mm long, 6 mm broad. *Petals* very small, scarcely 2 mm in total length, with dentate reduced yellow laminae 1–2 mm deep and 3 mm across and brownish-purple nectariferous projections or spurs scarcely 1 mm long. *Stamens* protruding, 5 mm long; anthers yellow, 2–2.5 mm long.

DISTRIBUTION. North Iran (Gilan province) and the adjoining Talysh district of Azerbaydzhan (Azerbaijan) near the Caspian Sea, in mountain woods.

E. pinnatum subsp. *colchicum*

This western subspecies of *E. pinnatum*, common in gardens, diverges from the eastern subspecies *pinnatum* chiefly in its longer petal-spurs, ovate inner sepals, and usually 3- or 5-foliolate leaves with larger and less spinous-, or even entire-, margined leaflets. Concerning the latter, Woronow noted (in Russian, in Herb. Fl. Ross., no. 1704): 'In many of the present specimens the edges of the leaflet are entirely without spines or have only one or two pairs on the lower part. Medwedow and Alboff distinguished these as a separate variety, var. *integrifolium*. In Caucasian forests I have often found both entire and spiny leaflets on one and the same leaf. This characteristic is purely individual and Alboff's variety cannot be maintained'. The plant generally cultivated has slightly broader inner sepals than wild specimens, including the plant introduced from Georgia by Roy Lancaster. It is known in gardens as *E. colchicum*, '*E. pinnatum*', and *E. pinnatum elegans*, the spires of bright yellow *Verbascum*-like flowers being very ornamental in April and the foot-high leaves remaining green all winter; it grows vigorously under almost any conditions. According to Morren (1846), Fischer, in a letter quoted by Hooker (London J. Bot. 1: 207, 1842) and by Lindley (in Bot. Reg. 32, sub. t.9, 1846), all of whom refer to it as '*E. pinnatum*', this plant was sent from Abkhasia by Count Worontzoff to N. de Hartwiss of the Nikita Botanic Garden, Crimea, who introduced it into European gardens about 1842, apparently under the provisional name *E. colchicum*. This name has been continuously used in gardens to the present day; but its first valid publication is uncertain, as botanists for a long time regarded the plant as identical with 'typical' *E. pinnatum* from the Caspian area and mentioned *E. colchicum* only as a horticultural

53. EPIMEDIUM PINNATUM

Fig. 90. The rich Caspian woodlands of northern Iran, where **Epimedium pinnatum** subsp. **pinnatum** occurs. Photograph: Brian Mathew.

synonym. Consequently most of the figures and literature on 'E. pinnatum' really refer to subsp. *colchicum*, a fact not appreciated by Franchet and Komarov, though they separate the two. Apparently Cosson was the first to regard them as distinct species.

The epithet *colchicus* refers to the region of Transcaucasia called Colchis (Kolxis of Strabo), in ancient Greek geography the almost triangular district of the west Caucasus (Georgia) at the south-east corner of the Black Sea (Pontus Euxinus), now represented by Abkhasia. This was the home of the enchantress Medea, according to mythology; here the Argonauts under Jason sought the Golden Fleece.

When E. pinnatum subsp. colchicum and E. perralderianum have been grown together for a long time, bees have hybridised them giving rise to a hybrid group of E. × perralchicum.

E. pinnatum subsp. **colchicum** (Boiss.) N.Busch, Fl. Caucas. Crit. 3: 209 (1903); Stearn in J. Linn. Soc., Bot. 51: 464 (1938); Coode in Davis, Fl. Turkey 1: 211 (1965). Type: 'Hab. in Caucasus Occid.', *Nordmann* (G-BOISS).

E. colchicum hort. ex. Maund, Bot. Gard. 12, t.276, no. 1102, *sine descript.* (1848).

E. pinnatum Fisch., sec. C.Morren in Ann. Soc. Roy. Agric. Gand 2: 139, t.61 (1846); Hook in Curtis's Bot. Mag. 75, t.4456 (1849); *auct. anon.* in Ann. Hort. 5: 377, *cum icone* (1850); C.Morren in Belgique Hort. 4: 35, t.6 (1854); Boiss., Fl. Orient. 1: 102 (1867) p.p.; Baker in Gard. Chron. II. 13: 683 (1880), p.p.; Franch. in Bull. Soc. Bot. France 35: 104 (1886), p.p.; Alboff, Prod. Fl. Colch.: 15 (1895); Kom. in Acta Horti Petrop. 29: 129 (1908), p.p. – *non* Fisch., *sensu stricto*.

E. pinnatum var. *colchicum* Boiss., Fl. Orient. 1: 102 (1867); Franch. in Bull. Soc. Bot. France 35: 104 (1886); Woronow in Sched. Herb. Fl. Ross. no. 1704 (1908).

E. colchicum hort. ex Ann. Hort. 5: 377 (1850); ex C.Morren in Belgique Hort. 4: 35, t.6 (1854); ex Boiss., Fl. Orient. 1: 102 (1867), *omnes pro syn.*

'*E. pinnati*'; Vilmorin, Fl. Pleine Terre: 226 (1863); ed. 2: 297 (1866); ed. 3: 375 (1870); 'Fischer mss.' ex Baker, in Gard. Chron. II. 13: 683 (1880) *pro var.*; Casson, Ill. Fl. Atlant. 1: 9 (1882) & Compend. Fl. Atlant 2: 57 (1887), *in obs.*; Trautvetter, Increm. Fl. Ross. 1, no. 274 in Acta Horti Petrop. 8: 65 (1883); Kom. in Acta Horti Petrop. 29: 129 (1908); Sosn. in Monit. Jard. Bot. Tiflis N.S. 1: 75 (1923); Bergmans, Vaste Pl.: 202 (1924); Grossheim, Fl. Kavkaza, 2: 124 (1930).

'*E. macranthum*' Garden 46: 356 *in icone*, '*E. pinnatum*' Goldring in textu (1894).

E. pinnatum var. *integrifolium* Medwedow & Alboff ex Alboff, Prod. Fl. Colch.: 15 (1895).

E. pinnatum f. *colchicum* Voss in Vilm. Blumengärtn. ed. 3, 1: 51 (1894).

E. pinnatum subsp. *colchicum* (Boiss.), *cum.* f. *integrifolio* (Alboff), N.Busch, Fl. Caucas. Crit. 3: 209 (1903).

E. pinnatum elegans hort. ex W.Miller in Bailey, Standard Cyclopedia of Hort. ed. 2: 1122 (1914); Farrer, Engl. Rock-Gard. 1: 327 (1919); Bergmans, Vaste Pl. 203, fig.46 (1924).

E. elegans hort. ex Bergmans, Vaste Pl. 203, fig.46 (1924) *pro syn.*

ILLUSTRATIONS. Curtis's Bot. Mag. 75: t.4456 (1849); Ann. Soc. Roy. Agric. Gand 2: t.61 (1846); Belg. Hort. 4: t.6 (1854); Phillips & Rix, Perennials 1: 35 (1993).

DESCRIPTION. *Flowering stem* leafless, 20–40 cm high. *Rhizome* long-creeping, about 5 mm thick. *Leaves* all basal, usually 3- or 5-foliolate; leaflets ovate to broadly ovate, the tip acute, the margin sparsely spinous-serrate or even entire (forma *integrifolium* Alboff), the base deeply cordate with rounded frequently overlapping lobes, dark green above, glaucescent beneath, frequently evergreen, up to 15 cm long, 11 cm broad. *Flowering stem* ending in a loose glandular or glabrous raceme of 15–20 (or more) flowers. *Flowers* about 1.8 cm across. *Outer sepals* oblong-ovate, blunt, 3–5 mm long, 1–3 mm broad. *Inner sepals* rounded, aureolin-yellow, the outer pair broadly ovate about 7 mm broad, the inner pair almost elliptic ('oblong') and slightly narrower. *Petals* small, up to 3.5 mm in total length, with dentate reduced yellow laminae about 2 mm deep and straight upcurved, brown or yellow (forma *unicolor* C.Morren. in Belg. Hort. 4: 35, *pro var.*, 1854), spurs about 2 mm long. *Stamens* protruding, 5–6 mm long; anthers yellow, 2.5–3 mm long.

DISTRIBUTION. Western Georgia and north-eastern Turkey (Çorum, Rize and Trabzon provinces), in woods at 50–300 m near the Black Sea coast.

E. pinnatum subsp. *circinatum*

This is an obscure plant known to me only from the 1923 publication of Sosnovsky. Examining the epimediums grown in the Colchic section of the Tiflis Botanic Garden under the name *E. colchicum*, and obtained from various parts of western Transcaucasia, D.Sosnovsky was able to distinguish three kinds, contrasting them as follows (translated from the Russian):

Locality whence obtained	(1) Dzhvari, Zugdidui, Novosenaki.	(2) Sochi	(3) Tiflis Bot. Gard., mixed with others
Sepal	sulphur yellow	bright yellow	bright yellow
Petal	in its widened part sulphur-yellow, the spur brown, straight, $1/3$ the length of sepal	entirely bright yellow, towards the tip convoluted like a snail-shell, $1/4$ the length of sepal	in its widened part yellow, the spur brown, slightly upcurved, shorter than in first kind
Stamen-filament	almost the same length as anther	half as long again as anther	as in first kind
Anther	yellow	green	green
Determined by Sosnovsky as	*E. colchicum*	*E. circinnato-cucullatum* sp. nov. (ad interim)	probably hybrid between 1 and 2

The first and second differed in definite floral characters, while the third, of which individuals grew scattered among patches of the others, had features belonging to one or the other of these and appeared to be a hybrid between them; the vegetative organs were similar in all three.

The second kind from Sochi, with yellow curled petal-spurs and green anthers, Sosnovsky provisionally described as a new species, *E. circinnato-cucullatum*; I prefer to regard it as another subspecies of the collective *E. pinnatum*.

Sochi, the type-locality of *E. circinnato-cucullatum*, is near the Black Sea in the northern Caucasus about 260 km north-west of Batum. According to Sosnovsky, specimens collected at Sochi by Medwedew belong to typical *E. colchicum*.

E. pinnatum subsp. **circinatum** (Sosn.) Stearn in J. Linn. Soc., Bot. 51: 466 (1938). Type: cult. Hort. Bot. Tiflis, ex Sochi (holotype TBI?, *n.v.*).
E. circinnato-cucullatum Sosn. in Monit. Jard. Bot. Tiflis II. 1: 75 (1923); Grossheim, Fl. Kavkaza 2: 124 (1930), ed. 2, 4: 79 (1950).

DESCRIPTION (after Sosnovsky). *Leaves* imparipinnate, 1–2-paired or biternate; leaflets ovate, deeply cordate, with shortly acuminate apex, margins more or less spinous-serrate.

Racemes simple. *Petals* golden yellow, nectariferous spurs golden yellow, cucullate, 4 times shorter than the petals, circinnate-subrevolute, concolorous. *Staminal filaments* 1.5 times as long as the anthers, green.

54. EPIMEDIUM PERRALDERIANUM

When in 1861 three French botanists, Ernest St.-C. Cosson (1822–1889), Jean Louis Kralst (1813–1892) and Henri de la Perraudière (1831–1861) ascended Mount Babor, east Kabylia, in north-eastern Algeria, for the first time, they found, much to their surprise, plants of an *Epimedium* growing at between 1200 and 1300 m under the oaks (*Quercus afares*) and cedars (*Cedrus atlantica*).

According to Trabut (in Bull. Soc. Bot. France 36: 61, 1889), the north slope of the Babor massif is 'entièrement couvert par un forêt très boisé de Cèdres, *Abies numidica*, *Acer obtusatum* [var. *africanum* Pax], *Populus tremula*, *Quercus mirbeckii* souvent courvert de galles, *Ilex aquifolium*, *Taxus baccata* et *Quercus afares*… Deux raretés habitent sous la futaie du versant nord: l'*Epimedium perralderianum* et le *Campanula trichocalycina*'. An essentially temperate flora. It is this region of Algeria 'qui recoit le maximum de pluies annuelles.'

This montane woodland is the last remnant of lowland forest which flourished along the North African coast during the period of heavy Mediterranean rainfall when the ice sheets of Europe were at their maximum. The *Epimedium* was not in flower but had unmistakable *Epimedium* leaves. Cosson brought a plant back to Paris and when it flowered, he described and illustrated it as a new species, named *E. perralderianum* in memory of his friend Henri René le Tourneux de la Perraudière who had died of fever during the expedition. The syllable *–ald* of vulgar Latin or old French became *–aud* in modern French. Cosson reversed this process when dedicating *E. perralderianum* to him. From Paris, *E. perralderianum* was distributed to other gardens. Only one clone being introduced, it is self-sterile.

Epimedium perralderianum is closely allied to *E. pinnatum* subsp. *colchicum* despite the vast distance and long period of time separating them, but may be distinguished by constantly having only three leaflets, rigid, with very spiny margins, more evidently veined on the upper surface (instead of three to seven leaflets). The spur of the minute petal is bent upwards at an angle of about 45° (instead of almost straight or only slightly upcurved). It is dwarfer than *E. pinnatum* subsp. *colchicum* grown under the same conditions.

E. perralderianum Coss. in Bull. Soc. Bot. France 8: 607, *nomen* (1862); 9: 167, *cum descript.* (1862); Baker in Gard. Chron. N.S. 13: 683 (1880); Hook.f. in Curtis's Bot. Mag. 106: t.6509 (1880); Cosson, Ill. Fl. Atlant. 1: 9, t.5 (1882) & Compend. Fl. Atlant. 2: 57 (1887); Franch. in Bull. Soc. Bot. France 33: 105 (1886); Battandier, Fl. Algér. Dicot.: 18 (1899); Kom. in Acta Horti Petrop. 29: 130 (1908); Stearn in J. Linn. Soc., Bot. 51: 467 (1938) & Europ. Gard. Fl. 3: 391 (1989). Type: 'in silvaticis umbrosis quercinis regionis montanae in provinciae Cirtensis Kabylia orientali, ad 1200–1500-met., in ditione Beni-Foughal ad fontem El-ma-Berk, *H. de la Perraudière* (holotype P).

Epimedium perralderianum

CHRISTABEL KING

Fig. 91. ***Epimedium perralderianum.*** Photographed by John Fielding in the garden of Elizabeth Strangman, in Kent, England.

E. pinnatum var. *perralderianum* (Coss.) H.R.Wehrh., Gartenstauden 1: 453 (1931) & in Bonstedt, Pareys Blumengärtn. 1: 621 (1930) as 'var. *peralderianum*'.

ILLUSTRATIONS. Curtis's Bot. Mag.: t.6509 (1880); Cosson, Ill. Fl. Atlant. 1: 9, t.5 (1882); Phillips & Rix, Perennials 1: 34 (1993).

DESCRIPTION. *Flowering stem* leafless, 15–30 cm high. *Rhizome* long-creeping, 2–3 mm thick. *Leaves* all basal, trifoliolate or rarely unifoliolate, never pinnate or biternate; leaflets ovate to broadly ovate, the tip acute, the margin undulate, rigid and very spinous-serrate, the base deeply cordate with rounded subequal lobes, at length firm and subcoriaceous, often bronzed when young, glaucescent and ultimately subglabrous beneath, from 2.5 cm long, 2 cm broad, to 6.5 cm long, 5.5 cm broad, or under cultivation even to 10 cm long, 7 cm broad. *Flowering stem* equalling the leaves, ending in a simple very glandular raceme of 9–25 flowers; pedicels 1–2.5 cm long. *Flowers* 1.5–2.3 cm across. *Outer sepals* greenish, oblong-ovate, blunt, 4–5 mm long, 1–2 mm broad. *Inner sepals* obovate, rounded, aureolin-yellow, 8–11 mm long, 5–9 mm broad. *Petals* small, 2.5 mm in total length, with dentate yellow laminae 2–3 mm deep, and brown upcurved spur 1–2 mm long. *Stamens* protruding, 5 mm long; anthers yellow, 2–3.5 mm long.

DISTRIBUTION. North Africa: Algeria, east Kabylia approx. 36°30'N, 5°25'E), growing in the shade of oaks and cedars (*Cedrus atlantica*) on the Chaîne des Babors at 1200–1500 m.

Appendix 1. *EPIMEDIUM* HYBRIDS

If two closely akin plant species chance to grow in the same locality, these may nevertheless remain distinct by being intersterile, by flowering at different months, attracting different pollinators, engaging only in self-pollination or by occupying different habitats; they thus avoid loss of specific identity through extensive interbreeding. Species of *Epimedium* mostly occupy different areas in nature with little overlap and have seemingly evolved in geographical isolation. However, on Mount Omei (Emei Shan), where both *E. acuminatum* and *E. fangii* occur, their meeting has resulted in the creation of a hybrid swarm, described as *E.* × *omeiense*, with some members so distinct from both parents as to give the impression of belonging to an independent species. Other natural hybrids, involving *E. diphyllum* and *E. grandiflorum*, occur in Japan. The only observations on the bees pollinating them have been made in Japan by Kazuo Susuki (1986).

In the first half of the nineteenth century there appeared in European gardens several distinct plants which a study of the known species established beyond doubt had no counterparts in the wild, although they were usually referred to as varieties of *E. alpinum*, *E. grandiflorum* or *E. pinnatum*. They were well illustrated at the time and plants still in cultivation can be confidently identified with the plants portrayed. These blend the characters of species long grown in gardens but which, with the exception of *E. diphyllum* and *E. grandiflorum*, never occur together in nature. Their history, intermediate characters and mingled flower pigments together made it certain they had arisen in gardens through hybridisation. The deliberate crossing of *Epimedium* species in the botanic garden in Ghent by André Donckelaar, although stated in his obituary by Dodart Spae in 1864, has hitherto been overlooked. To him we owe such good garden plants as *E.* × *rubrum*, *E.* × *versicolor* and some *E.* × *youngianum* cultivars.

Observation of plants in gardens led moreover to the conclusion in 1938 that clones of the species then in cultivation were self-sterile. Most species had been propagated vegetatively from a single introduced plant. There is no record, for example, of *E. alpinum*, which was introduced in the sixteenth century, ever producing seed until some time between 1830 and 1850 when it was hybridised with *E. grandiflorum*. Other species have likewise always been sterile when grown on their own. On the other hand, the existence of hybrids between species so geographically distant as European *E. alpinum* and Japanese *E. grandiflorum*, as also between *E. grandiflorum* and *E. pinnatum* subsp. *colchicum*, indicated that all species might well be interfertile when brought together. This has been proved by subsequent experience.

The earlier hybrids recorded are as follows, the dates being approximately when they and their parents were first introduced or noticed:

E. alpinum (1580) × *E. grandiflorum* (1830) = *E.* × *rubrum* (1854)

E. diphyllum (1830) × *E. grandiflorum* (1830) = *E.* × *youngianum* 'Youngianum' (1839), 'Roseum' (1849?), 'Niveum' (1866)

E. grandiflorum (1830) × *E. pinnatum* subsp. *colchicum* (1840) = *E.* × *versicolor* 'Versicolor' (1849), 'Sulphureum' (1849)

E. alpinum (1580) × *E. pinnatum* subsp. *colchicum* (1840) = *E.* × *warleyense* (1909)

E. perralderianum (1867) × *E. pinnatum* subsp. *colchicum* (1840) = *E.* × *perralchicum* (1932)

E. alpinum (1580) × *E. pubigerum* (1887) = *E.* × *cantabrigiense* (1950)

Their characteristics are, not surprisingly, intermediate and variable. Hybrids derived from *E. pinnatum* subsp. *colchicum*, e.g. *E.* × *versicolor* and *E.* × *warleyense*, occasionally have a leafless flower stem, broadly ovate inner sepals and long stamen-filaments; those from *E. grandiflorum* rather long pointed spurs to the petals developed at the base into distinct petaloid laminae; those from *E. diphyllum* obovate petals with pointed spurs of varying length or even no spurs and often paired leaflets; those from *E. alpinum* dark red or reddish inner sepals, etc. It is evident that crosses between unlike species can produce hybrids of high garden merit whereas crosses between closely allied species produce plants which are not much better, or no better, than their parents. Therefore there is nothing to be gained by crossing these closely related taxa. For example, *E.* × *perralchicum*, which arose naturally, is not really an improvement on *E. pinnatum* subsp. *colchicum* and *E. perralderianum*; and *E.* × *cantabrigiense*, also a natural hybrid, is perhaps an improvement on *E. alpinum* but not on *E. pubigerum*. On the other hand *E.* × *rubrum* (*E. alpinum* × *E. grandiflorum*) and *E.* × *warleyense* (*E. alpinum* × *E. pinnatum* subsp. *colchicum*) are good and distinctive garden plants, well separate from their unrelated parents. It is to be hoped that hybridisers of *Epimedium* will be severely critical and destroy any hybrids scarcely distinct from their parents or from hybrids already known. Too many hybrids benefit neither the nurseryman nor the buyer.

The Chinese *E. acuminatum*, *E. dolichostemon*, *E. leptorrhizum* and *E. davidii* were introduced into cultivation in the late twentieth century and have been grown side by side in gardens. Their nectariferous flowers are highly attractive to bumble-bees, and this has led to the production of abundant hybrid seed and the raising, sometimes as chance seedlings, of some very distinct and beautiful hybrids. Some of these hybrids of Chinese parentage are:

E. acuminatum × *E. dolichostemon* = *E.* 'Amanogawa', *E.* 'Kaguyahime'

E. acuminatum × *E. fangii* = *E.* × *omeiense* 'Emei' and 'Stormcloud'

E. dolichostemon × *E. leptorrhizum* = *E.* 'Enchantress' (1990)

E. davidii × *E. pubescens* = *E.* 'Kew Hybrid' (1995)

[Recently Mr Robin White of Blackthorn Nursery, Hampshire, has informed us that he has a hybrid between *E.* × *omeiense* and *E. membranaceum*, and no doubt others will emerge as more and more epimediums are cultivated together — ed.]

In Japan there has developed during the last twenty years an enthusiasm for cultivars of *Epimedium* derived from crossing variants of the Japanese *E. grandiflorum*, *E. sempervirens* and *E. diphyllum* and some of the Chinese species, apparently *E. acuminatum* and *E. membranaceum*. Approaching 200 cultivars have received names. All are beautiful but, since most gardeners have space for only a few kinds, such an abundance makes choice difficult and is likely to be commercially self-defeating. In a genus such as *Epimedium* it must be admitted that there can be too many cultivars.

No attempt has been made to include all the hybrids now known, but, as the origin of some of the early hybrids is obscure or has been misunderstood, accounts of these are given below. Some of the newer hybrids that have arisen between recently introduced Chinese species are also included.

A Key to Named Hybrids of *Epimedium*

An essential tool is a ×8 lens

1. Flower stem with 1 leaf or no leaf. Leaves with 2–9 leaflets 2
1. Flower stem with 2 (rarely 3) leaves, each with 3 leaflets 13
2(1). Petals very small, *c*. 4 mm long, very much shorter than the conspicuous petal-like inner sepals *c*. 8 mm long and 5–6 mm broad. Stamens protruding 3
2. Petals conspicuous, often about as long as or slightly longer than the inner sepals. Stamens largely covered by petals, rarely protruding 4
3(2). Inner sepals sulphur yellow. Leaves with 3–5 leaflets, the marginal spines 0.2–2.6 mm long. Flower stem always leafless (*E. perralderianum* × *E. pinnatum* subsp. *colchicum*) .... **E. × perralchicum** [Page 186]
3. Inner sepals at first coppery red, later dull reddish yellow. Leaves usually with 5 or 9 leaflets, rarely 3, the marginal spines *c*. 0.5 mm long. Flower stem leafless or leaf-bearing on the same plant (*E. alpinum* × *E. pinnatum* subsp. *colchicum*) .... **E. × warleyense** [Page 182]
4(2). Inner sepals concave, boat-shaped, dull red or carmine red, 5–12 mm long. Petals slipper-like, obtuse, very pale yellow, almost as long as the inner sepals. Inflorescence compound 5
4. Inner sepals flat, white, mauve, dull rose or yellow. Inflorescence usually simple 6
5(4). Flowers 1.5–2.5 cm across. Inner sepals crimson carmine, 10–12 mm long. Petals with basal lamina (*E. alpinum* × *E. grandiflorum*) .. **E. × rubrum** [Page 180]
5. Flowers *c*. 1 cm across. Inner sepals dull red, *c*. 5.5 mm long. Petals without lamina (*E. alpinum* × *E. pubigerum*) .... **E. × cantabrigiense** [Page 179]
6(4). Petals yellow, with cylindrical spurs. Flower stem 20–50 cm high, leafless or leaf-bearing on the same plant (*E. grandiflorum* × *E. pinnatum* subsp. *colchicum*) .... **E. × versicolor** (go to 7)

Epimedium* × *rubrum (upper), ***E.* × *cantabrigiense*** (lower left), ***E. alpinum*** (lower centre) MARK FOTHERGILL

6. Petals white or purplish, the spur absent or present and conical or slender and subulate, often varying on the same plant. Flowers stem 15–30 cm high, always leaf-bearing. Leaves with 2–9 leaflets, often without a median leaflet (*E. diphyllum* × *E. grandiflorum*) .... **E. × *youngianum*** (go to 9)

7(6). Inner sepals dull rose or copper-red, petals yellow. Leaves perishing in late autumn 8

7. Inner sepals and petals both yellow. Leaves normally remaining green all winter 9

8(7). Inner sepals dull rose or 'old' rose ...... ***E. × versicolor* 'Versicolor'** [Page 187]

8. Inner sepals coppery-red .... ***E. × versicolor* 'Cupreum'** [Page 187]

9(7). Leaves on flower stem usually with 9 leaflets, these deeply cordate with a very narrow sinus, the lobes touching or overlapping. Petals about as long as the inner sepals .... ***E. × versicolor* 'Sulphureum'** [Page 188]

9. Leaves on flower stem usually with 3 leaflets, these shallowly cordate with an open sinus, the lobes diverging at 60°–80°. Petals distinctly shorter than inner sepals .... ***E. × versicolor* 'Neosulphureum'** [Page 190]

10(6). Inner sepals and petals purplish ....
.... ***E.* × *youngianum* 'Roseum'** and **'Merlin'** [Pages 194, 195]

10. Inner sepals and petals white 11

11(10). Inflorescence longer than stem leaf and overtopping this. Leaflets obtuse
.... ***E. × youngianum* 'Niveum'** [Page 195]

11. Inflorescence shorter than stem leaf and overtopped by this. Leaflets acute or acuminate 12

12(11). Leaves with 9 long-acuminate leaflets ....
.... ***E. × youngianum* 'Youngianum'** [Page 191]

12. Leaves with 2 or 4 acute or shortly acuminate leaflets.......................................
.... ***E. × youngianum* 'Amy'** [See note page 194]

13(1). Petals with spurs *c*. 3 cm long, much longer than inner sepals (*E. acuminatum* × *E. fangii*) .... **E. × *omeiense*** (go to 14)

13. Petals with spurs 12–14 mm long, slightly longer to much shorter than inner sepals 15

14(13). Inner sepals dull rose, without any swellings or outgrowths. Petals red-streaked towards base .... ***E. × omeiense* 'Emei Shan'** [Page 184]

14. Inner sepals dull purple, often with a conical swelling or outgrowth. Petals dark purple towards base .... ***E. × omeiense* 'Stormcloud'** [Page 184]

15(13). Petals with yellow lamina enclosing stamens. Inner sepals reddish. Leaflets pubescent beneath (*E. davidii* × *E. pubescens*) .... ***E.* 'Kew Hybrid'** [Page 185]

15. Petals without distinct lamina. Stamens protruding. Inner sepals pale purple. Leaflets apparently glabrous or with minute appressed bristles 16

16(15). Inflorescence simple, with about 12 flowers, somewhat concealed by
the leaves (*E. dolichostemon* × *E. leptorrhizum*)........ **E. 'Enchantress'** [Page 184]
16. Inflorescence compound, with 12–40 flowers, held well above the foliage
(*E. acuminatum* × *E. dolichostemon*) .... **E. 'Kaguyahime'** [Page 185]

EPIMEDIUM × CANTABRIGIENSE [*E. alpinum* × *E. pubigerum*]

The collection of cultivated taxa of *Epimedium* grown for me under glass in the garden of St. John's College, Cambridge between 1930 and 1932 was later planted outdoors by Ralph Thoday the head gardener in a raised bed inside the college's wilderness garden. There, untended, they fended for themselves during the Second World War (1939–1945). In 1950 when Mr Thoday and I checked this collection, we found that vigorous growers like *E. alpinum* and *E. pinnatum* subsp. *colchicum* had ousted their weaker brethren but not sturdy *E. pubigerum* and among these survivors grew a clump of distinct plant different from them all. It was clearly a hybrid between *E. alpinum* and *E. pubigerum* combining the flowers of the one with the inflorescence of the other, which had evidently arisen during the wartime years of neglect. Mr Thoday propagated and generously gave this intermediate plant to other gardens under my unpublished name *E.* × *cantabrigiense*.

Grown under the same conditions the flowering stems of *E. pubigerum* rise to about 70 cm, those of *E.* × *cantabrigiense* to about 60 cm, those of *E. alpinum* to about 45 cm, its flowers usually somewhat hidden among the foliage. The leaves of the hybrid, though evergreen, like those of *E. pubigerum*, approximate more to those of *E. alpinum*. It is a pleasing but unspectacular ground cover plant, with parents too closely allied for it to be as distinctive as the hybrids of *E. alpinum* with *E. pinnatum* subsp. *colchicum* (i.e. *E.* × *warleyense*) and with *E. grandiflorum* (i.e. *E.* × *rubrum*).

The epithet *cantabrigiense*, referring to its city of origin, derives from Norman *Cantebrigie*, a corruption of the earlier Saxon *Grantebrycge*, the name for the Anglo-Saxon settlement by the bridge over the river Granta which grew into the university town (later city) of Cambridge.

Epimedium × cantabrigiense Stearn in Plantsman 1: 190 (1979); Europ. Gard. Fl. 3: 391 (1989).

ILLUSTRATIONS. Plantsman 1: 189 (1979); Phillips & Rix, Early Perennials 32 (1991).
DESCRIPTION. *Flowering stem* 30–60 cm high, bearing one biternate leaf. *Rhizome* short, with annual growth about 3–5 cm long, 5 mm thick. *Leaves* basal and cauline, usually biternate but with the lateral divisions sometimes having 7 leaflets, remaining firm and green through the winter; leaflets ovate or broadly ovate, the tip acute or shortly acuminate, the margin sparsely spinous, the base deeply cordate with the lateral leaflets uneven-sided as is usual in the genus, the terminal leaflet to 10 cm long and 7 cm broad, persistently pubescent beneath. *Inflorescence* compound, loose, many flowered, 17–30 cm

long, overtopping the stem-leaf, glandular with numerous red-tipped hairs. *Flowers* about 1 cm across. *Outer sepals* quickly falling, about 2 mm long. *Inner sepals* boat-shaped, blunt, dull red, about 5.5 mm long, 3.5 mm broad. *Petals* (nectaries) slightly shorter than the inner sepals, slipper-like, blunt, about 5 mm long, pale yellow with no laminae. *Stamens* protruding, about 3 mm long; anthers yellow, about 2 mm long.

PARENTAGE. *E. alpinum* L. × *E. pubigerum* (DC.) C.Morren & Decne.

EPIMEDIUM × RUBRUM [*E. alpinum* × *E. grandiflorum*]

By 1848 the university botanic garden in Ghent, Belgium, possessed *Epimedium alpinum, E. diphyllum, E. grandiflorum* and *E. pinnatum* subsp. *colchicum*. Philipp Franz von Siebold, on his return from Japan in 1830, had had planted in the Ghent garden *E. grandiflorum* together with other Japanese plants, and this species obviously attracted the attention of the Dutch gardener André Donckelaar (1783–1858) when he became head gardener at Ghent. He hybridized it with other *Epimedium* species then available. His cross of *E. grandiflorum* with *E. alpinum* produced the ornamental garden plant which Charles Morren named *E. rubrum* in 1854. Its origins were unknown to Joseph Hooker at Kew, as also to other authors until 1932, and he considered it as merely a large-flowered variety of *E. alpinum*. However from *E. alpinum* it differs not only in its larger and brighter flowers (about 1.5–2.5 cm across, instead of 0.9–1.3 cm as in *E. alpinum*) but also in its less vigorous growth of rhizome and the more tapering slightly curved spurs of the petals expanded upwards at the base to partly enclose the stamens.

Epimedium × rubrum is a good garden plant, attractive for its red young leaves as well as its crimson and pale yellow flowers. Moreover it is easy to cultivate in any good well-drained soil, which does not dry out in summer, preferably in half-shade.

Epimedium × rubrum C.Morren in Belgique Hort. 4: 33, t.6 figs.1–6 (1854); Baker in Gard. Chron. N.S. 13: 620 (1880); Stearn in J. Linn. Soc., Bot. 51: 515 (1938), Europ. Gard. Fl. 3: 391 (1989).
 E. purpureum Vilm., Fl. Pleine Terre, ed. 2: 299 (1866).
 E. alpinum var. *rubrum* (C.Morren) Hook. f. in Curtis's Bot. Mag. 93: t.5671 (1867).
 E. alpinum f. *rubrum* (C.Morren) Voss in Vilm. Blumengärtn. ed. 3, 1: 50 (1894).
 E. coccineum Silva Tarouca, Unsere Freiland-Stauden: 93 (1910).
 E. youngianum f. *rubrum* A.Lehm., Gartenzierpflanzen ed. 2: 111 (1937).

ILLUSTRATIONS. Belgique Hort. 4: t.6 (1854), 18: t.14 (1868); Bot. Mag. 93: t.5671 (1867); Gartenflora 11: t.373 (1862); Phillips & Rix, Early Perennials 34, 35 (1991).
DESCRIPTION. *Flowering stem* 25–35 cm high, bearing one (abnormally two) bi- or tri-ternate leaf. *Rhizome* elongated, ± 3 mm thick. *Leaves* basal and cauline, usually biternate, sometimes triternate or with the secondary petiolules 5-foliolate; leaflets ovate or narrowly ovate, the tip acuminate, sometimes 2- or 3-fid, the margin spinous serrate, the base deeply or shallowly cordate, at first pubescent beneath, later subglabrous, membranous in texture,

Epimedium × *warleyense*

often bright red when young, up to 14 cm long, 9 cm broad. *Inflorescence* compound, loose, glabrous or sparsely pilose, many-(10–23)flowered, shorter, equalling or overtopping the stem-leaf; lower peduncles often 5-flowered; pedicels 1–2 cm long. *Flowers* 1.5–2.5 cm across. *Outer sepals* ovate-oblong, grey, speckled red, 3–4 mm long, 1.5–3 mm broad. *Inner sepals* narrowly oblong-ovate, blunt, concave, light crimson-carmine or cochineal-carmine, 1–1.2 cm long, 4–5 mm broad. *Petals* about as long as or shorter than the inner sepals, slipper-like, cylindric, pale yellow or white, tinged with red, 1 cm long, with inflated blunt tips curving upwards and slight or distinct rounded basal laminae, to 4 mm high, tending to enclose the stamens. *Stamens* included or slightly protruding, 4 mm long; anthers 3 mm long.

PARENTAGE. *E. alpinum* L. × *E. grandiflorum* C.Morren; a hybrid raised by A.Donckelaar in the Ghent university botanic garden.

EPIMEDIUM × WARLEYENSE [*E. alpinum* × *E. pinnatum* subsp. *colchicum*]

Between 1890 and 1914 Miss Ellen Ann Willmott (1858–1934) by extravagant expenditure made her garden at Great Warley, Essex one of the most richly stocked and celebrated in England. Roses were her great love, as her book *The Genus Rosa* (1910–1914) testifies, but she assembled a collection of *Epimedium*. She employed an artist to paint these plants and in 1932 she lent me for a short time these beautiful paintings. One portrayed a plant hitherto unknown, evidently a hybrid between *E. alpinum* and *E. pinnatum* subsp. *colchicum* which had arisen naturally in the garden at Warley Place. After her death it passed into general cultivation.

Epimedium × *warleyense* stands out in the garden on account of its conspicuous 'coppery-red' inner sepals, anthocyanin derived from *E. alpinum* overlaying plastid yellow derived from *E. pinnatum* subsp. *colchicum*. It is a sturdy grower like both its parents.

The place-name 'Warley' is of medieval origin, recorded as 'Werle' in 1045 and 'Warleia' in 1086.

Epimedium × warleyense Stearn in J. Linn. Soc., Bot. 51: 521, figs.15–19 (1938); Europ. Gard. Fl. 3: 391 (1989).

ILLUSTRATIONS. Stoker, A Gardener's Progress, t.11 (1938); J. Linn. Soc., Bot. 51: 520 (1938); Phillips & Rix, Early Perennials 34 (1991).

DESCRIPTION. *Flowering stem* 20–55 cm high, leafless or with one 5- or 9-foliolate leaf. *Rhizome* elongated, 4–8 mm thick. *Leaves* basal and cauline, 9- or 5- (rarely 3-) foliolate; leaflets ovate to broadly ovate, the tip acute, the margin sparsely spinous-serrate, the base deeply cordate, with rounded or acute lobes, finely pubescent beneath, green when young, becoming subcoriaceous with age and often remaining green all winter, up to 13 cm long, 9.5 cm broad. *Inflorescence* simple or compound, glandular, 10–30-flowered, with sometimes a large secondary raceme in the leaf-axil;

lower peduncles often 5-flowered; pedicels 1–2 cm long. *Flowers* 1.5 cm across. *Outer sepals* green or purple-tinged, obovate, blunt, 3–4 mm long. *Inner sepals* ovate-oblong, blunt, 'coppery red' or 'orange pink' in effect, the yellow ground-colour being conspicuously tinged and veined with red ('rouge cerise'), especially in the opening flower, fading to a yellowish salmon colour, 8 mm long, 3 mm broad. *Petals* much shorter than the inner sepals, yellow ('jaune safran'), with blunt, occasionally red-streaked spurs scarcely 4 mm long and small bilobed (not erose) laminae about 3 mm across, 3 mm deep. *Stamens* protruding, 4.5 mm long; anthers 2.5 mm long, greenish. It flowers from early April into May.

PARENTAGE. *E. alpinum* L. × *E. pinnatum* Fisch. ex DC. subsp. *colchicum* C.A.Mey.: a hybrid which arose naturally in the garden of Miss E.A. Willmott, Warley Place, Brentwood, Essex.

EPIMEDIUM × OMEIENSE [*E. acuminatum* × *E. fangii*]

When different species of *Epimedium* are grown together in cultivation and bumble-bees (*Bombus* spp., etc.) abound, their nectar-seeking activities usually result in uncontrolled hybridisation. In the wild usually only one species occupies an area and then there is no hybridisation. But on Emei Shan (Mount Omei) the areas of *E. acuminatum* and *E. fangii* meet, resulting in a hybrid swarm discovered by Mikinori Ogisu. The hybrids ranged in form and colour from one to the other, and out of them Mikinori selected two very distinct forms described below as 'Emei' and 'Storm Cloud'.

Epimedium × omeiense Stearn in Curtis's Bot. Mag. 12: 22 (1995) (*E. acuminatum* Franch. × *E. fangii* Stearn). Type: Central China, Sichuan, Emei Shan (Mount Omei) at Jiuladong, (103°41'E, 29°28'N), *M.Ogisu 82001*; cultivated at Blackthorn Nursery by R.White, April 1994 (holotype K).

DESCRIPTION. *Flowering stem* to 60 cm long, with 2 opposite leaves. *Basal and cauline* leaves with 3 leaflets; leaflets ovate, to 9.5 cm long, 7 cm broad, the tip rounded or almost acute, the margin spinous with spines 0.5 mm long, the base deeply cordate with the lobes rounded or acute and diverging, those of the lateral leaflets moderately unequal, coriaceous, persisting green all winter, the underside glaucous and with minute erect hairs. *Inflorescence* compound, erect, loosely many-flowered (with up to 18 flowers), the lower peduncles 3-flowered; pedicels glabrous, to 4 cm long. *Flowers* large, *c.* 5 cm across. *Outer sepals* soon falling, 3–4 mm long, green or reddish. *Inner sepals* narrowly ovate, dull rose, 12 mm long, 5.5 mm broad. *Petals* much longer than the inner sepals; spur elongated, subulate, strongly curved, 3 cm long, pale yellow but profusely suffused with red streaks about a third or more towards the mouth, which has no fully developed lamina but two yellow side flanges. *Stamens c.* 4 mm long; filaments 1 mm long, pale yellow; anthers 3 mm long, greenish before dehiscence.

DISTRIBUTION. Central China, Sichuan province, Mount Omei.

Fig. 92. **Epimedium × *omeiense*** (*M.Ogisu* 82001). Photographed by John Fielding at Blackthorn Nursery, Hampshire.

Epimedium × omeiense 'Emei Shan'

Inner sepals dull rose, flat. *Petals* pale yellow but profusely suffused with red streaks towards the mouth of the spur.

Epimedium × omeiense 'Stormcloud'

Inner sepals dull purple often with a small cap-like projection. *Petals* dull purple.

A rather sombre plant, hence the cultivar name.

EPIMEDIUM 'ENCHANTRESS' (*E. dolichostemon* × *E. leptorrhizum*)

This charming hybrid was raised by Miss Elizabeth Strangman in the Washfield Nursery, Hawkhurst, Kent, and was named 'Enchantress'.

If a collective name is needed for hybrids of *E. dolichostemon* × *E. leptorrhizum*, the epithet *mago* would be suitable.

DESCRIPTION. *Flowering stem* about 25 cm long, with 2 opposite leaves. *Leaves basal and cauline* with 3 leaflets; *leaflets* narrowly ovate, to 10 cm long, 4 cm broad, the apex acuminate, the margin spinous with spines *c.* 1.5 mm long, the base deeply cordate with the lobes acuminate and diverging up to 50°, those of lateral leaflets markedly unequal, evergreen, the underside glabrous, conspicuously veined. *Inflorescence* simple, loose, few-flowered (with up to 12 flowers); 10–16 cm long; pedicels glabrous, the lower to 3 cm long. *Flowers* medium-sized, somewhat concealed beneath the leaves. *Inner sepals* concave, acute, lilac, 10 mm long, 3 mm broad. *Petals* slightly longer than the inner sepals, strongly curved, *c.* 11 mm long, purple at base, the rest of spur white. *Stamens* protruding, *c.* 5.5 mm long; filaments slender, *c.* 2.5 mm long, slightly yellow; anther 3 mm long, yellow.

EPIMEDIUM 'KAGUYAHIME' [*E. acuminatum* × *E. dolichostemon*]

The hybrid named 'Kaguyahime' arose as a chance seedling in the nursery of Mr S.Yamaguchi of Japan where *E. acuminatum* and *E. dolichostemon*, unrelated Chinese species, were growing close together. It is a beautiful plant with evergreen leaves bronzed in winter and many-flowered inflorescences of large pink and purple flowers. From *E. dolichostemon* it has inherited its compound inflorescence and strongly curved petals, but longer, thanks to the influence of *E. acuminatum*; the stamens have longer filaments than in *E. acuminatum* but shorter than in *E. dolichostemon*.

Should hybrids of *E. acuminatum* × *E. dolichostemon* need a collective botanical name, *amoenum* would be a fitting epithet.

DESCRIPTION. *Flowering stem* 22–50 cm long bearing 2 opposite, or occasionally 3, leaves. *Rhizome* compact. *Leaves* basal and cauline with 3 leaflets; *leaflets* mostly lanceolate, 5–10 cm long, 2–4 cm broad, the apex acuminate, the margin spinous with spines 1–1.5 mm long, the base moderately cordate, the terminal leaflet with rounded or acute lobes, the lateral leaflets with markedly unequal lobes, the inner acute, the outer acuminate, coriaceous, the underside glaucous and glabrous. *Inflorescence* compound, loose, many-flowered (with 12–40 flowers), 12–20 cm long, most peduncles 3-flowered; pedicels glabrous, 1–3.5 cm long. *Flowers* large, 2–2.5 cm across. *Outer sepals* dull purple with white margin, 3–4.5 mm long. *Inner sepals* elliptic, very pale purple, 1.2–1.4 cm long, 5–6 mm broad. *Petals* strongly curved, *c.* 6–10 mm long, dark purple. *Stamens* protruding, 5.5–6 mm long; filaments 2–5 mm long, whitish anthers 2.5 mm long, green before dehiscence.

EPIMEDIUM 'KEW HYBRID' [*E. davidii* × *E. pubescens*]

This hybrid arose in the Royal Botanic Gardens, Kew, where *E. davidii* and *E. pubescens* were growing close together, but it was not noticed until 1965. Referred to as 'Kew Hybrid', it manifests its parentage by possessing intermediate characters. Here at Kew, when in flower, *E. pubescens* grew to about 65 cm high, 'Kew Hybrid' about 40–50 cm and *E. davidii* about 37 cm; like *E. pubescens*, 'Kew Hybrid' has a compound inflorescence. The flowers, however, unlike those of *E. pubescens*, have reddish inner sepals and longer petals with a slender spur and distinct lamina. It is a low-growing, many-flowered plant but not of any great horticultural merit.

DESCRIPTION. *Flowering stem* 40–50 cm long, bearing 2 opposite leaves. *Rhizome* compact. *Leaves basal and cauline* with 3 leaflets; *leaflets* narrowly ovate, about 8 cm long, 4 cm broad, the apex acuminate, the margin spinous with spines 1–1.5 cm long, the base cordate, with the lobes diverging at about 55°, rounded, the lateral unequal, the underside pubescent all over with short erect hairs, evergreen. *Inflorescence* compound, loose, many-flowered (with 20–26 flowers), the lower peduncles 3-flowered; pedicels glandular-hairy, 1–2.5 cm long. *Flowers* small, 10 mm across. *Outer sepals* 2.5 mm long. *Inner sepals* lanceolate, acute, reddish, 3–4 mm long, 1–1.5 mm broad. *Petals* curved but not coiled, yellow, with tapering spur 7 mm long and lamina 3 mm high. *Stamens* protruding, 4 mm long; filament yellow, 1 mm long; anther 3 mm long, dull yellow.

EPIMEDIUM × PERRALCHICUM [*E. perralderianum* × *E. pinnatum* subsp. *colchicum*]

Some time between 1878 and 1902 George Ferguson Wilson (1822–1902) made a planting of *Epimedium perralderianum* and *E. pinnatum* subsp. *colchicum* in his wild garden at Wisley, Surrey, which later became part of the Royal Horticultural Society's Wisley Garden. By 1934 they had grown together into a wide mass which included forms that were neither one nor the other but combined their features in various ways. These were evidently hybrids. At least three seedlings had competed successfully with their parents, for three types could be distinguished when I designated in 1938 as Wisley Hybrid nos. 1, 2 and 3 under the collective name, an abbreviated formula, *E.* × *perralchicum* in *J. Linn. Soc., Bot.* 51: 515–516 (1938). They have evergreen leaves with 3 or 5 leaflets and profuse yellow flowers like their parents. The spur of the petals, though more curved upwards than the straight spur of *E. pinnatum* subsp. *colchicum*, is not bent so erect as the angular spur of *E. perralderianum*. The latter has marginal spines on the leaflets 0.5–1.9 mm long, those of *E. pinnatum* subsp. *colchicum* 0.65–0.9 mm long; the hybrids display much divergence in spine length.

One clone is known in the nursery trade as 'Wisley', another raised in Germany as 'Fröhnleiten'.

EPIMEDIUM × VERSICOLOR [*E. grandiflorum* × *E. pinnatum* subsp. *colchicum*]

The introduction into European gardens of *Epimedium pinnatum* subsp. *colchicum* about 1840 not only enriched them with an ornamental and easily cultivated plant; it provided a parent for robust hybrids, as André Donckelaar (1803–1858) at Ghent presumably surmised when he crossed it with *E. grandiflorum*. According to an obituary by Dodart Spae in 1858, he raised the plants known in gardens as *E. versicolor* and *E. sulphureum*. The name *E.* × *versicolor* is used here as a collective name for all hybrids of the parentage *E. grandiflorum* × *E. pinnatum*.

Of the three clones described below, 'Sulphureum' and 'Neosulphureum' have evidently derived their evergreen foliage and yellow flowers from *E. pinnatum*, while 'Versicolor' resembles *E. grandiflorum* in its deciduous foliage. The flower stem may be leafless as in *E. pinnatum*, or bear a single leaf, as in *E. grandiflorum*, both occurring together on the same plant. They thrive in any good, well-drained soil.

Epimedium × versicolor C.Morren in Ann Soc. Roy. Agric. Gand 5: 92, t.243 f.2 (1849); Stearn in J. Linn. Soc., Bot. 51: 518 (1938); Europ. Gard. Flora 3: 392 (1989).

DESCRIPTION. *Flowering stem* 20–25 cm long, leafless (as in *E. pinnatum*) or bearing one biternate, or rarely 3-foliolate, leaf (as in *E. grandiflorum*) on the same plant. *Rhizome* 3–5 mm thick. *Leaves* basal and cauline, usually biternate, sometimes triternate or 5- or 3-foliolate; leaflets ovate or narrowly ovate, acute or acuminate, the margin spinous, sparingly pubescent beneath, usually perishing in autumn in 'Versicolor' (as in *E. grandiflorum*), but some remaining green all winter in 'Sulphureum' and 'Neosulphureum' (as in *E. pinnatum* subsp. *colchicum*). *Inflorescence* usually simple, many flowered with up to 20 flowers, sometimes with 2–5-flowered lower peduncles. *Flowers* 2 cm across, in general form like those of *E. grandiflorum* but smaller and possessing shorter petal-spurs and broader inner sepals (thus approaching *E. pinnatum*). *Inner sepals* broadly ovate, subacute or blunt, 10–14 mm long, 5–8 mm broad. *Petals* shorter or slightly longer than the inner sepals, yellow, with distinct petaloid basal laminae 5–6 mm high and cylindric spurs slightly upcurved and globular at their tips. *Stamens* usually enclosed, 4.5 mm long; anthers yellow, 3 mm long.

Epimedium × versicolor 'Versicolor'
E. macranthum var. *versicolor* C.Morren in Ann. Soc. Roy. Agric. Gand 5: 91 (1849).
E. versicolor C.Morren, *op. cit.* 92: t.243, f.2 (1849); Franch. in Bull. Soc. Bot. France 33: 115 (1886); Kom. in Acta Horti Petrop. 29: 133 (1908).
E. discolor Vilm., Fl. Pleine Terre, ed. 2 (1866).
E. macranthum f. *versicolor* (C.Morren) Voss, Vilm. Blumengärtn. ed. 3, 1: 51 (1894).
E. versicolor cl. *versicolor* (C.Morren) Stearn in J. Linn. Soc., Bot. 51: 517 (1938).
E. macranthum var. *versicolor* (C.Morren) Bergmans, Vaste Pl. ed. 2: 314 (1939).

ILLUSTRATION. Ann. Soc. Roy. Agric. Gand 5: t.243, f.2 (1849).
DESCRIPTION. *Leaflets* usually 9, up to 18 cm long, 6 cm broad, conspicuously mottled or entirely red when young, later green, the basal lobes rounded, acute or acuminate and often diverging widely, membranous in texture and perishing in autumn. *Inflorescence* glandular, with 10–20 flowers. *Inner sepals* old rose. *Petals* yellow, with red-tinged spurs. This differs from *E. versicolor* 'Sulphureum' in the conspicuous red colouring of its young foliage, its more glandular inflorescence, its more acuminate leaflets of thinner texture, and its rose coloured sepals. It is also of less vigorous growth.

Epimedium × versicolor 'Cupreum'
E. versicolor cupreum C.Morren in Belgique Hort. 5: 34, t.6, fig.7 (1854); Franch. in Bull. Soc. Bot. France 33: 115 (1886); Kom. in Acta Horti Petrop. 29: 133 (1908); Wehrhahn, Gartenstauden 1: 455 (1930).

Fig. 93. ***Epimedium* × *perralchicum* 'Wisley'.** Photographed by John Fielding at Blackthorn Nursery, Hampshire.

E. macranthum f. *versicolor cupreum* Voss, Vilm. Blumengärtn. ed. 3, 1: 51 (1894).

ILLUSTRATION. Belgique Hort. 5: 34, t.6, fig.7 (1854).
DESCRIPTION. This is like 'Versicolor' but the inner sepals ('pétales' of C.Morren) are coppery-red ['la couleur des pétales est aussi remarquable que rare dans la végétation, c'est le rouge saumoné clair ou le rouge cuivré' (C.Morren, *loc. cit.*)].

Epimedium × versicolor 'Sulphureum'

E. macranthum var. *sulphureum* C.Morren in Ann. Soc. Roy. Agric. Gand 5: 91 (1849).
E. sulphureum C.Morren, *op. cit.* 92: t.243, f.3 (1849); Franch. in Bull. Soc. Bot. France 33: 115 (1886); Kom. in Acta Horti Petrop. 29: 133 (1908); Wehrhahn, Gartenstauden 1: 455 (1930).
E. citrinum Baker in Gard. Chron. II, 13: 683 (1880).
E. macranthum f. *sulphureum* (C.Morren) Voss, Vilm. Blumengärtn. ed. 3, 1: 51 (1894).
E. ochroleucum Farrer, English Rock Garden 1: 327 (1919).
E. pinnatum var. *sulphureum* (C.Morren) Bergmans, Vaste Pl.: 203 (1924), ed 2: 314 (1939).
E. versicolor var. *sulphureum* (C.Morren) Stearn in Kew Hand-list of Rock Garden Pl., ed. 4: 53 (1934).

ILLUSTRATION. Ann. Soc. Roy. Agric. Gand 5: t.243, f.3 (1849); Gartenflora 86: 53 (1937), as '*E. pinnatum*'.

Epimedium × *versicolor* **'Sulphureum'** (left), **'Versicolor'** (right)

ANN FARRER

DESCRIPTION. *Leaflets* 5–11 (usually 9), up to 8 cm long, 6 cm broad, green or sometimes red or brown mottled, the base deeply cordate, with usually rounded lobes, subcoriaceous in texture and sometimes remaining green all winter. *Inflorescence* usually glabrous, with 8–20 flowers; pedicels 5–25 mm long. *Inner sepals* pale yellow. *Petals* usually about as long as the inner sepals and brighter yellow.

This is a graceful plant with pendulous lemon-yellow flowers, common in gardens and often called *E. sulphureum*, '*E. pinnatum*', or *E. ochroleucum*. From *E. pinnatum* it differs in its paler flowers with longer spurred laminate petals and usually included stamens, and also in its frequently leafy flowering stem. Baker's *E. citrinum* (type ex hort Ware, K!) is identical with Morren's *E. sulphureum*. As Baker says, 'it is most like a yellow-flowered variety of *macranthum* [= *grandiflorum*], but differs by its short spur and the racemes being simple and sometimes produced direct from the rootstock'.

Epimedium × versicolor 'Neosulphureum'

E. versicolor var. *neo-sulphureum* Stearn in J. Linn. Soc., Bot. 51: 519, t.31 (1938).

DESCRIPTION. *Basal leaves* usually with 3 leaflets, less often 5 or 7; stem-leaves nearly always with 3 leaflets, occasionally 5 or 9; leaflets ovate or narrowly ovate, up to 9 cm long, 7 cm broad, acute or acuminate at the tip, the base cordate, with a rather open sinus sparsely pubescent beneath, broad. *Inflorescence* simple, loose, subglabrous, 7–16-flowered, overtopping the stem-leaf; pedicels 0.5–3 cm long. *Flowers* 2 cm across, pale yellow. *Outer sepals* purplish, 3–4 mm long. *Inner sepals* narrowly ovate, bluntish, flat, pale creamy yellow, 1–1.3 cm long, 5–7 mm broad. *Petals* shorter by 3–5 mm than the inner sepals, with pale lemon-yellow laminae 5–6 mm high, 6 mm broad, enclosing the stamens, and brownish-tinged slightly upcurved spurs 3–4 mm long. *Stamens* included, 4 mm long; anthers 3 mm long, yellow.

This attractive plant is very like *E.* × *versicolor* 'Sulphureum', but its leaves, especially those on the flowering stems, consist usually of 3 leaflets (in 'Sulphureum' nearly always of 9 leaflets), and the spurs of the petals are always distinctly shorter than the inner sepals. Beginning to flower in April, it finishes in mid or late May, a fortnight or more after 'Sulphureum'. It had been cultivated long before 1934 in the Royal Horticultural Society's Garden at Wisley, Surrey, and distributed thence to other gardens, but it is by no means so common as 'Sulphureum'. The young leaves are brownish.

EPIMEDIUM × YOUNGIANUM [*E. diphyllum* × *E. grandiflorum*]

Epimedium grandiflorum and *E. diphyllum* are both native to Japan and have hybridized there in the wild producing stable populations which have been specifically distinguished as *E. trifoliolatobinatum* (which see). They have also hybridized in gardens, producing the clones named below. Assuming *Aceranthus*, typified by *E. diphyllum*, to be generically distinct from *Epimedium*, Wehrhahn in 1930 proposed the name × *Bonstedtia* for this hybrid group in honour of Carl Bonstedt (b. 1866), who was from 1900–1931 Gartenoberinspektor in Göttingen. Phillipp Franz von Siebold had introduced the plant from Japan, probably a natural hybrid.

The name *E.* × *youngianum* is here used in a collective sense to cover all the garden hybrids of *E. diphyllum* × *E. grandiflorum*, this being the earliest binomial applied to a member of the group. The epithet commemorates the early nineteenth century Epsom nurserymen Messrs. Charles, James and Peter Young who grew a large collection of uncommon herbaceous plants and published in 1828 *Hortus Epsomensis* enumerating 4060 kinds then in cultivation. The nursery was situated on land leased from the celebrated English botanist Philip Barker Webb who sent the Youngs interesting plants or seeds collected on his travels in Spain and the Canary Islands. They supplied the Edinburgh Botanic Garden with the plant illustrated in *Curtis's Botanical Magazine* 66, t.3745 (1839) as '*E. musschianum*'; this name is of obscure application so in 1846 Fischer & Meyer renamed it (i.e. the plant featured in the *Botanical Magazine*) *E. youngianum*.

There are several clones with this parentage, some raised in Japan, some in Europe.

Epimedium × youngianum Fisch. & C.A.Mey., Sert. Petrop. sub t.1 (1846); Stearn in J. Linn. Soc., Bot. 51: 521 (1938) & Europ. Gard. Fl. 3: 392 (1989). Type: ex hort. Edinburgh, *Graham* (holotype K); and illustration in *Curtis's Bot. Mag.* 66: t.3745 (1846) as '*E. musschianum*'.

DESCRIPTION. *Flowering stem* 10–30 cm tall, bearing one 2–9-foliolate leaf. *Leaves* basal and cauline, 2–9-foliolate; leaflets ovate to narrowly ovate, thin in texture, and becoming subglabrous beneath. *Inflorescence* simple or with the lower peduncles 2-flowered, almost glabrous, few-flowered (with 3–12 flowers); pedicels 5–20 (rarely up to 40) mm long. *Flowers* campanulate, pendulous, white or rose, 1.6–2 cm across. *Inner sepals* narrowly ovate to lanceolate, blunt or subacute, horizontally spreading, 8–11 mm long, 3–5 mm broad. *Petals* broader and slightly shorter than the inner sepals, obovate, rounded, connivent, 7–10 mm long, 4–6 mm broad, spurless or with a short conical dorsal projection or a slender incurving subulate spur sometimes (in *E.* × *youngianum* 'Youngianum') up to 1 cm long, frequently varying in length and form on the same inflorescence. *Stamens* included, 3–4 mm long; anthers yellow, 2.5–3 mm long.

The following clones differ in habit, leaf and flower colour. In 'Youngianum' the leaf has usually nine acuminate leaflets as in *E. grandiflorum* while in 'Roseum' and 'Niveum' there is a tendency for the middle three to be suppressed and their leaves usually consist of six leaflets. However, leaves varying in form from the biternate leaf of *E. grandiflorum* through intermediates to the bifoliolate leaf of *E. diphyllum* may sometimes be found on one plant.

Epimedium × youngianum 'Youngianum'
"*E. musschianum*" sec. Graham in Edinburgh New Philos. J. 27: 190 (1839); Curtis's Bot. Mag. 66: t.3745 (1839); Maund, Bot. Garden, 12: t.281 (1848); vix Morren & Decne (1834).

Epimedium × *youngianum* **'Roseum'** (upper left), **'Niveum'** (right), **'Merlin'** (lower left)   CHRISTABEL KING

E. youngianum Fisch. & C.A.Mey., Sert. Petrop. sub t.1 (1846); Franch. in Bull. Soc. Bot. France 33: 115 (1886); Voss in Vilm. Blumengärten. ed. 3, 1: 51 (1894); 2: t.5, f.2 (1895); Kom. in Acta Horti Petrop. 29: 13 (1908).

E. youngianum var. *typicum* Makino in Bot. Mag. (Tokyo) 23: 142 (1909).

Bonstedtia youngiana (Fisch. & C.A.Mey.) H.R.Wehrh., Gartenstauden 1: 455 (1830).

Bonstedtia lilacina var. *youngiana* (Fisch. & C.A.Mey.) Kesselr. in Silva Tarousa & Schneider, Unsere Gartenstauden ed. 5: 99 (1934).

E. musschianum var. *vulgare* Koidz. in Acta Phytotax. Geobot. 5: 127 (1936).

ILLUSTRATIONS. Curtis's Bot. Mag. 66: t.3745 (1839); Maund, Bot. Garden 12: t.281 (1848); Voss, Vilm. Blumengärtn. ed. 3, 2: t.5, f.2 (1895); J. Linn. Soc., Bot. 51: t.31, f.2 (1938).

DESCRIPTION. *Flowering stem* 15–30 cm tall bearing one leaf. *Leaves* usually biternate (with 9 leaflets) but occasionally 3-foliolate (as figured in Bot. Mag. t.3745); leaflets narrowly ovate or sometimes ovate, ± 2–8 cm long, 1–5 cm broad, the tip abruptly long-acuminate, the margin sparsely or distinctly spinous, the base deeply cordate, with usually rounded lobes. *Inflorescence* with 3–8 flowers crowded towards its apex, much shorter than the stem-leaf; *pedicels* 5–10 mm long. *Flowers* (i.e. inner sepals and petals) white with a greenish tinge.

Hiding the few flowers under its foliage, and accordingly horticulturally unattractive, this historically important plant became extremely rare in gardens but has surprisingly

Fig. 94. ***Epimedium*** × ***youngianum* 'Niveum'**. Photographed by John Fielding in a garden near London.

survived although still rare. It flowers in April, earlier than 'Niveum' but is more often damaged by frost. It should be maintained in gardens for its interest value. The cultivar E. × *youngianum* 'Amy' usually has only 2 or 4 leaflets and these are less strongly acuminate, often merely acute at the apex.

Epimedium × youngianum 'Roseum'

E. violaceo-diphyllum C.Morren in Ann. Soc. Roy. Agric. Gand 5: 92 (1849).

E. hybridum C.Morren, *loc. cit.* t.243, f.1 (1849).

E. lilacinum Donckel. ex C.Morren, *loc. cit.* (1849); Vilm., Fl. Pleine Terre, ed. 2: 298 (1866); Franch. in Bull. Soc. Bot. France, 33: 113 (1886); Kom. in Acta Horti Petrop. 29: 132 (1908).

E. sinense, E. roseum, E. atroroseum Vilm., Fl. Pleine Terre ed. 2: 298 (1866).

E. concinnum Vatke in Regel, Gartenflora 21: 165, t.726 (1872); Baker in Gard. Chron. N.S. 13: 683 (1886); T.Ito in J. Linn. Soc., Bot. 22: 431 (1887).

E. macranthum f. *roseum* [Vilm.] Voss, Vilm. Blumengärtn. ed. 3, 1: 51 (1894).

E. macranthum var. *roseum* (Voss) W. Mill. in L.H.Bailey, Cycl. Amer. Hort. 2: 535 (1900).

E. youngianum var. *concinnum* (Vatke) Makino in Bot. Mag. (Tokyo) 23: 142 (1909); Makino & Nemoto, Fl. Japan ed. 2: 348 (1931).

× *Bonstedtia lilacina* (Donckel.) H.R.Wehrh., Gartenstauden 1: 455 (1930).

E. youngianum var. *roseum* (Vilm.) Stearn in Kew Hand-list of Rock Garden Pl., ed. 4: 54 (1934).

E. lilacinum var. *concinnum* (Vatke) Bergmans, Vaste Pl. ed. 2: 314 (1939).

ILLUSTRATIONS. J. Jap. Bot. 13: 811 (1937), J. Linn. Soc., Bot. 51: t.31, f.3 (1938).

DESCRIPTION. Flowering stem 10–30 cm tall, usually bearing one 2- or 6-foliolate leaf. *Leaves* very variable with 2, 6 or 7 leaflets, rarely 9; leaflets ovate, 2–5 cm long, 1–2.5 cm broad, the lateral leaflets usually with very unequal basal lobes, the middle leaflet sometimes peltate and scoop-like, the tip blunt (not abruptly long-acuminate), the margin sparsely spinous-serrate. *Inflorescence* loose, 4–12-flowered, about as long as or longer than the stem-leaf; pedicels 5–15 (rarely 40) mm long. *Flowers* (i.e. inner sepals and petals) purplish-mauve, varying in depth of colour.

This plant is known under a variety of names, the commonest being *E. lilacinum*. It is a small, pleasing plant well-known in gardens and is one of the epimediums briefly described in Vilmorin, *Les Fleurs de Pleine Terre*, and represented by specimens dating from 1863 in the Vilmorin-Andrieux herbarium near Paris. *Epimedium roseum*, described as having 'fleurs d'un rose clair', agrees with authentic specimens of *E. concinnum* Vatke (specimens in e.g. Jena, Leiden, Cambridge) grown in the Berlin botanic garden. Vatke's description and figure of his *E. concinnum* are so unsatisfactory that Maximowicz and Franchet thought it to be synonymous with *E. rubrum*! J.G.Baker of Kew, however, interpreted it correctly [Gard. Chron. 13: 683 (1886)].

Epimedium × youngianum 'Niveum'

E. niveum Vilm., Fl. Pleine Terre ed. 2: 298 (1866).

'*E. musschianum*' see Baker in Gard. Chron. N.S. 13: 683 (1880); Koenemann in Gartenwelt 3: 591 (1889); Farrer, English Rock Garden 1: 327 (1919).

E. macranthum f. *niveum* (Vilm.) Voss, Vilm. Blumengärtn. ed. 3, 1: 51 (1894).

E. macranthum var. *niveum* (Voss) W.Mill. in L.H. Bailey, Cycl. Amer. Hort. 2: 535 (1900).

E. youngianum var. *niveum* (Vilm.) Stearn in Kew Hand-list of Rock Garden Pl. ed. 4: 53 (1934).

ILLUSTRATION. Gartenwelt 3: 591 (1889); Silva-Terouca & Schneider, Unsere Freiland-Stauden ed. 5: 159, f.162 (1934); J. Linn. Soc., Bot. 51: t.31, f.4 (1938).

DESCRIPTION. Similar to 'Roseum' (see above) in that it is a low, neat plant, but possesses pure white flowers. It differs from 'Youngianum' in having blunt leaflets (which vary in number from 2 to 9) and a looser inflorescence (with pedicels 5–20 mm long) usually overtopping the stem-leaf ('Youngianum' has acuminate leaflets and inflorescences shorter than the stem). It occasionally happens that a specimen occurs in which all the flowers are spurless and the leaves are 2-foliolate; in such cases it cannot be distinguished from *E. diphyllum*, but usually there are spurred and spurless flowers present and the leaves have 2, 3, 6 or 9 leaflets, sometimes all together on one plant.

The small, dainty 'Niveum' has considerable garden merit, and it is easier to grow than *E. diphyllum*. It flowers from May into early June, later than 'Youngianum'. Although the name *E. niveum* was validated by Vilmorin in 1866 the epithet had already been used for this plant in the catalogue of the Belgian nurseryman Van Houtte well before this.

Another excellent white-flowered cultivar of *E.* × *youngianum* is 'Yenomoto', with noticeably larger flowers than those of 'Niveum'.

Epimedium × youngianum 'Merlin'

This clone is very similar to 'Roseum' but has slightly darker flowers. It was raised by Amy Baring (later Amy Doncaster) in her garden at Chandler's Ford, Hampshire, England.

Appendix 2. *EPIMEDIUM TAKHTAJANII* THE FOSSIL *EPIMEDIUM*

In 1965 a fossil species of *Epimedium* was described by E.F.Kutuzkina, based on a sample collected in the eastern Black Sea region on the river Pshekhi. It was named in honour of Prof. Armen Takhtajan. I am indebted to Claire Glaser for translating the description of *E. takhtajanii* and extracts from this are provided as follows:

A nearly complete impression of a single leaflet has survived. It is ovate with a long-tapering apex to an acuminate tip. The species of *Epimedium* distributed in the Caucasus, Europe and North Africa often have ovate leaflets but in comparing herbarium material of these with the fossil leaflet, there was not a single instance where their leaflets tapered so gradually from the lamina into a long, pointed apex — such as occurs in the Chinese species *E. pubescens* Maxim. and *E. acuminatum* Franch. There is a particularly striking similarity between it and the latter species, but *E. takhtajanii* is distinguished from *E. acuminatum* by the leaflet having smaller dimensions and shorter basal veins.

Epimedium takhtajanii Kutuzk. in Soviet Bot. 50, 8: 1114 (1965). Type: Krasnodar region, right bank of the river Pshekhi, in the vicinity of the town of Apsheronsk, Collection 972, sample 1 (LE).

DESCRIPTION. *Leaflets* petiolulate, petiole slender and *c.* 1 cm long. lamina with a symmetrically cordate base and straight [i.e. nearly parallel] sides in the lower part, gradually tapering to an acuminate apex; margin cartilaginous, toothed and cartilaginous, the teeth bristle-like at the tips. The central vein is straight, thicker than the rest, and there are two slender basal veins arising from the base, each forming an angle of 40° with the central vein; these lateral basal veins arch upwards in a series of loops and are joined at intervals by secondary veins which arise from the central vein at an angle of about 75°. From the outer side of these 'loops' other finer veins arise, running out to the margin. Under higher magnification a rather precise network of veins is visible.

Appendix 3. OBSCURE AND EXCLUDED TAXA

Epimedium angustifolium W.T.Macoun, List. Per. Expt. Farm. Gard. Ottawa (Bull. Cent. Exp. Farm. Ottawa, ser. 2, no. 5): 40 (1908). 'Half-hardy, 8–12 in. Fl. white. May 11 to May 22' (Macoun, loc. cit.). Perhaps *E.* × *youngianum* 'Niveum'.

Epimedium cavaleriei H.Lév., Cat. Pl. Yunnan: 18 *in adnot.* (1915) = *Stauntonia cavaleriana* Gagnep. (1908), described from the same gathering. A shrub belonging to the *Lardizabalaceae*.

Epimedium japonicum Makino in Iinuma, Somoku Dzus. ed. 3, II: 129 in obs. Jap. sub t.46 (1907), *non* Siebold ex Miq.
This is known to me only from Makino's brief note, published in Japanese and here translated: 'Herbaceous. One stem with three leaflets. Flowers as in the former [*E. diphyllum*], but pale pink. This should be another species of the genus and be called *Epimedium japonicum* Makino. The petals are spurred, but in the baikwa-ikariso [*E. diphyllum*] they are flat. Is it a form intermediate between the baikwa-ikariso [*E. diphyllum*] and the ikariso [*E. grandiflorum*]?'.
   The name *E. japonicum* does not occur in Makino & Nemoto's Flora of Japan, ed. 2 (1931); it may be a synonym of *E.* × *youngianum* 'Roseum'.

Epimedium lilacinum Donckel. ex C.Morren in Ann. Soc. Roy. Agric. Gand 5: 91 (1849).
The plant cultivated today as *E. lilacinum* (for which the name *E.* × *youngianum* 'Roseum' is here adopted) seems distinct from the original *E. lilacinum* (*E. diphyllum* × *E. violaceum*, *fide* D.Spae ex C.Morren, loc. cit.) as figured by Morren, the spurs of the latter being apparently stouter and the leaflets different. Morren describes *E. lilacinum* as having 'les feuilles des vrais épimèdes, la forme cordée, les lobes arrondis, les dents nombreuses et la seule ressemblance que j'y vois avec les feuilles de l'aceranthus [*Aceranthus diphyllus* = *E. diphyllum*] se borne à leur existence géminée…. Il n'y a rien dans ces fleurs qui rappelle l'*Aceranthus*: elles sont armées de vrais nectaires [petals] cucullifomes et l'éperon même est enpointe; les pétales [inner sepals] sont planes, grands, dépassent un peu les nectaires [petals] en longeur et le coloris est celui de '*Epimedium violaceum* [= *E. grandiflorum* f. *violaceum*], lilacine, lavée de blanc et de violet'.
   I have no doubt that this plant and our *E.* × *youngianum* 'Roseum' (*E. concinnum* Vatke) are hybrids of the same parentage, but they probably represent distinct clones.

Unfortunately no plant corresponding exactly to Morren's plate seems to be in cultivation today, the nearest being *E.* × *youngianum* 'Roseum'; it may be that the latter was incorrectly figured.

Epimedium musschianum C.Morren & Decne. in Ann. Sci. Nat. Bot. ser. II, 2: 353 (1834); Franch. in Bull. Soc. Bot. France 33: 111, 114 (1886).
E. macranthum var. *musschianum* (C.Morren & Decne.) Makino in Bot. Mag. (Tokyo) 23: 143, *quoad syn.* (1909).
'*E. parvulum* Baker' ex Koidz. in Acta Phytotax. Geobot. 5: 126, *pro syn.* (1936).

Because of its uncertain application, the name *E. musschianum* is best abandoned. It is unfortunate that Morren and Decaisne published no figure and left no type-specimen at Paris or Brussels since, as Franchet remarked, 'la longue description qu'ils en donnent ne dit absolument rien qui puisse mettre sur la trace de son identité'. They placed it between *E. macranthum* and *E. violaceum*, distinguishing it from other species 'par ses feuille simplement ternées, d'une couleur verte forcé, par ses fleurs d'un blanc sale, moins grandes que dans l'espèce precedente'. No species with such characters is known today, although it is possible Morren and Decaisne had before them an abnormal plant of *E. grandiflorum* or a state of the hybrid between *E. diphyllum* and *E. grandiflorum*. There is a specimen in the Paris Museum labelled '*E. musschianum*' which Franchet regarded as authentic– 'l'étiquette est de la main de Decaisne'. This specimen belongs to *E. sagittatum* (*E. sinense*); the handwriting on the label (ex herb. Lugd. Bat., i.e. Leiden) is not, however, Decaisne's, but that of the Leiden botanist Blume; the note of affirmation on this sheet is by Spach, and possibly Decaisne never saw it at all. The original description of *E. musschianum* definitely excludes *E. sagittatum*.

The name commemorates Jean Henri Mussche (1764–1834), head gardener at the Ghent Botanic Garden and author of the *Hortus Gandavensis* (1817); cf. N. Cornellissen, Quelques Souvenirs.......de Jean Henri Mussche (1835).

'**E. musschianum**' of Bot. Mag. t.3745, see *E.* × *youngianum* 'Youngianum'.

Epimedium musschianum var. **multifoliolatum** Koidz. in Acta Phytotax. Geobot. 5: 126 (1936). 'Folio caulino quarter ternato. Hab. Japonia occidentalis'.

Epimedium pteroceras C.Morren in Ann. Soc. Roy. Agric. Gand 1: 145–146, t.14 (1845); II: 140 in obs. (1846); Franch. in Bull. Soc. Bot. France 33: 114–15 (1886).

Appendix 4. *EPIMEDIUM* CULTIVARS

The following list is not complete but includes most of those that are likely to be encountered in nurseries and gardens at present. There are three groups: named cultivars that are known to be or presumed to be of hybrid origin, cultivars that are purely selections of a species, and cultivars of hybrids which have been formally described and given latinised epithets. Thanks go to Robin White, Darrell Probst and Teyl de Bordes for greatly assisting in the compilation and checking of this list — ed.

A. NAMED HYBRID CULTIVARS

'Amanogawa' (*E. acuminatum* × *E. dolichostemon*).	Evergreen leaves, mottled red at first; many large flowers; broad white sepals, spurred petals brown.
'Amber Queen' (*E. flavum* × *E. wushanense*)	Sepals faintly suffused pink, petal lamina amber with near-white spur tipped yellow.
'Arctic Wings' (*E. latisepalum* × *E. ogisui*)	Large, pure white flowers.
'Baoxing Mist' [Darrell Probst: "From a naturally-occurring hybrid swarm of *E. chlorandrum* × ?sp. nov. aff. *acuminatum*"].	Evergreen leaves. Light lavender and cream colored flowers with yellow spur tips.
'Black Sea' (*E. pinnatum* subsp. *colchicum* × *E. pubigerum*? [Darrell Probst: "More likely crossed with *E.* × *cantabrigiense* than straight *pubigerum*"]	Leaves evergreen, dark green, glaucous, turning glossy purple-black in winter; flowers creamy yellow.
'Buckland Spider' (*E. grandiflorum* × *E. koreanum*)	Deciduous; dusky pink flowers.
'Egret' (*E. franchetii* × *E. latisepalum*)	White sepals, primrose yellow petals with long white spur.
'Enchantress' (?*E. dolichostemon* × *E. leptorrhizum*)	Leaves dark green above, whitish below; flowers similar to *E. dolichostemon* but inflorescence taller.
'Fire Dragon' (*E. davidii* × *E. leptorrhizum*)	Sepals pale pinkish-purple, petals yellow with white spur.
'Flowers of Sulphur' (*E. flavum* × *E. ogisui*)	White sepals, yellow petals with long white spur.
'Honeybee' (*E. flavum* × *E. wushanense*)	Pale yellow with purple/brown shading to the lamina of the petals.
'Kaguyahime' (*E. acuminatum* × *E. dolichostemon*)	Evergreen leaves tinged pink in winter; pink sepals, spurred petals purple.

'Lilac Charm' (*E. acuminatum* × *E. latisepalum*)	Pale mauve outer sepals, red-purple petals.
'Little Shrimp'. [Darrell Probst: "exact match for what is known in the US as E. *alpinum* 'Shrimp Girl' but probably is a dwarf form of *E.* × *cantabrigiense*"]	Sepals coral pink, petals yellow.
'Pink Elf' (*E. brevicornu* or *E. pubescens* × ?) [Darrell Probst comments that he has had plants just like this that have arisen from hand pollinations between *E. pubescens* × *E. grandiflorum*]. `	Sepals heavily speckled purple, petals darker purple with paler spur.
'Sasaki'. This is not a valid cultivar name for a single clone. Japanese botanists recognize *E.* × *sasakii* as a group of mixed generation, naturally occurring hybrids between *E. sempervirens* red-flowered form and *E.* × *setosum*.	Plants are quite variable with a variety of flower colours and shapes between the two parents.
'Sohayaki'. Not a valid cultivar name. It is the Japanese common name for *E. grandiflorum* var. *higoense*	
'Starlet' (?*E. grandiflorum* selection or hybrid)	Leaflets edged brown-red, flowers lavender pink, tipped white.
'William Stearn' (*E. membranaceum* × *E.* × *omeiense*)	Dark red inner sepals, scarlet petals with yellow spurs.

B. NAMED SELECTIONS OF SPECIES

acuminatum 'Galaxy' (*Lancaster* 1962)	Sepals white, petals creamy white faintly tinged pink.
acuminatum 'Ruby Shan'	Inner sepals purple, spurs lighter
diphyllum 'Nanum'	Compact form only 10 cm high when in flower, flowers white.
diphyllum 'Roseum'	Flowers pink.
fargesii 'Pink Constellation' (*Ogisu* 93023)	Inner sepals pink, petals purple, both sharply reflexed.
franchetii 'Brimstone Butterfly' (*Ogisu* 87001)	Sepals rusty red-brown, petals yellow.
grandiflorum 'Akakage'	Large wine-red flowers.
grandiflorum 'Akebono'	Flowers white suffused with pale rose.
grandiflorum 'Album'	Flowers white.
grandiflorum 'Beni-chidori'	Flowers a little darker purple-red than those of 'Rose Queen', white-tipped spurs.
grandiflorum 'Crimson Beauty'. A synonym of 'Rose Queen'	
grandiflorum 'Crimson Queen' probably a synonym of 'Rose Queen'	
grandiflorum 'Elfenkönigin'	Flowers creamy-white.
grandiflorum 'Koji'	Flowers lavender.
grandiflorum 'La Rocaille' [A named form of the Japanese *E. koreanum*, but see Species 46, page 146]	Flowers pale yellow; a tall form, to 45 cm.

grandiflorum 'Lilacinum' [possibly a synonym of 'Tama no Genpei' — needs further checking]	Inner sepals pale pink, petals white.
grandiflorum 'Lilafee'	Deep purple sepals, paler petals.
grandiflorum 'Mount Kitadake' [also used for this are 'Mount Kitadake Red' and "red form from Mount Kitadake"].	Dwarf form with purplish flowers and the young foliage reddish.
grandiflorum 'Nanum'	White-flowered, small in habit and flower size, leaves dark-edged.
grandiflorum 'Nanum Freya'	Inner sepals purple, petals purple with paler spur.
grandiflorum 'Purple Prince'	Rich reddish-purple sepals and petals.
grandiflorum 'Queen Esta'	Purple sepals, paler petals stained purple. New foliage dark purple in spring.
grandiflorum 'Rose Queen'. This is a synonym for a very old cultivar in Japan correctly named 'Yubae' [Darrell Probst, pers. comm.].	Sepals and petals deep rosy purple.
grandiflorum 'Rubinkrone'	Deep rose, spurs white-tipped.
grandiflorum 'Saturn'. A clonal selection of *grandiflorum* var. *higoense* (var. not recognised by Stearn — ed.)	Flowers wholly white; dark edge to leaves which lasts longer than in 'Nanum'.
grandiflorum 'Saxton Purple'	Reddish-purple flowers, white-tipped spur.
grandiflorum 'Shiho'	Dark red flowers.
grandiflorum 'Sirius'	Flowers pale shell pink.
grandiflorum 'Sunset'	Darrell Probst regards this as nearly identical to 'Rose Queen' and is in fact probably the same clone.
grandiflorum 'Tama no Genpei'	Large flowers with lavender inner-sepals and creamy-white petals suffused with pale lavender.
grandiflorum 'Violaceum' = forma *violaceum*	Flowers large pale purple.
grandiflorum 'White Beauty'	Flowers white.
grandiflorum 'White Queen'	Large white flowers.
grandiflorum 'Yellow Princess'	Creamy sepals, primrose yellow petals.
leptorrhizum 'Mariko'	Pale magenta sepals, white petals.
rhizomatosum 'Golden Eagle'	Flowers yellow.
sempervirens 'Aurora'	Flowers lavender.
sempervirens 'Candy Hearts'	Flowers silvery pink.
stellulatum 'Wudang Star'	Sepals white, very pointed, petals small, yellow.
wushanense 'Caramel' (*Ogisu* 92009)	Pale green sepals suffused pale red, petals light yellow suffused orange with a purple rim to the spur opening.

C. CULTIVARS OF HYBRIDS WITH LATINISED EPITHETS

× *omeiense* 'Akame' (*Ogisu* 82001). This is the correct (earliest) cultivar name for 'Emei Shan'	Flower dark red flushed with yellow.
× *omeiense* 'Emei Shan' (*Ogisu* 82001). The correct cultivar name is 'Akame'; named in Japan before it reached the UK.	
× *omeiense* 'Stormcloud' (*Ogisu* 82002)	Flowers brownish-purple.
× *omeiense* 'Myriad Years'	Inner sepals whitish-pink.
× *perralchicum* 'Fröhnleiten'	Flowers bright golden yellow.
× *perralchicum* 'Weihenstephan'. [Darrell Probst: "The plants I've obtained of this from the UK were all true *E. perralderianum*, not the hybrid]. Apparently needs further investigation — ed.	Flowers bright yellow.
× *perralchicum* 'Wisley'	Flowers large, yellow.
× *rubrum* Cobblewood Form. The correct cultivar name is 'Sweetheart'.	
× *rubrum* 'Sweetheart' (first distributed as Cobblewood Form, but this is the correct cv. name)	Evergreen leaves. Flowers pinkish-purple, petals with white spur.
× *versicolor* 'Cupreum'	Sepals coppery-red.
× *versicolor* 'Neosulphureum'	Flowers pale yellow.
× *versicolor* 'Sulphureum'	Sepals pale yellow, petals brighter yellow.
× *versicolor* 'Versicolor'	Sepals old rose, petals yellow with red-tinged spurs.
× *warleyense* 'Orangekönigin'	Flowers pale orange.
× *youngianum* 'Beni-kujaku'	Flowers pale pink edged darker; petals & sepals similar in size, petals short-spurred.
× *youngianum* 'Capella'	Flowers deep rose.
× *youngianum* 'Lilacinum'	Flowers light pink.
× *youngianum* 'Merlin'	Flowers dusky lilac-purple.
× *youngianum* 'Niveum'	Flowers white.
× *youngianum* 'Roseum'	Flowers light pinkish-purple.
× *youngianum* 'Shikinomai' [may fall within *E. trifoliatobinatum* subsp. *maritimum* — see page 149]	Pure white flower; repeat flowering over a long period [Teyl de Bordes, pers. comm.].
× *youngianum* 'Tamabotan' (syn. 'Pink Ruffles') [Darrell Probst: "It was called 'Pink Ruffles' for a time in the US as Harold Epstein had lost the label it came with from Japan and came to think it was a seedling that had appeared in his garden"]	Sepals and petals suffused soft pale purple pink, the latter similar in shape to the sepals and with a very short spur, hence the flowers appear 'double'.
× *youngianum* 'Yenomoto'	Flowers wholly very pale pinkish-white.
× *youngianum* 'Youngianum' (syn. 'Typicum')	Flowers greenish cream.

8. TAXONOMIC TREATMENT OF *VANCOUVERIA*

Vancouveria C.Morren & Decne. in Ann. Sci. Nat. Bot. II. 2: 351 (834); Benth. & Hook. fil., Gen. Pl. 1: 44 (1862); Jepson, Fl. Calif. 1: 550 (1914); Abrams, Illustr. Fl. Pacific States, 2: 221 (1944): H.Loconte in Kubitzki, Fam. & Gen. Vascular Pl. 2: 150 (1993). Type species: *Vancouveria hexandra* (Hook.) C.Morren & Decne. (*Epimedium hexandrum* Hook.).

Sculeria Raf., Fl. Tellur. 2: 52, no. 188 (1937) non *Scouleria* Hook. (1829). Type: *Sculeria geminata* Raf. (*Epimedium hexandrum* Hook.).

Epimedium sect. *Vancouveria* (C.Morren & Decne.) Baill., Hist. Pl. 3: 56 (1871); Franchet in Bull. Soc. Bot. France 33: 4 (1886); Prantl in Engler & Prantl, Nat. Pflanzenfam. III, 2: 76 (1888).

Epimedium subgenus *Vancouveria* (C.Morren & Decne.) Kom. in Acta Horti Petrop. 29: 140 (1908).

GENERIC DESCRIPTION. *Herbaceous perennials*. *Rhizome* sympodial, horizontally creeping, elongated, slender, furnished with brown membranous bracts. *Leaves* normally basal, usually biternate, sometimes 3- or 5-foliolate; stipules united into a membranous ligulate sheath; petioles terete; leaflets pilose above (when young) or glabrous above, cordate at base, indented and rounded at the tip, spineless but often undulate or crisped at the margin, usually obscurely 3-lobed. *Flowering stem* normally leafless, sometimes bearing one leaf in *V. hexandra*. *Inflorescence* simple or compound and paniculate, few- or many-flowered, glabrous or glandular. *Flowers* trimerous with parts in threes, regular, pendulous, glabrous or glandular, white or yellow; aestivation imbricate; parts free, opposite. *Outer sepals* ('sepals', A.Gray) 6–9, small, unequal, the inner larger, glabrous or glandular, soon falling. *Inner sepals* ('petals', A.Gray) 6, petaloid, spathulate, reflexing, c. 4–9 mm long. *Petals* ('nectaries', A.Gray), 6, quite flat and lobed at the tip in *V. planipetala* but with a narrowly oblong stalk expanded and folded over at the tip as a nectariferous pouch in *V. hexandra* and *V. chrysantha*, slightly shorter than the inner sepals, reflexing. *Stamens* 6, connivent; filaments glabrous or glandular, without appendages; anthers about as long as the filaments, dehiscing by two oblong valves which separate from the connective but remain attached at the top, curling upwards. *Ovary* superior, glabrous or glandular with parietal placentation; ovules two or several. *Style* slender, with a slightly dilated stigma. *Capsule* splitting to the base as two valves. *Seeds* smooth, almost black, about 3 mm long, semi-orbicular when viewed laterally before removal of the conspicuous aril.

Vancouveria chrysantha (left), *V. hexandra* (centre), *V. planipetala* (right)

MARK FOTHERGILL

The genus comprises three species distributed in the Pacific States of USA, from northern Washington through Oregon to Mendocino County in central California. They are woodland plants, requiring very much the same treatment in cultivation as *Epimedium* (see Chapter 6, page 35).

The name *Vancouveria* commemorates the English seaman George Vancouver (1757-1798), who, as Captain of His Britannic Majesty's Ship *Discovery*, led a survey of the Pacific coast of North America in 1792. 'Vancouver' is an anglicised form of 'van Couverden' (from Couvorden in Holland), the Vancouver family being of Anglo-Dutch descent like other East Anglian families who left the Netherlands during the war of independence from Spain; Vancouver himself was born at King's Lynn, Norfolk, and is buried at Petersham in Surrey. The dedication of the genus by Morren and Decaisne to Captain Vancouver was no doubt suggested by Douglas' type-material having been collected near Fort Vancouver on the Columbia River, as also by Hooker's statement that 'it was discovered by Mr Menzies during the Voyage of Discovery of Captain Vancouver'.

Key to the species of *Vancouveria*

1. Flowers about 1–1.3 cm long. Petals bent over at apex forming a nectar-bearing hood. Stamens and ovary with glandular hairs 2
1. Flowers about 6–8 mm long. Petals flat, lobed at apex but not hooded. Stamens and ovary glabrous. Leaves evergreen: leaflets with thickened margin. Inflorescence paniculate with 20–50 flowers .... 3. ***V. planipetala***

2. Leaves dying in early winter; leaflets thin, not thickened at margin. Pedicels glabrous. Inflorescence many-flowered (with up to 45 flowers). Flowers white .... 1. ***V. hexandra***
2. Leaves evergreen; leaflets coriaceous, with thickened margin. Pedicels glandular. Inflorescence few-flowered (with 6–18 flowers). Flowers yellow ...... 2. ***V. chrysantha***

1. VANCOUVERIA HEXANDRA

Archibald Menzies, surgeon and naturalist on Captain Vancouver's ship *Discovery*, collected specimens of *Vancouveria hexandra* in 1792, probably in May near Olympia, Thurston County, Washington, but his material lay unstudied in the Banksian Herbarium, (now in the Natural History Museum of London), until the 1820s when W.J.Hooker began preparing his great *Flora Boreali-Americana* (1829–1840). In 1825 David Douglas collected it near the Hudson Bay Company's Fort Vancouver on the Columbia River and on his material Hooker based the description and illustration he published in 1829 as *Epimedium hexandrum*. Morren and Decaisne founded their genus *Vancouveria* on this in 1834. Since then the species has been collected all along the Pacific coast from Washington to California but never in Canada.

The date of its introduction into European gardens is uncertain but was sometime before 1886; it has proved an easy plant to grow under shady conditions in light moist soil. Reginald Farrer described it in 1919 as 'a most beautiful *Epimedium*, with the ample leafleted leafage of the race, soft bright-green in colour, and much more graceful, wide, airy and light in habit. The flowers, which are small and creamy, come up in summer on stems of 10 inches or a foot, in the most delicate and dainty loose showers, so that each little star seems to float pendulous on the air by itself'. Before flowering, it looks like an elegant ground-covering fern.

In 1914 E.L.Greene, an arch-splitter described by Marcus Jones as 'the bane of botany', published some supposed new species: *V. brevicula*, *V. parvifolia* and *V. picta*. When I examined the types of these in about 1932 I found no reliable characters to distinguish them from *V. hexandra* and that has certainly been the opinion of later botanists.

The epithet *hexandra*, from the Greek *hexa-* (six-), *aner*, *andros* (Man), refers to its six stamens.

Vancouveria hexandra (Hook.) C.Morren & Decne. in Ann. Sci. Nat. Bot. II. 2: 351 (1834); Torrey & Gray, Fl. N. Amer. 1: 52 (1838); Calloni in Malpighia 1: 266 (1887); Jepson, Fl. Calif. 1 (5): 551 (1914); Stearn in J. Linn. Soc., Bot. 51: 446 (1938); Abrams, Illustr. Fl. Pacific States 2: 221, 222 (1944); Munz & Keck, Calif. Fl.: 111, 112 (1959); Hitchcock & Cronquist, Vasc. Pl. Pacific Northwest 2: 415, 418 (1964); Rickett, Wild Flowers U.S. 5(1): 207 (1971); Hickman, Jepson Manual Higher Pl. Calif.: 364 (1993); M. Moore, Medicinal Pl. Pacific West: 156 (1993).

Epimedium hexandrum Hook., Fl. Bor.-Amer. 1: 30 (1829); Franch. in Bull. Soc. Bot. France 33: 112 (1886) p.p.; Komarov in Acta Horti Petrop. 29: 140 (1908) p.p. Type. 'Fort Vancouver' etc. 1825, *Douglas* (holotype K).

Sculeria geminata Raf., Fl. Tellur. 2: 52 (1837). Type as for *Epimedium hexandrum* illustrated in Hook., Fl. Bor.-Amer. 1: t.13 (1829).

V. brevicula Greene ex House in Muhlenbergia 9: 91 (1914). Type: Oregon, Coos County, head of Coos River, *House* 4832 (US).

V. picta Greene in Repert. Spec. Nov. Regni Veg. 13: 322 (1914). Type: Oregon Coast mountains, 1893, *Holzinger* (US).

V. parvifolia Greene in Repert. Spec. Nov. Regni Veg. 13: 323 (1914). Type: California, Humboldt County, Humboldt Bay, *Chandler* 1162 (US).

ILLUSTRATIONS. Hook., Fl. Bor.-Amer. 1: t.13 (1829); Garden 30: 263 (1886); J. Linn. Soc., Bot. 51: 447, t.24 (1938); Abrams, Illustr. Fl. Pacific States, 2: 222 (1944); Hitchcock & Cronquist, Vasc. Pl. Pacific Northwest, 2: 418 (1964); Rickett Wild Fl. U.S. 5 (1): t.67 (1971); M.Moore, Medicinal Pl. Pacific West: 157 (1993).

DESCRIPTION. *Flower stem* 10–40 cm high, leafless (or abnormally and rarely bearing one leaf), rising well above the foliage, pilose below. *Rhizome* long-creeping, 1–3 mm thick but stouter at the tip. *Leaves* normally basal, bi- or tri-ternate, clothed when young with profuse white hairs; leaflets narrowly ovate to broadly ovate or almost oblate, often three-lobed, very variable in size and shape, up to 7.5 cm broad, 7 cm long, the tip rounded and indented, the margin not cartilaginously crisped or spiny, the base cordate with rounded lobes, always thin and membranous in texture, perishing in autumn, bright green above, glaucescent beneath, at first pilose on both surfaces but becoming almost glabrous. *Inflorescence* loose, usually many-flowered with 6–45 flowers, the lower peduncles often 3- (sometimes 6-) flowered and up to 12 cm long, the upper one-flowered; pedicels glabrous, 1–3 cm long, two or three sometimes arising together. *Flowers* white, becoming yellowish on drying, 1–1.3 cm long, pendulous, the reflexing sepals and petals and connivent stamens (as in other species) giving them a shuttlecock appearance. *Outer sepals* dotted with very short red glandular hairs, unequal, the outer very short, the inner up to 5 mm long. *Inner sepals* spathulate, narrowing to a clawed base, slightly cuspidate at the tip, the margin undulate, 8–9 mm long, towards the tip 3–4 mm broad. *Petals* (nectaries) long-clawed, shorter than the inner sepals, the narrowly oblong stalk (or lamina) 5 mm long and 1 mm broad and expanded and folded over at the tip as a quadrate flap to form a rounded nectariferous pocket 1.5 mm high, 2 mm broad. *Stamens* 4 mm long, sparingly glandular with very short red glands; filaments 2 mm long. *Ovary* very glandular. *Capsule* about 1.5 cm long, 1–4-seeded; seeds black, about 3 mm long, 2 mm broad, almost covered by the aril.

DISTRIBUTION. Western United States, along the Pacific Coast hills and mountains from northern Washington to central California. Washington: Counties Chehalis, Pierce,

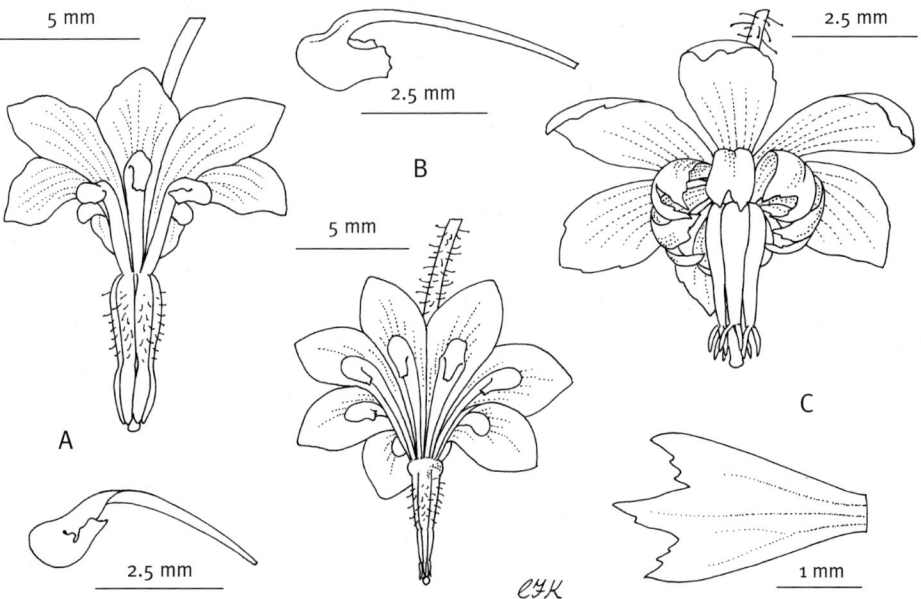

Fig. 95. *Vancouveria hexandra* (A), *V. chrysantha* (B), *V. planipetala* (C), showing whole flower and petal in each case. Drawn by Christabel King.

Thurston, Clarke, Skamania, Klickitat. Oregon: Counties Multnomah, Washington, Hood River, Clackamas, Marion, Benton, Douglas, Coos, Curry, Josephine, Jackson, Klamath. California: Counties Del Norte, Siskiyou, Humboldt, Mendocino.

2. VANCOUVERIA CHRYSANTHA

The yellow flower colour of *Vancouveria chrysantha* sets it apart from *V. hexandra*, which it resembles in flower form, and from *V. planipetala*, which has rather similar evergreen leaves. Unlike these two species of wide distribution, this is restricted to southern Oregon, in western Josephine County and Curry County, with a few localities in Del Norte County, northern California along the Oregon border. It grows on rocky hillsides.

Introduced into English gardens in 1935, *E. chrysantha* has become established in cultivation but, despite its conspicuous flowers, remains a very uncommon plant. The epithet *chrysantha* means 'golden flowered'.

Vancouveria chrysantha Greene in Bull. Calif. Acad. Sci. 1(2): 66 (1885); Howell, Fl. Northwest Amer. 28 (1897); Gabrielson, West Amer. Alpines: 254 (1932); Rowntree, Hardy Californians: 15819 (1936); Stearn in J. Linn. Soc., Bot. 51: 450 (1938); Peck, Man. Higher Pl. Oregon: 319 (1944); Abrams, Illustr. Fl. Pacific States, 2: 221, 222 (1944); Munz & Keck, Calif. Fl.: 111 (1959); Rickett, Wild Flowers U.S. 5 (1): 207 (1971); Hickman, Jepson Manual Higher Pl. Calif.: 364 (1993). Type: Oregon, coast mountains, fortieth parallel, 1884, *Howell* (US).
V. hexandra var. *aurea* Rattan, Anal. Key West Coast Bot.: 17 (1887).
V. hexandra var. *chrysantha* (Greene) Greene in Pittonia 2: 100 (1890).
Epimedium chrysanthum (Greene) Kom. in Acta Horti Petrop. 29: 141 (1908).

ILLUSTRATIONS. J. Linn. Soc., Bot. 51: t.24 (1938); Abrams., Illustr. Fl. Pacific States 2: 222 (1944); Rickett, Wild Flowers U.S. 5 (1): t.67 (1971).
DESCRIPTION. *Flower stem* 20–40 cm long, leafless, pilose below. *Rhizome* long-creeping, 2 mm thick, but stouter at the tip. *Leaves* basal, biternate or 3- or 5-foliolate, the petioles and nodes more profusely clothed than in other species with slender spreading hairs; leaflets ovate to almost oblate (i.e. much broader than long), variable in size and shape, but up to 4 cm long, 4 cm broad, obscurely 3-lobed, the margin cartilaginously thickened and crisped, subcoriaceous in texture, and 'almost evergreen', dark green and glabrous above, glaucescent and usually very pubescent beneath with numerous soft hairs. *Inflorescence* usually simple, loose, few-flowered, with 4–15 flowers, very glandular the lower peduncles sometimes 2- or 3-flowered; *pedicels* glandular up to 3 cm long. *Flowers* 'soft yellow' (Gabrielson) or 'brilliant rich yellow' (Rowntree), darkening on drying, 1–1.3 cm long. *Outer sepals* glandular, unequal, the inner up to 4 mm long. *Inner sepals* spatulate, narrowing to a clawed base, 7–8 mm long, towards the rounded tip 3.5 mm broad. *Petals* shorter than inner sepals, the narrowly oblong stalk about 4 mm long and expanded and

folded over at the tip to form a rounded nectariferous pocket 2 mm high, 1 mm broad. *Stamens* 4 mm long, sparingly glandular; filaments 2 mm long. *Ovary* glandular.

DISTRIBUTION. USA, Pacific Coast, in the Siskyou mountains of south-western Oregon (Josephine County and Curry County) and northern California (Del Norte County).

3. VANCOUVERIA PLANIPETALA

Although with smaller but more profuse flowers than the other two species of *Vancouveria*, *V. planipetala* is a pleasing if unobtrusive plant with elegant firm evergreen leaves, in Farrer's words 'especially beautifully goffered round the leaflets'. It was introduced into English gardens apparently about the end of the nineteenth century and distributed by the nurserymen Backhouse of York who specialised in rock garden plants. In California it is sometimes known as REDWOOD IVY or INSIDE-OUT FLOWER. Whereas *V. hexandra* with its thin foliage only just manages to extend southward into northern California, *V. planipetala* with more coriaceous foliage and a wide range in California only just manages to extend northward into southern Oregon. In their area of overlap occurs *E. chrysantha* which combines the floral structure of *V. hexandra* with the foliage of *V. planipetala* but differs from both in its yellow flower colour. This led me to suggest in 1938 that it might have originated from their distant hybridization.

Until 1938 this species was known as *V. parviflora* Greene (1890). However, the overlooked elaborate description and illustration of *E. planipetala* published in 1887 by Silvio Calloni (1850–1931), together with the type specimen (in Geneva) collected by Bolander in Marin County, had made clear that the name *V. planipetala* had priority over E.L.Greene's *V. parviflora*. This has now been generally accepted.

In 1914 Greene described three supposed new species as *V. vaseyi*, *V. concolor* and *V. crispa* but examination revealed no reliable characters by which they could be specifically distinguished.

Vancouveria planipetala Calloni in Malpighia 1: 263 (1887); Stearn in J. Linn. Soc, Bot. 51: 452 (1938); Abrams, Illustr. Fl. Pacific States 2: 221 & 222 (1944); Munz & Keck, Calif. Fl.: 112 (1859); Rickett, Wild Flowers U.S. 5 (1): 207 (1971); Peck, Man. Higher Pl. Oregon: 319 (1941); Hickman, Jepson Manual Pl. Calif.: 364 (1993). Type: California, Marin County, 'paper mill red-woods' 1867, *Bolander* (holotype G).

V. parviflora Greene in Pittonia 2: 100 (1890); Parsons & Buck, Wild Fl. Calif.: 88 (1897);
    Armstrong, Field Book West. Wild Fl.: 152 (1915); Jepson, Fl. Calif. 1 (5): 551 (1914).
    Type: not stated.

Epimedium planipetalum (Calloni) Citerne, Berbéridées et Erythrospermées: 21 (1892).

V. chrysantha var. *parviflora* (Greene) Jepson, Fl. West. Mid. Calif.: 204 (1901).

E. parviflorum (Greene) Kom. in Acta Horti Petrop. 29: 141 (1908) as *E. parvifolium*.

E. hexandrum f. *planipetalum* (Calloni) Himmelb. in Denkschr. Kaiserl. Akad. Wiss., Math.-Naturwiss. Kl. 89: 744 (1914).

V. crispa Greene in Repert. Spec. Nov. Regni Veg. 13: 321 (1914). Type: California, Mendocino County, 1902 *Congdon* (holotype US).

V. vaseyi Greene in Repert. Spec. Nov. Regni Veg. 13: 321 (1914). Type: California, place unknown, 1875, *Vasey* (holotype US).

V. concolor Greene in Repert. Spec. Nov. Regni Veg. 13: 322 (1914). Type: California, Sans Mateo County, *A.D.Elmer* 4922 (holotype US).

ILLUSTRATIONS. Malpighia, 1: t.6 (1887); Parsons & Buck, Wild Fl. Calif.: 89 (1897); Armstrong, Field Books West. Wild Fl.: 153 (1915); J. Linn. Soc., Bot. 51: 433, t.26 (1938); Abrams, Illustr. Fl. Pacific States 2: 222 (1944); Rickett, Wild Flowers U.S. 5 (1): t.67 (1971).

DESCRIPTION. *Flower stem* 18–32 cm long, leafless, rising above the foliage, pilose below, 18–32 cm high. *Rhizome* long-creeping, 2–4 mm thick, the leafy tips rising above the ground. *Leaves* basal, usually biternate, sometimes ternately compound with the three divisions 5-foliolate, or rarely simply 5-foliolate; leaflets broadly ovate, often broader than long and three-lobed, very variable in size but up to 6 cm broad, 5 cm long; margin cartilaginously thickened and slightly or much crisped, subcoriaceous and evergreen, dark glossy green and glabrous above, dull grey-green and glaucescent beneath with short sparse hairs. *Inflorescence* loose, paniculate, many-flowered with about 20–50 flowers, the lower peduncles up to 15-flowered; pedicels glandular, 1–2 cm long. *Flowers* white or lavender-tinged, 6–8 mm long (including stamens and reflexed sepals). *Outer sepals* glabrous, unequal, the inner up to 3 mm long. *Inner sepals* spathulate, narrowing to a clawed base, the margin erose, 4 mm long, towards the tip 2 mm broad. *Petals* shorter than the inner sepals, oblanceolate, flat and notched at the tip with a slight median lobe and larger yellow lateral lobes but not saccate, 3 mm long, 1 mm broad. *Stamens* 2 mm long, glabrous; filaments about 1 mm long. *Ovary* glabrous. *Capsule* 4–6 mm long, one- or two-seeded; seeds black, 3–4 mm long, 2 mm broad.

DISTRIBUTION. Western United States, from the Rogue River, Oregon to Santa Lucia mountains of California. Oregon: Jackson County. California: Counties Del Norte, Humboldt, Trinity, Mendocino, Sonoma, Napa, Marin, San Mateo, Santa Cruz, Santa Clara, Monterey.

PART II — REVIEW OF OTHER HERBACEOUS BERBERIDACEAE

ACHLYS

Achlys DC., Syst. Veg. 2: 35 (1821); Hooker, Fl. Bor.-Amer. 1: 30 (1829); Bentham & Hooker fil., Gen. Pl. 2: 45 (1862); Prantl in Engler & Prantl, Nat. Pflanzenfam. 3 (2): 76 (1888); Takeda in Bot. Mag. (Tokyo), 29: 168 (1913); Abrams, Illustr. Fl. Pacific States 2: 219 (1944); Stearn in Walters *et al.*, Europ. Gard. Fl. 3: 394 (1989); H.Loconte in Kubitzki, Fam. Gen. Vasc. Pl. 2: 150 (1993). Type species: *Leontice triphylla* Sm. (= *Achlys triphylla* (Sm.) DC.)

GENERIC DESCRIPTION. *Herbaceous glabrous perennials* to about 30 cm high. *Rhizomes* slender, long-creeping. *Stems* leafless. *Leaves* basal with 3 sessile fan-shaped, palmately veined leaflets. *Flowers* minute, numerous in a slender spike 2.5–5 cm long. *Sepals and petals* absent. *Stamens* 6–15, usually 9, with long clavate white filaments; anthers opening by 2 up-rolling valves. *Ovary* with 1 basal ovule. *Stigma* sessile, broad. *Fruit* a minute achene 3–4 mm long.
DISTRIBUTION. Western North America, Japan and Korea.

A genus of three closely allied taxa which can be treated as three microspecies (*triphylla* and *californica* in western North America, *japonica* in Japan and Korea), or as one tetraploid species (*californica*) and one diploid species (*triphylla*) with two subspecies (*triphylla* and *japonica*) or as one species with three subspecies.

Achlys is a woodland plant in nature, and in cultivation requires a half-shaded position in humus-rich soil; the leaves are its main attraction, apart from its botanical interest as the only wind-pollinated genus of *Berberidaceae*. In North America it is known as DEER FOOT or VANILLA LEAF.

The genus is named after the Greek goddess Achlys, a goddess of hidden places — probably an allusion to the wooded habitat of this plant.

Key to subspecies of *Achlys triphylla*

1. Middle leaflet with 3 entire triangular lobes or sometimes deeply 3-cleft. Inflorescence interrupted. $2n = 12$. Japan, Korea .... subsp. ***japonica***
1. Middle leaflet with 3–9 shallow lobes. Inflorescence continuous. North America 2

2. Middle leaflet with mostly 3–5 lobes. Diploid ($2n = 12$) .... subsp. ***triphylla***
2. Middle leaflet with mostly 7–8 lobes. Tetraploid ($2n = 24$) .... subsp. ***californica***

A. triphylla (Sm.) DC., Syst. Nat. 2: 35 (1821).

1. subsp. **triphylla**

Achlys triphylla JOANNA LANGHORNE

Leontice triphylla Sm. in Rees, Cyclop. 20: art. Leontice no. 5 (1812). Type: 'gathered by Mr Archibald Menzies, on the west coast of North America' (holotype LINN).

A. triphylla (Sm.) DC., Syst. Nat. 2: 35 (1821); Hooker, Fl. Bor.-Amer. 1: 30 (1829).

A. triphylla var. *typica* T. Ito in J. Linn. Soc., Bot. 22: 435 (1881).

ILLUSTRATIONS. Hooker, Fl. Bor.-Amer. 1: t.12 (1829); Bot. Mag. (Tokyo) 29: 171, t.7, fig. 30–36 (1913); Abrams, Illustr. Fl. Pacific States, 2: 222 (1944); Rickett, Wild Flowers U.S. 5: 207, t. 67 (1971).

DISTRIBUTION. Western North America, from British Columbia through Washington and Oregon to Mendocino County, California, predominantly in the mountain regions.

2. subsp. **californica** (Fukuda & H.G.Baker) Brayshaw, Buttercups, Waterlilies & Relatives in British Columbia (Roy. Brit. Columb. Mus. Mem. 1): 5, 179 (1989). Type: USA, California, Humboldt Co., 15 May 1963, Fukuda & Baker (holotype UC).

A. californica Fukuda & H.G.Baker in Taxon 19: 341 (1970).

ILLUSTRATION. Taxon 19: 343 (1970).

DISTRIBUTION. Western North America, from British Columbia through Washington and Oregon to Sonoma County, California, generally closer to the coast than subsp. *triphylla*.

3. subsp. **japonica** (Maxim.) Kitam. in Acta Phytotax. Geobot. 20: 202 (1962); Fukuda in Taxon 16: 315 (1967). Type: Japan, 'in silvis subalpinis principatus ivambu Nippon borealis deterit op. juv. *Tschonoski* a. 1865' (holotype LE).

A. japonica Maxim. in Bull. Acad. Imp. Sci. Saint-Pétersbourg 12: 61 (1868); Takeda in Bot. Mag. Tokyo 29: 171 (1913); Ohwi in Fl. Jap.: 465 (1965).

Fig. 96. *Caulophyllum robustum* photographed by Kazuo Mori in Japan at Hara-mura, Nagano-ken.

A. triphylla var. *japonica* (Maxim.) T.Ito in J. Linn. Soc., Bot. 22: 435 (1887).

ILLUSTRATIONS. J. Linn. Soc., Bot. 22: t.21 (1887); Bot. Mag. (Tokyo) 29: 171, t.7, fig.1–29 (1913); Satake *et al.*, Wild Flowers of Japan 2: t.90 (1981).

DISTRIBUTION. Japan, in northern Honshu and Hokkaido; Korea.

CAULOPHYLLUM

Caulophyllum Michx., Fl. Bor. Amer. 1: 204 (1803); H.Loconte & Blackwell in Rhodora 87: 463 (1983). Type: *Caulophyllum thalictroides* (L.) Michx. (*Leontice thalictroides* L.).

GENERIC DESCRIPTION. *Herbaceous perennials* with short, thick rhizomes, to 150 cm in height. *Stem* with 1–2(–3) leaves. *Leaves* compound, ternately or pinnately divided; leaflets broadly obovate, long-stalked, up to *c*. 15, each 2–5-lobed, to 8 cm long, 5 cm wide. *Inflorescence* a raceme or panicle. *Flowers* small (to c. 1 cm diameter), numerous, greenish, yellowish-green or purplish-green; outer sepals 3–4; inner sepals 6, petal-like; petals 6, much shorter than inner sepals, thick fan-shaped, hooded, producing nectar. *Stamens* 6; anthers dehiscing by up-rolling flaps. *Style* short; stigma gland-like, minute. *Fruit* almost non-existent because the seeds burst through the ovary wall early and ripen naked; seeds blue and berry-like, 5–8 mm long.

DISTRIBUTION. Eastern North America, eastern Asia.

A genus of three closely allied taxa, two from the eastern United States and one from eastern Asia. In North America *C. thalictroides* is called BLUE COHOSH or PAPOOSE ROOT. They are naturally plants of mountain woodlands and in cultivation require partially shaded conditions in humus-rich soil. They may be propagated by division in spring just before active growth commences.

In North America it is reported that *Caulophyllum* has been used in the past by native peoples to induce uterine contractions in the hastening of childbirth; but it is also alleged to be poisonous.

The name *Caulophyllum* is derived from Greek *kaulos*, stem, and *phyllon*, leaf, as the stem of this plant appears to form the stalk for a solitary leaf

Key to the species of *Caulophyllum*

1. Sepals pale yellow, green or purple, up to c. 5 mm long; style less than 0.7 mm 2
1. Sepals uniformly purple, 6 mm or more long; style 1–1.5 mm long E.N. America .... 3. **C. giganteum**
2. Sepals green, yellow or purple; plant up to 80 cm tall. E.N.America .... 1. **C. thalictroides**
2. Sepals pale yellow; plant up to 150 cm tall. Eastern Asia .... 2. **C. robustum**

1. **Caulophyllum thalictroides** (L.) Michx., Fl. Bor. Amer. 1: 204 (1803); H.Loconte & Blackwell in Rhodora 87: 463 (1983). Type: USA, 'habitat in Virginia' (LINN).
Leontice thalictroides L., Sp. Pl. 1: 312 (1753).

ILLUSTRATIONS. Rickett, Wild Flowers of the United States 1: pl.46 (1966); Plantsman 4: 13 (1982); Phillips & Rix, Perennials 1: 38 (1991).

DISTRIBUTION. Widespread in Eastern United States and south-eastern Canada, from New Brunswick to South Carolina and Tennessee.

2. **Caulophyllum robustum** Maxim., Primit Fl. Amur.: 33 (1859). Syntypes: Russia, south-eastern Siberia, Amur region, Kitsi, 8 June 1855; Borbi, 3 June 1855; Adi to Tottjcho, 1 June 1855; near Zjanka, 7 July 1855; Myllki, 17 May 1855; Chúngar-Mündung, 11 July 1855; Geong-Rückens, 25 May 1855; Onmoy, 21 May 1855; Dshare, 18 July 1855, *Maximowicz* s.n.; Chorroko, 9 July 1855, *Maack* s.n. (? all specimens in LE).

ILLUSTRATION. Somoku Dzusetsu, Icon. Pl. Nippon 7: t.25 (1910).
DISTRIBUTION. South-eastern Siberia, China, Korea, Japan, Sakhalin.

3. **Caulophyllum giganteum** (Farw.) H.Loconte & W.H.Blackw. in Phytologia 49: 483 (1981). Type: USA, Michigan, Oakland Co., Farmington, 1917, *Farwell* 4450 (BLH 38877). *C. thalictroides* var. *giganteum* Farw. in Michigan Acad. Sci. Rep. 20: 178 (1918).

ILLUSTRATIONS. None traced
DISTRIBUTION. Widespread in the eastern United States and northwards into south-eastern Canada (Ontario and Quebec).

DIPHYLLEIA

Diphylleia Michx., Fl. Bor.-Amer. 1: 204, t.19–20 (1803); Benth. & Hook. fil., Gen. Pl. 1: 44 (1862); Prantl in Engler & Prantl, Nat. Pflanzenfam. 2, ii: 75 (1888); Ernst in J. Arnold Arbor. 45: 16 (1964); T.S.Ying et al. in J. Arnold Arbor. 65: 70 (1984); Stearn in Walters *et al.*, Europ. Gard. Fl. 3: 395 (1989); H.Loconte in Kubitzki, Fam. Gen. Vasc. Pl. 2: 1251 (1993). Type species: *Diphylleia cymosa* Michx.

GENERIC DESCRIPTION. *Herbaceous perennials* 25–100 cm high. *Rhizomes* creeping, short, with leaf scars as in *Polygonatum*. *Stem* with 2, well separated leaves. *Leaves basal* with centrally peltate blade; cauline leaves 2, alternate, with marginally peltate blades, the lower one with a long petiole, the upper one with a short petiole or sessile; blade orbicular or reniform, lobed and toothed, palmately veined. *Flowers* in a terminal umbel-like cyme, usually many-flowered. *Sepals* 6, soon falling. *Petals* 6, flat, white. *Stamens* 6; anthers opening from the base by two uprolling valves; pollen globose, spinous. *Ovary* with few [2–(6)] parietal ovules; stigma almost sessile, peltate. *Fruit* a globose dark blue berry, with 2–10 seeds.
DISTRIBUTION. Eastern North America, Japan, China.

Diphylleia cymosa

SYDENHAM EDWARDS

Ranzania japonica   STELLA ROSS-CRAIG

A genus of three woodland species, *D. cymosa* in eastern North America, *D. grayi* Maxim. in Japan and *D. sinensis* H.L.Li in China, needing in cultivation a moist humus-rich soil in half-shade. The information below is taken from Ying, T.S., Terabayshi, S. & Boufford, D., A monograph of *Diphylleia* in *J. Arnold Arbor.* 65: 57–94 (1984). The generic name refers to the two leaves on the stem.

Key to the species of *Diphylleia*

1. Leaves glabrous or sparsely pubescent; inflorescence glabrous .... 1. **D. cymosa**
1. Leaves pubescent; inflorescence pubescent 2
2. Upper leaf with petiole 2.5 cm or more long; flowers usually 15 or more; petals less than 9 mm long, 6 mm broad .... 2. **D. sinensis**
2. Upper leaf sessile or with petiole less than 1 cm long; flowers usually 12 or less; petals more than 8 mm long, 6 mm broad .... 3. **D. grayi**

1. **Diphylleia cymosa** Michx., Fl. Bor.-Amer. 1: 204, t.11–12 (1803); T.S.Ying *et al.* in J. Arnold Arbor. 65: 71, t.13 (1984). Lectotype: USA, North Carolina, *Michaux* (P).

ILLUSTRATIONS. Michaux, Fl. Bor.-Amer. 1: t.11–12 (1803); Curtis's Bot. Mag. 40: t.1666 (1814); Gray & Sprague, Gen. Fl. Amer. 1: t.33 (1848); Britton & Brown, Illustr. Fl. North. U.S. 2: 91 (1897), (2nd ed.) 2: 129 (1913); Rickett, Wild Fl. U.S. 2: 217 (1966); J. Arnold Arbor. 65: t.13 (1984).
DISTRIBUTION. South-eastern United States, south-western Virginia, western North Carolina, eastern Tennessee, north-eastern Georgia and Orange County, South Carolina.

2. **Diphylleia sinensis** H.L.Li in J. Arnold Arbor. 28: 442 (1947); T.S.Ying *et al.* in J. Arnold Arbor. 65: 78, t.14 (1984). Type: China, Sichuan province, 1908, *E.H.Wilson* 814 (holotype GH).
D. cymosa subsp. *sinensis* (H.L.Li) T.Shimizu in Hikobia Suppl. 1: 450 (1981).

ILLUSTRATION. J. Arnold Arbor. 65: t.14 (1984).
DISTRIBUTION. China, Gansu (Kansu), Shaanxi (Shensi), Hubei (Hupeh), Sichuan (Szechwan) and Yunnan provinces.

3. **Diphylleia grayi** F.Schmidt in Mém. Acad. Imp. Sci. Saint-Pétersbourg 109 (1868); T.S.Ying *et al.* in J. Arnold Arbor. 65: 82, t.15 (1984). Syntypes: Sakhalin, Arkai, August 1860, *Glehn* s.n., 27 May 1861, *Glehn* s.n.; Dui, June 1860 & 21 July 1860, *Glehn* s.n.; Estaing Bay, *Brylkin* s.n.; Kussunai, 25 April 1860, *Brylkin* s.n.; Manne, August 1860, *collector unknown* (? all in LE).
D. cymosa subsp. *grayi* (F.Schmidt) Kitam. in Acta Phytotax. Geobot. 20: 202 (1962).

ILLUSTRATIONS. Takeda & Tanabe, Kozan Shokobutsu Shash., figs.134–140 (1931); J. Arnold Arbor. 65: t.15 (1984).
DISTRIBUTION. Japan, from Hokkaido south to south-western Honshu; Sakhalin.

NOTE. The specific epithet commemorates the celebrated American botanist Asa Gray (1810–1888) of Harvard University who called attention in 1859 to the relationship of the Japanese flora to that of eastern North America [in *Mem. Amer. Acad.* 6: 377–453 (1859)].

RANZANIA

Ranzania T.Ito in J. Bot. 26: 302 (1888). Type: *Ranzania japonica* (T.Ito) T.Ito. (= *Podophyllum japonicum* T.Ito).

GENERIC DESCRIPTION. *Herbaceous perennials* with slender rhizomes, glabrous, 20–50 cm in height. *Stem* with 2 opposite leaves. *Leaves* divided into 3, shortly stalked, leaflets; leaflets with 3–5 lobes, palmately veined, 8–12 cm long, 8–12 cm wide. *Flowers* 1–6 in a cluster between the leaves, drooping on pedicels 4–8 cm long, pale purple, 2–5 cm diameter; outer sepals 3, small; inner sepals 6, much larger than outer; petals smaller than inner sepals, flat, each with 2 glands at the base producing nectar. *Stamens* 6; anthers dehiscing from the base by up-rolling flaps. Ovary with numerous lateral ovules and with a large, sessile stigma. *Fruit* a berry.

DISTRIBUTION. Japan (Honshu).

A Japanese genus of one species, *R. japonica*, which is a woodland plant in its natural habitat. In cultivation it can be grown successfully in partial shade under deciduous trees or shrubs in a humus-rich, freely draining soil. It may be propagated by division in spring.

The genus is named after the celebrated Japanese naturalist Ono Ranzan (1729–1810).

Ranzania japonica (T.Ito) T.Ito in J. Bot. 26: 302 (1888). Type: Japan, monte Tagakushi, *Ito* (LE?).
Podophyllum japonicum T.Ito in Maxim., Mélanges Biol. Bull. Phys.-Math. Acad. Imp. Sci. Saint-Pétersbourg 12: 417 (1886).
Achlys triphylla var. *japonica* T.Ito in J. Linn. Soc., Bot. 22: 437, t.21 (1887).
Yatabea japonica (T.Ito) Yatabe in Bot. Mag. (Tokyo) 5: 281, t.28 (1891).

ILLUSTRATION. Bot. Mag. (Tokyo) 5: 281, t.28 (1891); Bull. Alpine Gard. Soc. Gr. Brit. 5: 223 (1937); Teresaki, Ic. Fl. Jap.: 764, 765 (1933).
DISTRIBUTION. Japan, Honshu.

JEFFERSONIA

Jeffersonia Barton in Trans. Amer. Philos. Soc. 5: 340 (1793); Bentham & Hooker fil., Gen. Pl. 1: 44 (1862); Prantl in Engler & Prantl, Nat. Pflanzenfam. 3, 2: 75 (1888); Hutchinson in Bull. Misc. Inform., Kew 1920: 242 (1920); Ernst in J. Arnold Arbor. 45: 17 (1964); Stearn in Walters *et al.*, Europ. Gard. Fl. 3: 394 (1989); H.Loconte in Kubitzki, Fam. Gen. Vasc. Pl. 2: 150 (1993). Type species: *J. binata* Barton (= *Podophyllum diphyllum* L. = *J. diphylla* (L.) Pers.)

Jeffersonia dubia

STELLA ROSS-CRAIG

THE GENUS EPIMEDIUM
PLATE 21

Jeffersonia diphylla

SYDENHAM EDWARDS

Plagiorhegma Maxim., Prim. Fl. Amur.: 34 (1859); Hutchinson in Bull. Misc. Inform., Kew 1920: 242 (1920). Type species: *P. dubium* (*J. dubia* (Maxim.) Baker & S.Moore).

GENERIC DESCRIPTION. *Herbaceous, low-growing perennials* with short rhizomes. *Stems* leafless, 1-flowered. *Leaves* basal, long-stalked; blade entire, lobed, or divided into 2 stalkless, palmately-veined leaflets. *Sepals* 4, soon falling. *Petals* 8, flat. *Stamens* usually 8; anthers opening from the base with 2 up-rolling flaps. *Ovary* with many lateral ovules; style short with small, 2-lobed stigma. *Capsule* opening either by an almost horizontal, incomplete slit below the top (thus appearing to have a lid (*J. diphylla*), or by an oblique, downward slit from the top (*J. dubia*). *Seeds* numerous, each with a small, lacerate aril.
DISTRIBUTION. Eastern North America, eastern Russia, northern China, North Korea.

A genus of two species, one from eastern North America, the other from Far Eastern Russia, China (Manchuria) and North Korea. They are naturally woodland plants and are best grown in a light or sandy soil with abundant leaf-mould or peat, in partial shade. Propagation is by careful division in spring or by seed.

The generic name commemorates Thomas Jefferson (1745–1826), the Third President of the United States of America who drafted the 1776 Declaration of Independence. He was a keen horticulturist.

Key to the species of *Jeffersonia*

Leaf-blade divided into 2 fan-shaped leaflets. Flowers white .... 1. ***J. diphylla***
Leaf-blade not divided but reniform or orbicular. Flowers lavender-blue
   (or rarely white) .... 2. ***J. dubia***

1. **Jeffersonia diphylla** (L.) Pers., Syn. Pl. 1: 418 (1803); Britton & Brown, Illustr. Fl. N. States & Canada 2: 92 (1897), 2nd ed. 2: 129 (1913); Ernst in J. Arnold Arbor. 45: 17 (1964). Type: Virginia (LINN 667.1).
Podophyllum diphyllum L., Sp. Pl. 1: 505 (1753).
Jeffersonia binata Barton in Trans. Amer. Philos. Soc. 5: 334 (1793).
Jeffersonia bartonis Michx., Fl. Bor.-Amer. 1: 237 (1803).

ILLUSTRATIONS. Curtis's Bot. Mag. 37: t.1513 (1812); Loddiges, Bot. Cab. 11: t.1036 (1825); Gray & Sprague, Gen. Fl. Amer. 1: t.34 (1848); Britton & Brown, Illustr. Fl. 2: 92 (1897), 2nd ed. 2: 129 (1913); Addisonia 5: t.176 (1920); Rickett, Wild Fl. US 1: 157, t.46 (1966).
DISTRIBUTION. Eastern North America, from southern Ontario, Iowa, New York State, Virginia, South Wisconsin to Alabama and Tennessee.

2. **Jeffersonia dubia** (Maxim.) Baker & S.Moore in J. Linn. Soc., Bot. 17: 377 (1879); Komarov in Acta Horti Petrop. 22: 322, 335 (1903); B.Fedtschenko in Komarov, Fl. URSS 7: 540 (1937); Airy Shaw in Curtis's Bot. Mag. 164: t.9681 (1948). Type: Eastern Asia, 'Am untern Amurin Nadelwäldern bei Pachale........', 29 May 1855, 1 June 1856, *Carl von Ditmar* (?LE).

Plagiorhegma dubium Maxim., Prim. Fl. Amur. 34, t.2 (1831); Hutchinson in Bull. Misc. Inform., Kew 1920: 242 (1920).

ILLUSTRATIONS. Maxim., Prim. Fl. Amur.: t.2 (1859); Bull. Misc. Inform., Kew 1920: 242 (1920); Beih. Bot. Centralbl. 337, 378 figs.62–66 (1928); Fl. URSS. 7: 540, t.37 (1937); Curtis's Bot. Mag. t.9681 (1948).

DISTRIBUTION. Far Eastern Russia (Ussuri region), Manchuria, North Korea.

BONGARDIA

Bongardia C.A.Mey., Verz. Pfl. Cauc.: 174 (1831). Type species: *Bongardia chrysogonum* (L.) Endl. (*Leontice chrysogonum* L.)

DESCRIPTION. *Herbaceous, tuberous perennial* 20–60 cm high. *Tuber* large, subglobose. *Leaves* all basal, 10–40 cm long, often spreading ± horizontally, pinnately divided; leaflets 7–17, in pairs or sometimes whorls of 3 or 4, sessile, obovate or cuneate, usually 3–6-lobed or toothed at the apex, sometimes bifid to the base, up to 3 cm long. *Inflorescence* a loose panicle with ascending branches. *Flowers* yellow, long-stalked, to 2 cm across; sepals 6, suborbicular, concave, caducous; petals 6, flat, conspicuous, each with a small nectar-producing pore at the base, 8–12 × 3.5–5 cm, irregularly crenate. *Stamens* 6, anthers opening from the base by 2 up-rolling valves. *Ovary* with 5–6 basal ovules. *Stigma* almost sessile, folded or lobed. *Fruit* an ovoid or ellipsoid, papery capsule to 1.5 cm long, opening by short, acute flaps at the top; seeds 1–2, large, black, pruinose.

DISTRIBUTION. Eastern Mediterranean region to Central Asia.

A genus of probably two species distributed in the eastern Mediterranean region and South-west Asia, from Turkey to Central Asia and Pakistan where they grows in fields, open stony hillsides and waste places.

Bongardia chrysogonum is an attractive yellow-flowered tuberous plant with glaucous-green pinnatisect leaves which are often marked with a reddish zone near the base of each leaflet. It is suitable for alpine house cultivation, or a well-drained position in full sun; the tubers must be kept fairly dry and warm in summer when the plant is dormant. Propagation is by seed.

It is recorded that 'baked or boiled the tubers are edible' (*Flora of the USSR,* English ed. 7: 421, 1970).

The generic name *Bongardia* commemorates August Gustav Bongard (1786–1839), a German botanist employed in Russia.

Key to the species of *Bongardia*

Inflorescence 3–4-flowered; capsule 5–6 mm long, 3 mm wide; leaves with 7–9 leaflets. Pakistan .... 2. **B. margalla**
Inflorescence with many flowers; capsule to 15 mm long and 10 mm wide; leaves with up to 17 leaflets. Mediterranean region to Baluchistan & C.Asia .. 1. **B. chrysogonum**

Fig. 97. *Jeffersonia dubia.* Photograph: Brian Mathew.

Fig. 98. *Jeffersonia dubia* **'Alba'** growing in garden in Surrey, England. Photograph: Brian Mathew.

Bongardia chrysogonum

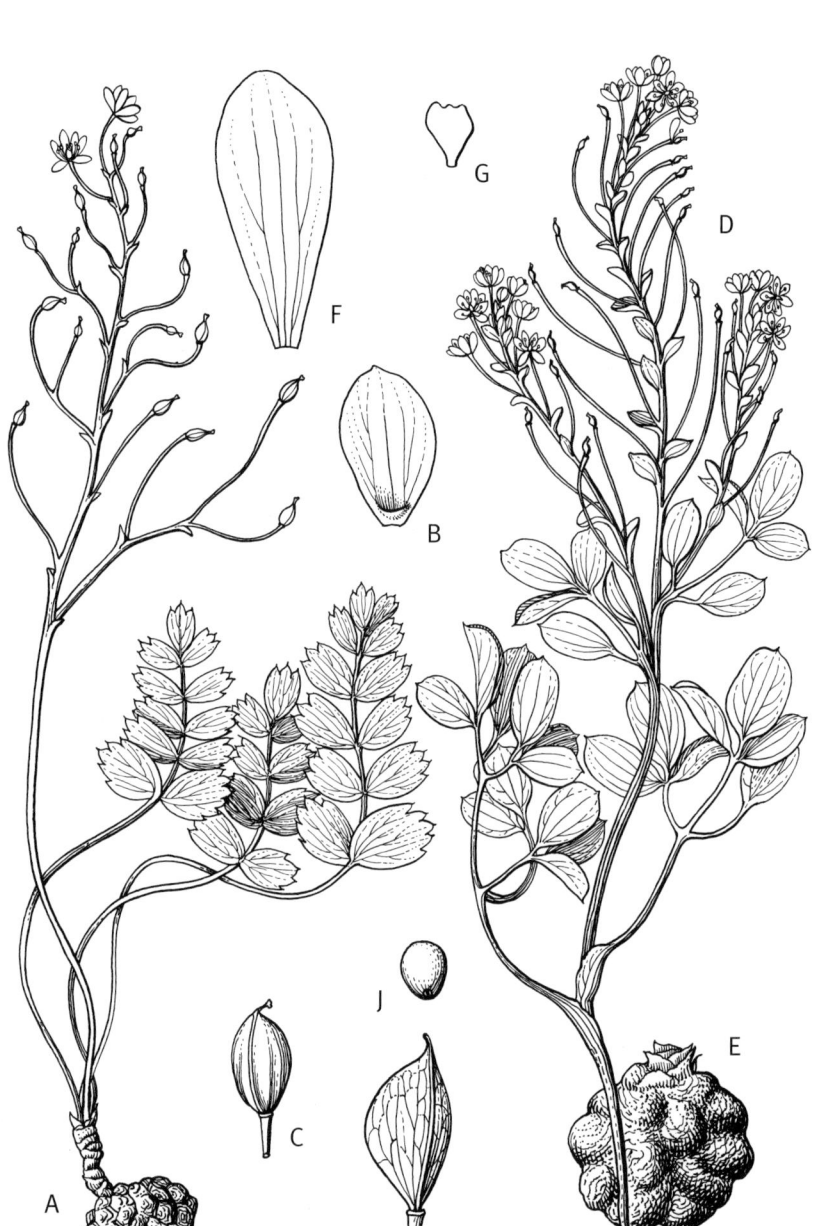

Fig. 99. **Bongardia chrysogonum.** A, habit, × 1; B, petal, × 6; C, fruit, × 1. **Leontice leontopetalum.** D, habit, × 1; E, tuber, × 1; F, petal, × 6; G, nectary, × 6; H, fruit, × 1; J, seed, × 1. From *Flora of Iraq* Vol. 4, Plate 136 (1980). Drawn by Laura M. Ripley.

1. **Bongardia chrysogonum** (L.) Spach, Hist. Veg. (Phan.) 8: 65 (1839). Type: Greece (holotype LINN).
Leontice chrysogonum L., Sp. Pl.: 447 (1753).
Bongardia rauwolfii C.A.Mey., Verz. Pfl. Cauc.: 174 (1831), *nom. illegit*. Type: Azerbaydzhan, 'in campis, collibus et agris provinciae Baku'.
B. olivieri C.A.Mey., Verz. Pfl. Cauc.: 86 (1831). Type: Aegean Islands, 'in insula Chio', Olivier.

ILLUSTRATIONS. Curtis's Bot. Mag. t.6244 (1876); Polunin, Flowers of Greece and the Balkans, pl. 8 (1980); Everett, New York Botanical Gardens Illustr. Encyclop. Hort. 2: 447 (1981); Rix & Phillips, The Bulb Book: 48 (1981); Plantsman 4: 3 (1982).
DISTRIBUTION. Eastern Mediterranean region, eastwards to Central Asia.

2. **Bongardia margalla** R.R.Stewart ex Qureshi & Chaudhri in Pakistan Syst. 3(1): 21 (1987). Type: Pakistan, Rawalpindi District, Margalla Pass, *R.R.Stewart* 9544 (holotype RAW).

ILLUSTRATION. Pakistan Syst. 3(1): 21 (1987).
DISTRIBUTION. Pakistan (Swat and Rawalpindi districts).

GYMNOSPERMIUM

Gymnospermium Spach, Hist. Nat. Veg. Phan. 8: 66 (1839); Takhtajan in Bot. Zhurn. (Moscow & Leningrad) 55: 1191–1193 (1970). Type species: *Gymnospermium altaicum* (Pall.) Spach (= *Leontice altaica* Pall.).

GENERIC DESCRIPTION. *Perennial glabrous herbs* with tubers. *Leaves* basal and cauline, the basal ones long-stalked, the cauline sessile to shortly petiolate, ternate, often with the leaflets palmately divided into 4–7 lobes. *Flowers* in a short, unbranched raceme. *Sepals* 6, petal-like. *Petals* much shorter than sepals, convolute, nectar-producing. *Stamens* 6, anthers opening from base by 2 up-rolling flaps. *Ovary* with 2–4 basal ovules; style long, stigma minute. *Fruit* early dehiscent, the seeds ripening in an exposed state.
DISTRIBUTION. South-eastern Europe and western Asia to Central Asia and western China.

A small genus of about six species, inhabiting open situations on hillsides and mountain slopes. Their tubers lie deep underground during the hot, dry summer months, so in cultivation they require a warm dry rest period when dormant. Thus, in cool temperate climates with summer rainfall they are best grown in a cool glasshouse or bulb frame for protection, although they are generally very frost-hardy. The two most commonly cultivated species, *G. altaicum* and *G. albertii*, flower in spring and are often to be found near melting snow so they receive a lot of moisture at this time; it is essential, therefore, that in cultivation plenty of moisture is available, but with good drainage to avoid stagnant conditions. Propagation is by seed sown in late summer or autumn.

Gymnospermium albertii — MATILDA SMITH

The name *Gymnospermium* is derived from *gymnos*, naked, and *spermum*, -seeded, referring to the way in which the fruit bursts open very soon after development and the seeds continue to mature in a naked state.

Key to the species of *Gymnospermium*

1. Leaves 2–4-ternate and ultimate leaflets lobed, resulting in many 'segments' per leaf. E.China .... 10. **G. kiangnanensis**
1. Leaves usually 1-ternate and leaflets either undivided or divided into 3–5(–7) lobes 2

2. Sepals only 5 mm long; N.China .... 9. **G. microrrhynchum**
2. Sepals 7–15 mm long 3

3. Ovary carried on a distinct stipe, at least 1 mm long 4
3. Ovary sessile, or if subsessile, the stipe less than 0.5 mm long 8

4. Leaflets on long stalks (petiolules) 1–6 cm long 5
4. Leaflets subsessile or on petiolules at most 8 mm long. Tadjikistan ... 6. **G. darwasicum**

5. Sepals 7–8 mm long. Afghanistan .... 8. **G. silvaticum**
5. Sepals usually more than 10 mm long 6

6. Petiolules to 1.3 cm long; flowers wholly yellow, not coloured differently on exterior. Tadjikistan .... 7. **G. vitellinum**
6. Petiolules 1.5–6 cm long; flowers usually stained or veined brown or violet on the exterior 7

Fig. 100. ***Gymnospermium albertii*** in the wild in the Tien Shan near Chimgan, Uzbekistan. Photograph: Brian Mathew.

7. Terminal leaflets truncate or emarginate, mucronate. Albania .... 3. **G. shquipetarum**
7. Terminal leaflets usually obtuse or rounded at apex. C.Asia .... 5. **G. albertii**

8. Nectaries (petals) saccate at base and with 2 recurved apical teeth 9
8. Nectaries not saccate, and with 2 erect apical teeth. Caucasus .... 4. **G. smirnovii**

9. Stamens much exceeding the nectaries. Ukraine, Crimea, Romania ... 2. **G. odessanum**
9. Stamens only just exceeding the nectaries. Altai region .... 1. **G. altaicum**

1. **Gymnospermium altaicum** (Pall.) Spach, Hist. Nat. Veg. Phan. 8: 67 (1839). Type: 'habitat in apricis montium Altaicorum Sibiriae', *Pallas, Patrin*; 'circa Zmeof', *Fischer* (?LE). *Leontice altaica* Pall. in Acta Acad. Sci. Imp. Petrop. 11: 255 (1783).

ILLUSTRATION. Curtis's Bot. Mag. t.3245 (1833).
DISTRIBUTION. Kazakhstan (Altai mountains).

2. **Gymnospermium odessanum** (Fisch. ex DC.) Tahkt. in Bot. Zhurn. (Moscow & Leningrad) 55(8): 1192 (1970). Type: 'in collibus calcareis circa Odessam', *Beaupré* (not traced).
Leontice altaica var. *odessana* Fisch. ex DC., Syst. Nat. 2: 26 (1821).
Leontice odessana (Fisch. ex DC.) G.Don in Gen. Syst. 1: 119 (1831).
Gymnospermium altaicum subsp. *odessanum* (Fisch. ex DC.) E.Mayer & Pulević in Willdenowia 13: 278 (1984).

ILLUSTRATION. None traced.
DISTRIBUTION. Romania, southern Ukraine, Moldava, Crimea.

3. **Gymnospermium shquipetarum** Papar. & Qosja in Bul. Shkencacet Nat. 30(2): 95 (1976). Syntypes: Albania, 'pr. Krujae (m. Skënderbè)', 1000 m, 23 March 1974 (fl.), 5 May 1974 (fr.), *Tartari s.n.* (Tirana Univ., Fac. Of Science); 'in montuosis pr. Elbasanis m. Shalqinii', 1400 m, 27 March 1974 (fl.), *Tartari s.n.* (Tirana Univ., Fac. of Science).
G. scipetarum Papar. & Qosja ex E.Mayer & Pulević in Willdenowia 13(2): 278 (1982).

ILLUSTRATION. Bul. Shkencacet Nat. 30(2): 95 (1976)
DISTRIBUTION. Albania.

4. **Gymnospermium smirnovii** (Trautv.) Tahkt. in Bot. Zhurn. (Moscow & Leningrad) 55(8): 1198 (1970). Type: Georgia, near Tbilisi, 'prope Lagodechi Kachetiae australis, *M.N.Smirnow* (holotype LE).
Leontice smirnovii Trautv. in Trudy Imp. S.-Petersburgsk. Bot. Sada 7: 405 (1880)

ILLUSTRATIONS. None traced.
DISTRIBUTION. Georgia, Caucasus.

5. **Gymnospermium albertii** (Regel) Tahkt. in Bot. Zhurn. (Moscow & Leningrad) 55(8): 1192 (1970). Type: Central Asia, Uzbekistan, between Tashkent and Samarkand, 'in Turkestaniae montibus alatavicus occidentalis', *A.Regel* (holotype LE).

Leontice albertii Regel in Gartenflora 30: 293 (1880).

ILLUSTRATIONS. Curtis's Bot. Mag. t.6900 (1886); Plantsman 4: 2 (1982).
DISTRIBUTION. Uzbekistan, ?Kirgizia, ?Tadjikistan

6. **Gymnospermium darwasicum** (Regel) Tahkt. in Bot. Zhurn. (Moscow & Leningrad) 55(8): 1192 (1970). Type: Central Asia, Tadjikistan, 'in Buchara orientale in chanato Darwas prope Schikai', *A.Regel* (holotype LE).
Leontice darwasica Regel in Trudy Imp. S.-Petersburgsk. Bot. Sada 8: 692 (1884).

ILLUSTRATION. Trudy Imp. S.-Petersburgsk. Bot. Sada 8: t.14 (1884).
DISTRIBUTION. Central Asia, probably confined to Tadjikistan.

7. **Gymnospermium vitellinum** M.Král in Preslia 53(1): 67 (1981). Type: Tadjikistan, picked material, probably from around Hissar (but actual location unknown), *M.Král s.n.* (indigenous collector unknown), 25 April 1976 (holotype PRC).

ILLUSTRATIONS. None traced.
DISTRIBUTION. Tadjikistan.

8. **Gymnospermium silvaticum** (Freitag) Tahkt. in Bot. Zhurn. (Moscow & Leningrad) 57(10): 1277 (1972); Browicz in Fl. Iranica 101: 9 (1973). Type: Afghanistan, Nangarhar province, Dara-e-Nur N Jalalabad, 2000–2300 m, *Freitag* 4748 (holotype GOET, isotype W).
Leontice sylvatica Freitag in Bot. Jahrb. Syst. 91: 473 (1972).

ILLUSTRATION. Bot. Jahrb. Syst. 91: 474, fig.3 (1972).
DISTRIBUTION. Eastern Afghanistan.

9. **Gymnospermium microrrhynchum** (S.Moore) Tahkt. in Bot. Zhurn. (Moscow & Leningrad) 55(8): 1182 (1970). Type: China, 'in provincia Schin King' [Sinkiang = Xinjiang], *J.Ross* (holotype K).
Leontice microrrhyncha S.Moore in J. Linn. Soc., Bot. 17: 377 (1889) and var. *venosa*.
L. microrrhyncha S.Moore f. *venosa* (S.Moore) Kitag., Neo-Lineam. Fl. Manshur.: 317 (1979). Type: China 'in montibus prope Kwandien', *J.Ross* (holotype K).

ILLUSTRATION. J. Linn. Soc., Bot. 17: t.16, fig.3 & 4 (1889).
DISTRIBUTION. Northern China.

10. **Gymnospermium kiangnanensis** (P.L.Chiu) H.Loconte in Canad. J. Bot. 67(8): 2315 (1989). Type: China, Zhejiang (Chekiang), Hangzhou, 19 April 1974, *Chiu* 95 (holotype Herb. Hort. Bot. Hangzhou).
Leontice kiangnanensis P.L.Chiu in Acta Phytotax. Sin. 18: 96 (1980).

ILLUSTRATION. Acta Phytotax. Sin. 18: 96, fig.1 (1980).
DISTRIBUTION. Eastern China.

Fig. 101. **Gymnospermium albertii** flowering as the snow melts at Chimgan, Uzbekistan. Photograph: Brian Mathew.

Fig. 102. **Gymnospermium microrrhynchum** (upper) and **G. microrrhynchum** var. **venosum** (lower). From: *Journal of the Linnean Society, Botany* 17: t.16 (1889).

LEONTICE

Leontice L., Sp. Pl. 1: 312 (1753). Lectotype: *Leontice leontopetalum* L.

GENERIC DESCRIPTION. *Perennial herbs* with tubers. *Leaves* basal and cauline, the basal long-stalked, the cauline ones shortly petiolate to sessile, 2- to 3-ternate or pinnate, with numerous, shortly stalked or sessile, entire leaflets. *Flowers* numerous in axillary and terminal bracteate racemes, often forming a panicle. *Sepals* 6, flat, conspicuous and petal-like. *Petals* smaller, convolute and nectar-producing. *Stamens* 6, anthers opening by 2 up-rolling flaps. *Ovary* with 2–4 basal ovules; style short, stigma truncate. *Fruit* much inflated, bladder-like, net-veined, breaking open irregularly at the top when dry, or indehiscent.
DISTRIBUTION. South-eastern Europe, western Asia, North Africa.

A small genus of 3 species, inhabiting open fields and hillsides, often in places which become sun-baked in summer. In cultivation they require plenty of moisture in the winter-spring growing period, then a warm dry rest period in summer. In cool, moist temperate climates they are best grown in a cool glasshouse or bulb frame where conditions can be controlled. Propagation is possible by seed, although this takes several years to produce a flowering plant.

It is recorded that the tubers of *L. leontopetalum* subsp. *ewersmannii* 'contain a significant quantity of starch, from which liquor is distilled' [*Flora of the USSR* Vol. 7, English edition: 419 (1970)].

The name *Leontice*, from *leon*, a lion, refers to the supposed resemblance of the leaf pattern to the paw print of a lion.

Key to the species of *Leontice*

1. Fruits indehiscent .... 3. **L. incerta**
1. Fruits irregularly dehiscent at apex 2
2. Inflorescence branched; plant to 80 cm in height when in flower 3
2. Inflorescence unbranched; plant to 20 cm in height when in flower
   .... 2. **L. armeniacum**
3. Leaflets obovate or suborbicular; fruits to 3 cm long, carried on
   spreading pedicels .... 1a. **L. leontopetalum** subsp. **leontopetalum**
3. Leaflets lanceolate or elliptical; fruits less than 2.5 cm long, on
   ascending pedicels .... 1b. **L. leontopetalum** subsp. **ewersmannii**

1. **Leontice leontopetalum** L., Sp. Pl. 1: 312 (1753).

1a. subsp. **leontopetalum.** Type: Italy, Crete (Herb Linn 433/1).

ILLUSTRATIONS. Holmboe, Veg. Cyprus: 228 (1914); Huxley & Taylor Wild Flowers of Greece and Aegean: plate 57, 58 (1977).
DISTRIBUTION. Bulgaria, Egypt, Greece (incl. Crete), Iran, Iraq, Israel, Lebanon, North Africa (Algeria), Syria, Turkey.

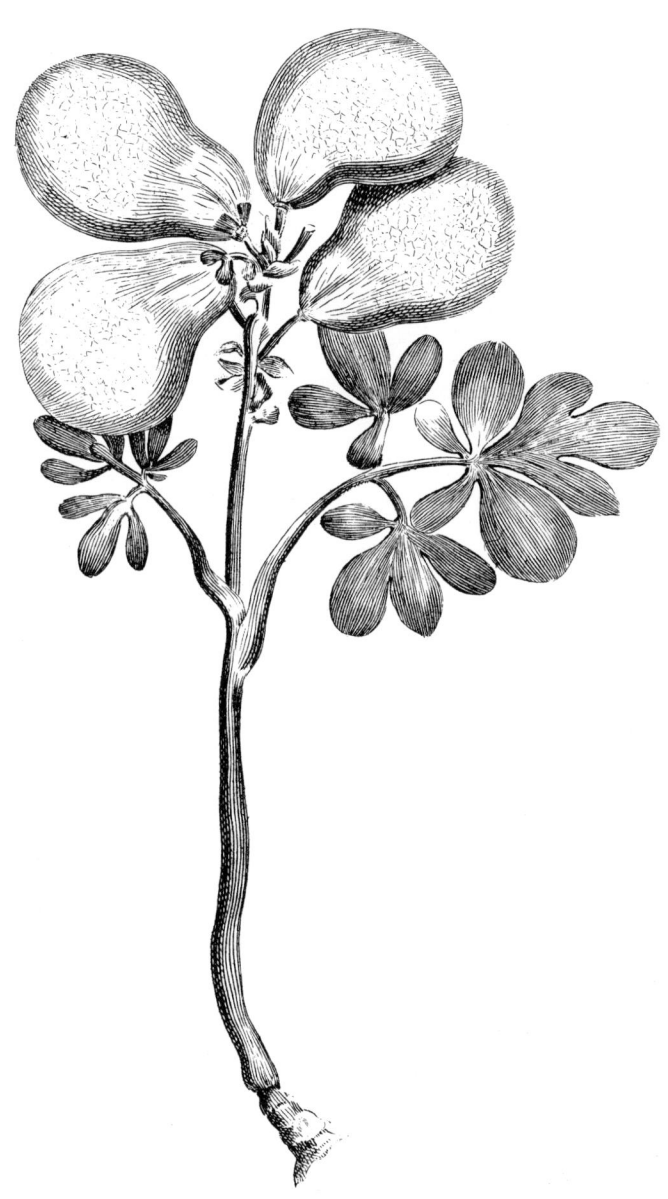

Fig. 103. **Leontice incerta.** From P. Pallas, *Reise durch Verscheidene Provinzen des Russischen Reichs* 3(2): 726, t.5 (1776).

Fig. 104. **Leontice armeniaca.** Photographed in Iran near Arak, north-west of Esfahan, by Brian Mathew.

1b. subsp. **ewersmannii** (Bunge) Coode in Notes Roy. Bot. Gard. Edinburgh 26: 42 (1964). Type: Uzbekistan, 'im sande und Lehmsande des Jaman-Kisil-Kum und des Batkak-Kum', 21–23 April 1842, *Ewersmann* (holotype LE, isotype P).

ILLUSTRATIONS. Flora USSR 7: t.37, fig.4 (1937); Rix & Phillips, The Bulb Book: 39 (1981).
DISTRIBUTION. Central Asia, Western Iran, Pakistan, Syria, south-eastern Turkey.

2. **L. armeniaca** Boivin in Bélanger, Voy. Indes Orient. Ic. (1846). Type: illustration in Bélanger, *op. cit.*: Pl. 2A.
L. minor Boiss., Fl. Orient. 1: 100 (1867). Syntypes: 'Aderbidjan', *Bélanger*; 'inter Salmas et Khoi', *Szovitz*; 'monte Elbrus prope Passgala', *Kotschy* 99; 'in Persia australi ad Ispahan, *Aucher* 4036; 'circa Persepolin', *Kotschy* 804.
L. leontopetalum subsp. *armeniaca* (Boivin) Coode in Notes Roy. Bot. Gard. Edinburgh 26: 42 (1964).
?*L. apiifolia* Fisch. in herb.

ILLUSTRATION. Rix & Phillips, The Bulb Book: 39 (1981).
DISTRIBUTION. Syria, Jordan, Iraq, Iran, Transcaucasia.

3. **Leontice incerta** Pall., Reise Russ. Reich. 3: 726 (1776). Type: Kazakhstan, 'in praeruptis limosis circa Lacum Inderiensem', *Pallas* (not traced: iconotype: Pallas, Reise Russ. Reich.: t.5, fig.2 (1776)).
L. vesicaria Pall. in Acta Acad. Sci. Imp. Petrop. 1779: 257 (1783). Types: 'in argillosis Turcomaniae', *Karelin, Lehmann.*

ILLUSTRATION. Pallas, Reise Russ. Reich.: t.5, fig.2 (1776).
DISTRIBUTION. Central Asia (?Kazakhstan, Turkmenistan, Uzbekistan).

Chromosome counts in Berberidaceae (excluding *Podophyllum*)

Taxon	Gametic	Somatic count	Reference
Achlys japonica		12	Matsuura & Suto, 1935
A. japonica		12	Fukuda, 1967; Kuroki, 1967
A. triphylla subsp. *triphylla*		12	Fukuda, 1967
A. triphylla subsp. *californica*		24	Fukuda, 1967
Bongardia chrysogonum		14	Tören, 1950
Caulophyllum robustum	8	16	Matsuura & Suto, 1935
C. robustum		16	Kuroki, 1965
C. robustum		16	Kawano & Ihara, 1967
C. thalictroides		16	Langlet, 1928
C. thalictroides var. *thalictroides*		16	Moore, 1963
C. thalictroides var. *giganteum*		16	Moore, 1963
Diphyleia cymosa		12	Langlet, 1928
D. grayi		12	Miyaji, 1930
D. grayi	6	12	Matsuura & Suto, 1935
D. grayi		12	Soeda, 1942
D. grayi		12	Kurita, 1956
D. grayi		12	Kuroki, 1967
D. grayi	(18 in endosperm)	12	Noda & Fujimura, 1970
D. grayi		12	Noguchi & Kawano, 1974
D. sinensis		12	Ma, S.-B. & Hu, Z.-H., 1996
Epimedium alpinum		12	Maude, 1939
E. diphyllum		12	Suzuka, 1950
E. diphyllum		12	Kurita, 1956
E. diphyllum		12	Kuroki, 1967
E. grandiflorum (macranthum)		12	Langlet, 1928
E. grandiflorum f. *violaceum*		12	Miyaji, 1930
E. koreanum		12	Kuroki, 1967
E. pinnatum		12	Langlet, 1928
E. pinnatum		12	Miyaji, 1930
E. rubrum		12	Langlet, 1928
E. youngianum (musschianum)		12	Langlet, 1928
Gymnospermium albertii		16	Kosenko, 1978
G. darwasicum		16	Kosenko, 1978
G. odessanum		16	Kosenko, 1978
G. smirnowii		16	Kosenko, 1979
Jeffersonia diphylla (binata)		12	Langlet, 1928
J. dubia		12	Miyaji, 1930
J. dubia		12	Langlet, 1928
J. dubia		12	Kurita, 1956
Leontice ewersmannii		32	Kosenko, 1977
L. incerta		32	Kosenko, 1977
L. leontopetalum		16	Tören, 1962
Ranzania japonica		14	Miyaji, 1930
R. japonica		14	Kurita, 1956
Vancouveria hexandra		12	Langlet, 1928

CYTOLOGY REFERENCES

Fukuda, I. (1967). The biosystematics of Achlys. *Taxon* 16: 308–316.

Kawano, S. & Ihara, M. (1967). Chromosome morphology of Caulophyllum robustum and its systematic implications. *J. Jap. Bot.* 42: 129–135.

Kosenko, V.N. (1977). [Comparative karyological study of Leontice ewersmannii Bunge and L. incerta Pall.]. *Bot. Zhurn.* (Moscow & Leningrad) 62: 7.

Kosenko, V.N. (1978). [Comparative karyological study of Gymnospermium altaicum (Pall.) Spach and G. darwasicum (Regel) Takht.]. *Bot. Zhurn.* (Moscow & Leningrad) 63: 8.

Kosenko, V.N. (1979). Comparative karyological study of representatives of the family Berberidaceae. *Bot. Zhurn.* (Moscow & Leningrad) 64(11): 1539–1553.

Kurita, M. (1956). Karyotype studies in Berberidaceae 1. *Mem. Ehime Univ., Sect. 2, Nat. Sci., Ser. B* 2(3): 19–24.

Kuroki, Y. (1965). Chromosome counts in three species of Berberidaceae. *Mem. Ehime Univ., Sect. 2, Nat. Sci., Ser. B* 5(2): 19–24.

Kuroki, Y. (1967). Chromosome study in seven species of Berberidaceae. *Mem. Ehime Univ., Sect. 2, Nat. Sci., Ser. B* 5(4): 27–33.

Matsuura & Suto (1935). Contributions to the idiogram study in phaneogamous plants 1. *J. Fac. Sci. Hokkaido Imp. Univ., Ser. 5, Bot.* 5(5): 33.

Ma, S.-H. & Hu, Z.-H. (1996). A karyotypic study on Podophylloideae. *Acta Bot. Yunnan.* 18(3): 325–330.

Moore, R. J. (1963). Karyotype evolution in Caulophyllum. *Canad. J. Genet. Cytol.* 5: 384–388.

Noda, S. & Fujimura, T. (1970). Karyotypes in root-tip cell and endosperm nucleus of Diphylleia grayi. *Kromosomo* 79–80: 2548–2551.

Noguchi, J. & Kawano, S. (1974). Brief notes on the chromosomes of some Japanese plants. *J. Jap. Bot.* 49(3): 76–86.

Soeda, T. (1942). On the chromosomes of Diphylleia grayi. *Jap. J. Genet.* 18: 47–48.

Tören, J. (1950). Les caractères morphologiques, anatomiques et cytologiques de Bongardia chrysogonum. *Istanbul Üniv. Fak. Mecm., B* 15: 239–263.

Tören, J. (1962). Recherches sur les Berberidaceae de la Turquie. 2. Caracteres cytologiques du Leontice leontopetalum L. *Istanbul Üniv. Fak. Mecm., B* 27: 3–4.

PART III — THE GENUS PODOPHYLLUM

THE GENUS PODOPHYLLUM

by Julian M.H. Shaw

The genus *Podophyllum* comprises about 14 species with one species in eastern North America and the remainder distributed in Asia along the Himalayan range and in central and southern China and Taiwan. They are woodland plants requiring moist shady conditions.

According to Joseph Pitton de Tournefort (1656–1708), a French plantsman Morin first coined the name *Anapodophyllum* (Duck-foot-leaf) for the plant we now know as *Podophyllum peltatum*. This name Tournefort adopted in his *Elemens de Botanique* (1694, p. 204) and *Institutiones Rei Herbariae* (1700, p. 239). Linnaeus, who thought that a botanist ought to be able to memorise generic names, either replaced or shortened what he called 'ell-long' names. Thus he replaced *Lepidocarpodendron* by *Protea* and *Anapodophyllum* by *Podophyllum*. He castigated in his *Critica Botanica* (1737), "By ell-long I mean those of such a length that they almost take up a whole line on the page and which almost strangle us in the effort to pronounce them."

In modern taxonomy the name *Podophyllum* dates from Linnaeus, *Species Plantarum* (1753) and the associated *Genera Plantarum* (1754). Here he included two species, both North American: *Podophyllum peltatum* (Tournefort's *Anapodophyllum*) and *P. diphyllum* (now excluded as *Jeffersonia diphylla*).

MORPHOLOGY OF *PODOPHYLLUM*

RHIZOME AND ROOT

The rhizome of *P. peltatum* is very distinct from other *Podophyllum* species due to its elongated internodes, which develop from stolons that typically arise from lateral buds. The growth of the rhizome in *P. peltatum* has been studied extensively by Foerste (1884), Holm (1899) and more recently by Frye (1977) and DeMaggio & Wilson (1986).

In the seedling stage the rhizome is vertical, producing scale leaves (cataphylls) and a single green aerial leaf. After the fourth year an axillary bud appears in one of the cataphyll bases which develops into a horizontal shoot giving rise to the characteristic rhizome with elongated internodes. At this point the terminal bud of the vertical rhizome becomes dormant, but may later develop if the terminal bud of the horizontal rhizome is damaged.

In an adult plant, new rhizome growth commences in late April from a small lateral bud at the base of the aerial shoot bud, proceeding rapidly until mid June. It appears as a white stolon, with paired cataphylls about every 2 cm, which forms the internodal region, while the terminal bud forms the nodal region of the rhizome. Up to six of these stolons, between 2 and 20 cm in length, may arise at the base of each aerial stem, although a single stolon is most frequently produced. Only the youngest node in a rhizome, which is usually the terminal bud, produces an aerial shoot, while the older nodes remain dormant yet viable for up to six years before senescence and decay occur. According to Foerste (1884) the internodes produced from older nodes are very short, producing in time several nodes which are almost adjacent, whereas Frye (1977), after several years of intensive fieldwork, reports frequently encountering both forms of growth.

Kumazawa (1930, 1936) has investigated the rhizome morphology of *P. hexandrum*, *P. pleianthum* and *Diphylleia* species. He reports that in each of these the main axis of the rhizome is replaced each year by an axillary bud which grows horizontally and terminates in an erect dormant bud. The sympodial axis formed develops in one direction horizontally, producing a single aerial stem scar each year. These scars, equivalent to the nodal regions of the *P. peltatum* rhizome, are very close to each other. The rhizome of *P. pleianthum* is identical in appearance to that of *Diphylleia*. Long internodes similar to those of *P. peltatum* occur in juvenile plants of some species of section *Dysosma*, suggesting that the retention of elongated internodes in adult *P. peltatum* is a juvenile character.

The histology of the rhizomes of *P. hexandrum* and *P. peltatum* has been described in detail by Wallis and Goldberg (1937a, 1937b). They observe that the anatomy of the *P. hexandrum* rhizome resembles that of the enlarged nodal sections of *P. peltatum*, but differs considerably from that of the elongated narrow internodal regions. Cortical bundles are present in the rhizome of *P. peltatum* and rare in *P. pleianthum* (Kumazawa, 1936), but are entirely absent from *Diphylleia* (Kumazawa, 1930).

Adventitious roots arise from the nodal regions of the rhizome in *P. peltatum* and from the underside of the rhizome in the Asiatic species; in *P. peltatum* the roots are contractile. Plants of *P. pleianthum* in cultivation produce roots near the soil surface which occasionally produce small adventitious buds that develop into plants and become separate from the parent, sometimes several metres distant, thus providing an unexpected means of clonal propagation. Similar regeneration from large root fragments of *P. hexandrum* has been observed. This suggests that some wild populations of Asiatic species may be clonal.

CATAPHYLLS

The cataphylls are bract-like structures associated with rhizome growth and enclosing the dormant overwintering bud. Internode elongation between cataphylls formed late in the growing season does not occur, and these cataphylls subsequently enclose the dormant bud (DeMaggio & Wilson, 1986). The cataphylls of this dormant apical bud are large and conspicuous, and their scars are visible at the base of an aerial shoot. The number of terminal bud cataphylls varies but in *P. peltatum* it is usually three outer ones enclosing the bud and three to five internal 10–20 mm long cataphylls that overlap one another by one third of their width (DeMaggio & Wilson, 1986). These bud cataphylls produce a small peltate blade at the apex.

Cataphylls are present around the apical bud of all *Podophyllum* species and particular taxonomic importance is attached by Chatterjee (1952, 1953) to the appearance of the largest cataphyll, the remains of which often adhere to the base of the aerial flowering stem. Examination of many specimens has shown this to be of little value.

AERIAL STEM

The term aerial stem is used here to refer to the stalk between the rhizome and the juvenile leaf lamina or as far as the node at which the aerial stalk branches. Petiole is used for the stalk between an aerial stem node and an adult leaf lamina. Some authors refer to the whole aerial stalk as the petiole (Ellis & Fell, 1962, 1963), while others studiously avoid applying the term stem to anything but the subterranean rhizome itself.

The most frequently observed aspect of *Podophyllum* morphology has been the irregular arrangement (as opposed to concentric bands) of the vascular bundles within the annual aerial stem. The stem in transverse section has an appearance like that of a monocot, with the exception that a cambium is present in each bundle (Harvey-Gibson & Horsman, 1919). The vascular bundles in the petioles are also very irregularly distributed (Holm, 1899). *Podophyllum* is similar in this respect to *Diphylleia*, *Leontice* and some species of *Papaver* (*Papaveraceae*) and *Ranunculaceae*, notably *Actea* and *Thalictrum* (Holm, 1899; Worsdell, 1908; Ellis & Fell, 1962), which has led in the past to their classification among the 'anomalous dicotyledons' (Holm, 1899).

Several mechanical aspects of the extension of the aerial stem in *P. peltatum* were studied by Himmel (1927). Continuous flexing of the aerial stem brought about a 12 per cent increase in stiffness in the plane of flexing, and a 40 per cent increase in stiffness at right angles to the plane of flexing. This may be partly due to a cross sectional change in shape of the aerial stem from round to elliptic. Himmel (1927) also demonstrated that individual vertically emerging aerial stems could lift a weight varying from 175 to 500 g, with an average of 300 g. Dormer (1972) reviewed some of these results and made an interesting observation concerning the geotropic response of the aerial stem. In this particular experiment, Himmel constructed an apparatus that released lead shot to weigh down the apex of a horizontally positioned aerial stem in response to its efforts to assume a vertical position. The combined results from six aerial stems tested separately resulted in 102 particles of lead shot being dropped on to the aerial stem apices in the first hour. For subsequent hours the numbers were 42, 43, 59, 62, 54, 43. If plotted against time a second smaller peak appears after the third hour. Dormer (1972) suggests this may indicate the presence of two mechanisms of geotropic response. An initial rapid reaction and a second, which comes in to play when an obstacle has to be overcome.

One of the most useful taxonomic characters from the aerial stem is its colour. The newly emerged aerial stems of *P. hexandrum* contain a maroon pigment which immediately distinguishes them. This gradually becomes less intense with age, but the stem retains a pale reddish-orange colour, which is preserved in herbarium specimens. This pigmentation is also characteristic of P. aurantiocaule.

LEAF

There are three noticeable trends with regard to leaf shape in *Podophyllum*:
1. In sections *Dysosma* and *Paradysosma*, the depth of the leaf lobes increases with latitude and altitude, with the simplest forms in Taiwan and the most dissected in western China.
2. Radial symmetry in juvenile leaves may change to bilateral symmetry in adult foliage.
3. There is a transition from peltate to palmate leaves from sections *Dysosma* and *Paradysosma* to sections *Podophyllum* and *Hexandra*.

The leaf shape appears to be one of the more stable and hence useful characters in the genus. Many herbarium specimens consist of leaves only, and in some cases these cannot be positively determined. Its usefulness varies between species and some of the differences are subtle. Young plants produce juvenile leaves with a different shape and degree of lobing from the adult form. Adult foliage is produced by most species as flowering draws near, except in the case of *P. hexandrum*, some individuals of which appear to retain the juvenile shape. The change in leaf shape usually consists of a reduction in lobing on the adaxial leaf margin, that is the side of the leaf blade facing the centre of the plant. This change is made to accommodate the inflorescence and is particularly noticeable on plants with the inflorescence inserted on the petiole just below the lamina. The entire aerial shoot, consisting of typically two leaves and some flowers along with petioles and aerial stem, is formed the previous year at the end of active growth, usually mid to late summer. This annual shoot is packed in to a single axillary bud which appears to be the rhizome apex. Imagine the mechanical problems of folding and packing two telescopic umbrellas tied together by the handles so that they are held close together with a bunch of flowers tied to the stem of one but positioned so that the flowers are packed above the tip of the tallest umbrella and then opening them simultaneously without damaging anything. This is roughly the sort of feat performed effortlessly by the average *Podophyllum* plant. As one contemplates this feat it becomes apparent why a change of shape in leaf is desirable with the onset of maturity. The zone of leaf retardation corresponds with the upward fold across the leaf in bud on the adaxial side. Lobe shape is fairly consistent within each taxon and a very useful character. The lobes may be divided near the apex and the subdivisions are termed lobules.

Several authors describe the occurrence of two stem types in *P. peltatum*, one bearing single leaves, the other two or rarely three (Bigelow, 1840). Some even excavated rhizome systems to demonstrate that only one species is involved (Krochmal et al., 1974; Sohn & Policansky, 1977). This shoot variation, like the dimorphic foliage discussed above, is linked to fertility, for only the multiple-leafed shoots normally bear flowers. Benner and Watson (1989) have shown that the size of the rhizome differs between juvenile and adult leafed shoots, and discuss the factors known to influence which type of shoot is produced by the terminal bud.

Peltate leaves are not very frequently encountered amongst flowering plants, although they are more common in some families than others. Occasionally they may characterize a genus such as *Umbilicus* or *Tropaeolum*. Various attempts have been made to explain the distribution of peltate leaves among flowering plants. Uittien (1928) attempted to correlate leaf form with inflorescence type and came to the conclusion that peltate leaves occur in families with either opposite foliage or rosettes of leaves, or a cymose inflorescence as does *Podophyllum* (Kumazawa, 1936). A number of genera that contain one or more species with peltate leaves display a variation of leaf shapes similar to *Podophyllum*, in which regard *Dorstenia bernimiana* provides an interesting parallel with *P. hexandrum*.

LEAF MARGIN

The small marginal teeth are a useful character and helpful to separate the generic sections. In *P. peltatum* they are coarse with 4–5 teeth per 2.5 cm of margin and in *P. hexandrum* they are usually finer, with 6–12 teeth per 2.5 cm and Stearn (1989) makes good use of them in the *European Garden Flora* to distinguish between these species. Plants of the *P. aurantiocaule* group usually have a mixture of different sized teeth on the leaf margins along with hairs. Section *Dysosma* produces small often very regularly spaced teeth which are also a useful character. Some confusion has been caused in the past by calling these teeth cilia, which is usually taken to mean hairs. They are, however, vascularised extensions of the lamina surface usually supplied by three veinlets with a glandular apex called a papilla by Hickey (1979) and described in detail by Ellis and Fell (1963).

INFLORESCENCE

The position of the inflorescence is a character more stable in some species than others. Some species invariably produce the flower at the junction of the petioles including *P. peltatum*, some wild plants of *P. pleianthum* and typical *P. delavayi*. In other species such as *P. hexandrum* it is very variable, even on an individual plant, whereas the majority of taxa produce the flowers along the petiole of the upper leaf, some about midway (*P. aurantiocaule* subsp. *furfuraceum*) and some close under the leaf lamina such as *P. versipelle*. Several factors appear to influence this. One seems to be introgression and is perhaps the reason for some populations of *P. pleianthum* producing an inflorescence along the petiole. Another is environmental; this particularly seems to affect cultivated plants of *P. pleianthum* which, while apparently stable in the wild, often resort to producing flowers from the upper petiole or even on a petiole-like stalk without a leaf. The control of this character and its inheritance would make an interesting study. In this account I have attempted to make use of this character since it is easily observed and, if used with caution, appears useful. In the *Solanaceae* it has been found to be useful taxonomically, (Child & Lester, 1991) and there also sympodial shoot generations are produced in which the position of the inflorescence appears to vary from a nodal to an internodal position.

As with *Podophyllum* the basic cause of this variation is the nature of the sympodial shoot system itself. Because the shoot system is sympodial I have refrained from using the term axillary for flowers that arise at the junction of the petioles.

SEPALS AND PETALS

The sepals are arranged in two whorls of three, that often differ in size. Some authors prefer to speak of perianth segments, which are either sepaloid or petaloid (Rao & Hajra, 1993). Section *Dysosma* is especially prone to producing flowers in which the inner three sepals are petaloid. Terabayashi (1983) has reported 9 sepals in 3 whorls (in addition to 6 petals in 2 whorls) in specimens of *P. pleianthum*. The sepals are fugaceous, and in *P. peltatum* and *P. hexandrum* they often drop off as a cap without opening.

The petals are a conspicuous feature of the flower in all species of *Podophyllum*, and useful characters are provided by shape and colour. In the genus the number of petals is usually six or occasionally nine, in two or three series, but this may vary from five to eight in *P. peltatum*. The occurrence of extra petals apparently depends upon sufficient free space being available between existing petal primordia around the meristem at the time of initiation (DeMaggio & Wilson, 1986). Dedoublement, where a single petal is deeply lobed into two parts is sometimes observed in petals of the inner whorl (Terabayashi, 1983). The petals of *P. peltatum* arise from primordia that are initiated in acropetal sequence during early August, however growth ceases after they attain a length of about 2 mm. They remain dormant until mid-May of the next season when within 2 weeks they grow to a length of 12 mm; by anthesis the petals are 2 cm or more in length (DeMaggio & Wilson, 1986). This occasionally results in individual flowers with extremely reduced petals, due to a failure of the petals to expand as dormancy breaks. Development appears to follow a similar pattern in *P. hexandrum*, since collections have been encountered during this study with unexpanded petals.

STAMENS

Six stamens are usually present in a *Podophyllum* flower with the exception of *P. peltatum*, which may produce up to 18, with 12 being the norm. This unusually large number of stamens apparently results in a departure from the acropetal sequence of initiation of the floral organs, that is; sepals, petals, stamens and gynoecium. In *P. peltatum* only three stamen primordia are formed initially, the remaining stamens appearing later after the gynoecium (DeMaggio & Wilson, 1986). This situation is known in other species outside *Podophyllum*, such as *Portulaca grandiflora* (*Portulacaceae*), which DeMaggio and Wilson regard, along with *P. peltatum*, as having basipetal inception of a second group of stamens. The sequence of stamen initiation appears to be reflected in their arrangement. In *P. peltatum* there are two whorls, the outer consisting of 3 stamens and the inner whorl of 3 groups of 3 or 4 stamens each (Terabayashi, 1983). Some stamens of the inner whorl are often abnormally reduced.

A wild Himalayan population of *P. hexandrum* was reported by Royle (1834) in which the number of stamens varied from 6 to 10 per flower, with some filaments bearing 2 or 3 anthers. It was also observed that plants of *Paris polyphylla* (*Trilliaceae*) at this locality displayed similar anomalies. Accordingly it may be inferred that local environmental factors may have been responsible.

The stamens in all species of *Podophyllum* consist of a filament of variable length with a flattened base supporting four anther loculi, two on each side of the connective. In section *Dysosma* the connective is swollen and often extended beyond the thecae to form a mucro of variable length. Anther dehiscence is extrorse, in common with other members of the *Berberidaceae*. Although Woodson (1928) first reported introrse anthers in *P. pleianthum* and used this as a feature to separate *Dysosma* from *Podophyllum*, in which he is followed by Hu (1937), subsequent workers with the advantage of live material have confirmed the extrorse dehiscence of the anthers in *P. pleianthum* (Kumazawa, 1936; Terabayashi, 1983). Dehiscence in *Podophyllum* is by longitudinal slits which is not common in the *Berberidaceae*, where it is known only in *Nandinia*, *Podophyllum*, and some species of *Berberis* (Stearn, 1978; Benson, 1979). Other berberidaceous genera effect anther dehiscence by apically hinged flaps. In *Diphylleia* these are formed by a rupture along the basal and lateral edges of the abaxial anther loculi (Terabayashi, 1983). The longitudinal slits of *P. peltatum* and *P. pleianthum* form along the suture of the abaxial loculus wall with the connective. Additionally the internal wall between the loculi in each pair also ruptures permitting release of the pollen. However in *P. hexandrum* the longitudinal slits form along the suture between the loculi of each pair on opposite sides of the anther (Terabayashi, 1983). Anther length is variable and can be diagnostic, therefore it is discussed in detail under *P. aurantiocaule* and *P. hexandrum*.

The anther connective is often swollen and may be produced above the thecae to form a mucro of variable length. In the sections *Podophyllum*, *Hexandra* and *Paradysosma*, which have predominantly white flowers, the mucro may be absent or, if present, reduced to only 0.2–0.4 mm. It is particularly noticeable in the illustration of *P. peltatum* in Krochmal *et al.* (1974). The greatest development of an anther mucro occurs in section *Dysosma*, especially in *P. delavayi*. In well grown specimens it is conspicuous and may reach 5 mm or more in length, tapering to a fine point. Plants apparently derived from *P. delavayi* by hybridisation with members of the *P. pleianthum* complex produce a shorter mucro, and in some herbarium specimens of *P. pleianthum* it is less than 1 mm long. The length of the mucro together with that of the petals is stressed by Ying (1979) to separate taxa within his concept of *Dysosma*; he even provides a petal and anther key to species. However a number of specimens encountered during this study do not fit into the scheme he proposes. There does seem to be considerable variation in both petal and mucro length within the species of section *Dysosma*, especially in the *P. delavayi* group. In section *Dysosma* the anther connective and mucro are usually pigmented and swollen with the abaxial surface somewhat pitted in dried specimens, apparently due to the collapse of large cells. Investigation of live and dried material has revealed that this surface is probably secretory and appears to function as an osmophore.

Carpeloid stamens were first reported from *P. peltatum* by Halsted in 1894. Sawyer (1926) investigated material from a colony of *P. peltatum* in Iowa where numerous stamens were found that bore both thecae containing pollen and also ovule-like structures. On some stamens the anther connective was expanded in an open fan with a thickly convoluted stigmatic margin. Some ovules were found to contain a mature embryo sac. There are no reports of seeds being produced from carpeloid stamens.

POLLEN

Recent studies on the structure of pollen in the genus *Podophyllum* (including *Dysosma* and *Sinopodophyllum*), have been conducted by Ying (1979) using optical microscopy and by Kosenko (1980) and Nowicke and Skvarla (1981) using both scanning and transmission electron microscopy. These last two studies have included some species of *Podophyllum* and *Diphylleia* as part of family-wide pollen surveys in the *Berberidaceae*. A short SEM study of *Diphylleia* pollen by Takahashi is incorporated in Ying *et al.* (1984).

Pollen in all material examined of *Podophyllum* and *Diphylleia* is 3-colpate, a feature shared with all the other herbaceous genera of *Berberidaceae* except *Ranzania* (6-pantocolpate) and also with *Nandinia* (Terabayashi, 1982). There are however some striking differences, notably the pollen of *Diphylleia* which is spinose, resembling that of some *Compositae* (Kosenko, 1980; Nowicke & Skvarla, 1981; Ying *et al.*, 1984), and that of *P. hexandrum* with its verrucate surface and its being released in tetrads, while all other *Berberidaceae* release pollen in monads (Ying, 1979; Nowicke & Skvarla, 1981). The pollen grains are two-celled when shed (Davis, 1966).

Ferguson and Skvarla (1982) have reviewed pollen morphology in relation to pollinators in the *Papilionoideae* (*Leguminosae*). They note that the occurrence of pollen with a coarse rugulate or verrucate surface sculpturing is related to pollination by birds or bats, whereas species with small insect pollinators have pollen with a simple reticulate or perforate surface. To some extent this parallels the findings of Poole (1981) from *Zimmermannia* pollen, which suggests that ecological requirements such as pollinators are reflected in the differing pollen forms encountered. Also the amount, distribution and consistency of pollenkit on the pollen surface is known to change with the pollination system. Consequently there is considerable scope for investigation in *Podophyllum* as the genus appears to provide an interesting case in which the changes in pollen characters, from reticulate monads to verrucate tetrads, partly coincide with the change in breeding system and pollination. The aggregation of pollen into tetrads is probably connected with the higher ovule number present in *P. hexandrum*. The suspicion that change in pollinators exaggerates taxonomic gaps, raised by Arroyo (1981), emphasises the advisability of maintaining the species of *Podophyllum* within a single genus until the fascinating and complex function of pollen morphology and pollenkit distribution is better understood in this group.

GYNOECIUM

In *Podophyllum* each flower produces a single ovoid ovary, surmounted by a short style and an irregularly lobed peltate stigma. More than 40 (up to 130 have been observed in *P. hexandrum*), bitegmic, hemitropous or anatropous (in *P. hexandrum*), pseudo-crassinucellar ovules are borne on a globose parietal placenta (Davis, 1966; Corner, 1976; Terabayashi, 1983).

The shape of the ovary varies slightly in the genus. It tends to be elliptic in the *P. pleianthum* complex, but some individuals from western China, near the range of *P. delavayi* produce conspicuously elongated elliptic ovaries with an elongated style. As a consequence of its ovary shape, *P. delavayi* itself is the only *Podophyllum* with a stipitate fruit, the others abruptly join the pedicel. Section *Paradysosma* produces a more or less spherical or shortly elliptic ovary. *P. hexandrum* is variable, with some ovaries that are spherical (*Yu* 19871 from Yunnan) with a tendency to become more elliptical in the west. The style is very short. In *P. peltatum* the ovary is elliptic-globose with no visible style.

As with other genera of the *Berberidaceae*, the ovary of *Podophyllum* has been the subject of considerable study and debate. Smith (1923) regards it as modified from free central placentation. Saunders (1925, 1928) and Chapman (1936) view the ovary as composed of two unequal carpels, but disagree over these two carpels as being derived from either two different types of carpel (Saunders) or from the expansion of a lower carpel accompanied by the contraction of the higher carpel of "exactly similar basic structure" (Chapman, 1936).

Chapman (1936) notes the rare occurrence of bilocular ovaries in *Podophyllum* and infers from this the probability that the typical unilocular type has been derived from a bilocular form, drawing on the parallel development of bilocular ovaries in *Epimedium* to support this idea and seeking to ally *Podophyllum* with *Epimedium* rather than with *Diphylleia*, for which a relationship to *Nandinia* is postulated.

Recent investigations by Guedes (1977) conclude that the *Podophyllum* gynoecium is made up of a single carpel congenitally closed along all its length. The placenta is borne from the upper half of the line of fusion of the carpel margins. Based on material of *P. peltatum*, *P. veitchii* (probably *P. delavayi*) and *P. versipelle* (probably subsp. *boreale*), Guedes decided that there was no indication of a second sterile carpel. He notes that the teratological appearance of second carpels is known in other families that are typically monocarpelate and describes a characteristic reorganisation of the vascular traces associated with this event, concluding that it is quite unnecessary to invoke the explanation of a reduced hypothetical second carpel. Noting that abnormal pistils with a single locule and two placentas occur widely in the family and intermediate conditions are found in *Berberis* and *Diphylleia*, Terabayashi (1983) agrees, observing that these may be derived through multiplication from normal monomerous pistils. *P. delavayi* was observed to differ slightly from *P. peltatum* and *P. versipelle* in the origin of the vascular supply to the placenta and ovules, otherwise a uniform picture is presented for the species studied so far (Guedes, 1977).

Terabayashi (1985) published a detailed review of floral anatomy in the *Berberidaceae*. He groups *Diphylleia* with *Dysosma* and *Podophyllum* as genera with parietal placentation, noting in these the occurrence of two kinds of vascular bundles in the placental region. The bundles on the locular side of the ovary, which sometimes display the xylem oriented inversely, provide the traces to the ovules along the lower to middle parts of the ovary, while the ventral bundles supply the traces to the ovules along the upper half of the ovary.

The ovary in *Podophyllum* differs from others of the *Berberidaceae* in the large and well developed vascular network. The ovary wall is traversed by several ventral bundles, some lateral bundles and a dorsal bundle which supply the style and stigma. The ovary venation consists of dichotomously branched veins, with anastomoses in the second and third order veins. The well developed vascular system is thought to be related to the swollen placenta and large number of ovules (30–50). Terabayashi (1983) considers this to be a specialised condition when compared to the normal situation in which the ovular traces are derived directly from ventral bundles in the placental region as in *Diphylleia* with 10–15 ovules. The fruit of *Podophyllum* is the largest known in the *Berberidaceae*.

Ovule orientation is described by Terabayashi (1983) as hemitropous (amphitropous) in the genus *Podophyllum*, while Corner (1976) notes it to be anatropous in *P. hexandrum*.

The ovary locule is open in *P. peltatum* due to the presence of a transmitting tract lined with glandular cells. These cells extend from the stigmatic surface through the style to a point about half way to the loculus. The glandular cells are epidermal with a papilla projecting into the tract. Their function is little understood, but they are thought to be involved in pollen grain recognition, determining the path of the pollen tube and supplying nutrients to the pollen tube (DeMaggio & Wilson, 1986).

Embryological investigations have been reviewed by Davis (1966). The embryo sac is of a Polygonum-type and embryogeny conforms to the Onograd type, but the pattern of early cleavages is irregular in both *P. hexandrum* and *P. peltatum* (Clark, 1923; Lublinerowna, 1925a, 1925b, 1925c; Davis, 1966).

FRUIT

The *Podophyllum* fruit is a berry without any tendency to predetermined regular dehiscence. McLean and Ivimey-Cook (1956) describe it as a fleshy capsule. As would be expected from the variation in ovary shape, the shape of the mature fruit varies even within a species. The fruit of *P. peltatum* is typically yellow when ripe with a strong fruity odour. There are many records of fruits with several (2 to 6) carpels (Gray, 1868; Porter, 1877; Clute, 1915, 1918; Plitt, 1931). Clute (1915) reported finding a "large number of examples in an area several rods square" and concluded that these may represent one clone.

Fruit colour in *P. peltatum* has also occasioned comment. There are reports of orange (Steyermark, 1952), red (Raymond, 1948) and maroon (Deam, 1940; Steyermark, 1952, 1963), or white and rose coloured (Rafinesque, 1828) fruits which is discussed in the taxonomic account. Fruit shape in *P. hexandrum* varies between populations as does the shade of colour. The size is related to the number of seeds contained.

SEED

A thick fleshy placental aril covers the seeds in *Podophyllum*. In *P. hexandrum* the short funicle becomes fleshy along with the placental tissue and invests each seed with an almost complete coat (Corner, 1976). Here the aril, often a conspicuous structure on the seeds of herbaceous *Berberidaceae*, appears to have been transferred almost entirely to the placental tissue.

Only the seeds of *P. peltatum*, *P. hexandrum*, *P. pleianthum* and *P. versipelle* seem to be known. It is immediately apparent that the seed of *P. hexandrum* is distinguished by its dark brown testa and more globose shape from those of *P. peltatum* and *P. pleianthum* which resemble each other, being somewhat reniform with a straw-pale brown testa.

Corner (1976) views the structure of the testa as being of taxonomic value. In *P. hexandrum* it is composed of rather thick-walled, slightly lignified cells, appearing in transverse section as a palisade, in which the cells elongate longitudinally with oblong facets. This is a feature shared with *Jeffersonia* as opposed to fibriform cells with strongly thickened outer walls as found in *Epimedium*. Seed surface patterns might well yield valuable characters, giving insight into the relationship between sections *Paradysosma* and *Hexandra*. Unfortunately the seed of section *Paradysosma* remains unknown at present.

GERMINATION AND SEEDLINGS

Germination is epigeal in both *Diphylleia* and *Podophyllum*, and follows the same pattern, termed pseudomonocotyledonous germination by McLean and Ivimey-Cook (1956). Pseudomonocotyledonous germination is also known in some species of Celastraceae, Cruciferae, Cucurbitaceae, Papaveraceae, Primulaceae, Ranunculaceae and Umbelliferae (Dickson, 1882; Holm, 1899; Kumazawa, 1930; McLean & Ivimey-Cook, 1956).

Recourse to live plant material revealed that germination proceeds as follows: The radicle emerges from the seed first and when anchored, the hypocotyl and cotyledonary tube emerge and arch, the apex of the loop pushing upwards above the soil, bearing the folded cotyledons still inside the testa and elevating them as it straightens out. As the cotyledons expand the testa is forced off. At the seedling stage the two cotyledons are fused along their base with a small pit in the centre of the join. This small pit is the apex of a cotyledonary tube which extends below ground and terminates at the plumule. The hypocotyl is very short and during the first few months growth of the root system proceeds underground. When the plumule begins to develop it ruptures the cotyledonary tube near its base and emerges above the ground opposite the cotyledons atop their tube, which then wither away. The plumule consists of a small irregularly orbicular peltate leaf, or in the case of *P. hexandrum* a small subtrifoliate leaf. Germination usually occurs in early spring after the seed has over-wintered, or rarely in a mild autumn in the case of cleaned seed of *P. hexandrum* at Nottingham.

There is experimental evidence that the fruit pulp contains a germination inhibitor, which in the case of *P. peltatum* has also been found to inhibit germination of *Avena* spp. (*Graminae*), (Krochmal *et al.*, 1974). Fruit borne inhibitors are also present in *P. hexandrum* (Nautiyal in Kaushik, 1988; Nautiyal *et al.*, 1987), and these may perform the adaptive role of delaying germination during mild autumn weather until the following spring, thus reducing loss of seedlings in a severe winter.

In addition to cleaning to remove the inhibitors in the fruit pulp, *P. hexandrum* seed requires a 32 day post harvest ripening period before germination commences. Maximum germination of 64 per cent has been achieved experimentally at 20°C in the light and 4 per cent at 20°C in the dark (Nautiyal in Kaushik, 1988). Sown out of doors at Jaunsar, India, seeds of *P. hexandrum* were found to lie dormant for two winters, germinating in the spring of the second or even third year after a dormancy of at least two years (Troup, 1915). Seed left outside in the fallen fruit at Nottingham (UK) germinated in the following spring. Similar results are reported by Badhwar and Sharma (1963) at Chakrata, with about 45 per cent germination in April from an autumn sowing. Seed stored indoors through the winter and sown the following spring gave 7.9 per cent germination from an April sowing and <7 per cent from a May sowing (Badhwar & Sharma, 1963), demonstrating that seed stored in paper packets is sensitive to desiccation.

Several workers have been unable to germinate seeds of *P. peltatum* (Badhwar & Sharma, 1963; Ellis & Fell, 1963; Krochmal *et al.*, 1974). Seeds of this species are extremely sensitive to desiccation and is rarely if ever offered in seedlists. Krochmal *et al.* (1974) tested embryos of *P. peltatum* for viability with tetrazolium dye and found freshly collected seed to be 80 per cent viable. Seeds stored at room temperature for $2^{1}/_{2}$ years were apparently all dead when tested. At Nottingham, it has proved possible to maintain viability of stored seed of both *P. hexandrum* and *P. peltatum* for at least 2 years by storing cleaned seed moist in a refrigerator at about 4°C, a technique used by commercial seed companies with seed of *Lapageria rosea* (Philesiaceae). Rust and Roth (1981) reported germination of 45 per cent in a greenhouse using cold-wet stored seed of *P. peltatum*, and obtained the highest germination (87.5 per cent) and seedling survival rates from seed that had been ingested and voided by eastern box turtles (*Terapene carolina*). This seed also germinated 5–7 days earlier than non-ingested seed planted adjacent to them. Seed sown at a depth of about 1 cm under natural conditions out of doors averaged 24.4 per cent germination when protected by mesh against large predators, whereas seed placed on the soil surface ranged from 0.7 to 7.1 per cent, depending on the degree of predator exclusion (Rust & Roth, 1981).

Cotyledon shape provides a potentially useful taxonomic character. The cotyledons of all accessions of *P. peltatum* and *P. hexandrum* seen are elliptic with entire margins, occasionally with a slight notch at the apex. In *P. pleianthum*, the cotyledons are slightly larger and bear marginal teeth each with a small papillate apex, similar to the marginal projections of adult foliage. It remains for the cotyledons of section *Paradysosma* and *P. delavayi* to be seen before the value of this character can be fully ascertained in separating sections within the genus. Kumazawa (1930) is the only author to report serrated margins for the cotyledons of *P. peltatum*, which may be due to error or, if true, he may have encountered a hybrid plant (with *P. pleianthum*).

Terabayashi (1987) recognises seven groups of genera in the *Berberidaceae* based on cotyledonary morphology, seedling vasculature and germination behaviour. He groups *Diphylleia* with *Podophyllum* (including *Dysosma*) in a separate tribe *Podophylleae*.

POLLINATION AND BREEDING SYSTEM IN *PODOPHYLLUM*

SECTION *PODOPHYLLUM*

Podophyllum peltatum is the most intensively investigated species. The flower appears to be a typical bumble bee flower with a wide open bowl shape on a deflexed nodding pedicel. However, each clone seems to differ in the floral scent produced and degree of self compatibility. The floral scent can best be detected by placing a cut flower in a small clean glass bottle which is then sealed and left for half an hour or so. After this time the scent has had time to collect and may be detectable by carefully sniffing the contents of the jar. One clone thus sampled had such a strong ginger scent that it brought tears to the eyes. Others report clones with orange scent, or scentless or too sweet.

Mathews (1912) thought that bumble bees and early bees cross-pollinated the flowers. Pollinators observed include at least two *Bombus* species, the honey bee *Apis mellifera* (Swanson & Sohmer, 1976a, 1976b), *Andrena* sp. and *Halictus* sp. (Rust & Roth, 1981). In addition, Laverty and Plowright (1988) reported eight 'bumble bee species'. Of interest here is the discovery of high concentrations of the flavonoid, kaempferol 3-diglucoside in the petals of *P. peltatum* (Schilling & Calie, 1982). Because this flavonoid is widespread amongst early spring ephemeral flowers in eastern North America and exhibits a peak of UV absorbency in the range of 345–350 nm, which is near the optimum sensitivity of *Apis mellifera* in the UV spectrum, Schilling and Calie suggest that it may render white flowers conspicuous to pollinators. Very similar petal flavonoids are found in the related *Diphylleia cymosa*, *Caulophyllum thalictroides* and *Jeffersonia diphylla*.

The honey bee usually visits flowers within a single clone (thus effecting geitonogamy), while *Bombus* species are more likely to move from clone to clone (xenogamy). This observation led Rust and Roth (1981) to conclude that scarcity of pollinators in early spring and their concentration on a single clone "reduce the probability of pollination in this obligate outcrossing species." These findings have been confirmed by Whisler and Snow (1992), who reported that seed production was limited by a lack of pollinators and consequent pollen shortage at the stigma. In support, they noted that fruits resulting from hand pollinations contained about twice as many seeds as did those derived from naturally pollinated flowers. A possibility, as yet unconsidered by investigators, is that the original pollinator of *P. peltatum* is now extinct.

On the other hand, Bernhardt (1975) reported that plants in flower at Brockport, New York, received very little attention from insects. Solitary *Apoidea* and *Bombus* queens visited the flowers but did not forage for pollen. Bernhardt concluded that *P. peltatum* reproduced entirely without insect vectors, and consequently that fruit production was the result of successful agamospermy.

In contrast, Laverty and Plowright (1988) found that clonal colonies of *P. peltatum* which were within 45 metres of *Pedicularis canadensis* (Scrophulariaceae) plants (which produce copious nectar and are heavily visited by bumble bees) showed significant increases in fruit set and seed production compared with similar groups of *P. peltatum* more distant from *Pedicularis* plants. This suggests nectar rich plants in the vicinity of *P. peltatum* colonies may assist in attracting pollinators rather than excluding them. Later Laverty (1992) coined the term magnet species for these plants which attract pollinators on behalf of others, which may lack suitable scent and are nectarless, as is *P. peltatum*.

Bernhardt (1975) based his conclusions on the achievement of an apparently similar high degree of fruit set in all four of his experimental categories; control plants, plants with anthers excised at anthesis, plants isolated by enclosure in polythene bags; and those both isolated and emasculated. Anthesis commenced in the Brockport population on 20 May, and ended on 2 June, a period of 14 days inclusive, while his study finished on 15 June, when the developing fruits were collected and examined. At the most these fruits could have been developing for only 32 days, and the majority would have been developing for less time. Swanson and Sohmer (1976b) emphasize the importance of assessing the fertility of crosses by seed production rather than fruit set, as ovaries not pollinated, or pollinated with incompatible pollen, often expand before senescing. They recommend that at least four weeks after floral withering are required to accurately assess fruit set and six to eight weeks to determine seed production in *P. peltatum*. Therefore, while it is not impossible that Bernhardt found an apomictic population, since, "a low frequency of apomixis is known in several species, and it is quite possible that such a low incidence is widespread in many Angiosperms" (Davis & Heywood, 1963), it cannot be regarded as a well-founded claim until confirmed by actual seed set.

Moreover, attempted interspecific crosses made at Nottingham, involving *P. peltatum* with pollen from *P. hexandrum* have resulted in apparently mature fruits, which have even ripened, but when opened in September have been empty. Rust and Roth (1981) also report greenhouse pollination of eight flowers which all led to fruit development, however only the two from interclonal crosses produced seed. Parthenocarpic fruit development is known to be a secondary effect of pollination and its occurrence is in agreement with the view expressed by Whisler and Snow (1992) that gametophytic incompatibility probably operates in *P. peltatum*.

The anthers have been observed touching the stigma, by Mathews (1912), who considered autogamy likely. Swanson and Sohmer (1976a, 1976b) found *P. peltatum* to be protandrous. The anthers shed pollen on to the stigma at anthesis or even before, according to Rust and Roth (1981), and the stigma reaches optimum receptivity two days after anthesis. Self-incompatibility in some clones prevents self-fertilization.

In most populations studied, clones were self-incompatible and consequently cross-pollination between clones (xenogamy) was required for successful seed production (Swanson & Sohmer, 1976a, 1976b; Rust & Roth, 1981; Policansky, 1983; Laverty & Plowright, 1988; Whisler & Snow, 1992). In contrast to Bernhardt (1975), no evidence of agamospermy has been found in any of these studies. However, Policansky (1983) reported a degree of self-compatibility in the populations at Weston, Massachusetts and Oak Ridge, Tennessee, indicating an apparent tendency for the self-incompatibility mechanism to breakdown. This is corroborated by Whisler and Snow (1992), who discovered 13 colonies of *P. peltatum* (out of 49 colonies studied), which produced seed from self-pollination. While the self-compatible colonies had a higher fruit set (ave. 55 per cent) than the self-incompatible ones (ave. 10–25 per cent), the number of seeds per self-pollinated fruit (ave. 5) was lower than that of those resulting from outcrosses (ave. 50). Inbreeding depression or partial self-incompatibility may limit seed set in these plants (Whisler & Snow, 1992). Pollen viability also varies between populations. Swanson and Sohmer (1976a) found pollen sterility ranged from 0 to 21.5 per cent in different populations examined.

SECTION *DYSOSMA*

The plants of section *Dysosma*, mostly exhibit sapromyophily, that is carrion fly pollination. Typical features associated with this system are: a dark dull purplish colour, an absence of nectar and nectar guides, motile hairs or tail-like appendages, sexual organs hidden in the interior of the flower, small openings through which the flies crawl into the flower and an odour resembling decaying protein. All these features occur in some species of *Podophyllum*. The petals of *P. pleianthum* are dark reddish-purple and slightly convex so as to form a spherical chamber around the stamens and gynoecium, with longitudinal slits where they fail to overlap completely, thus providing the small openings through which flies like to crawl. The flowers generate a noticeable putrid pungency which certainly attracts flies! In the case of *P. delavayi* there are further refinements. The petals are longer, so that, in addition to forming a chamber, the slender tips form a funnel-like entrance. The conspicuous hairs on the outer sepals, pedicels and stem, and the clustering of flowers together likely all enhance the flowers attractiveness from the fly's point of view. The odour from the flowers is probably of prime importance in attracting carrion flies since the flowers in many species are hidden from the view of an insect flying overhead by the large peltate leaves. The odour is very noticeable, even in a large greenhouse when only one or two flowers are open. In order to establish the source of the odour, a flower of *P. pleianthum* was separated into parts, which were then stored for a few hours in small clean glass phials. By this simple method it became apparent that the large purple petals produce a strong odour reminiscent of over-ripe Stilton cheese. The large anther connectives also produced a different strong acrid smell like acetic acid. Dual scent production is also known from some other sapromyophilous plants such as the inflorescence of *Biarum*. It is not known whether the two scents combine to enhance the overall attractiveness of the flower or, as seems more likely, the scent from the petals initially attracts, while that from the anther connectives initiates another type of behaviour conducive to pollination. The organs that produce these scents have been termed osmophores by Vogel (1990) and they appear to be situated in the petals and anther connective mucros in *P. pleianthum*. Some features associated with osmophores such as papillose epidermal calls and increased numbers of stomata have been observed during studies by scanning electron microscopy of an anther connective and mucro. Sapromyophilous flowers are usually known only from highly specialised families of plants, such as *Orchidaceae* and *Araceae*. Their occurrence in the *Berberidaceae* is therefore remarkable, and probably unique in the *Ranunculales*. There is a great deal more to be discovered about this phenomenon in

Podophyllum. Field work on pollinators would be interesting, and examination of the populations of *P. delavayi* with different flower colours may reveal subtle differences in the pollination system as it has done with different ecotypes of *Arum palaestinum*. Indiscriminate pollination by carrion flies, which are not species specific, helps to explain the widespread occurrence of hybrids within the genus, and especially section *Dysosma*.

SECTION *PARADYSOSMA*

Podophyllum aurantiocaule is basically an unknown entity when it come to breeding system. However, it does display some unusual features which probably relate to pollination. These include the brightly coloured red stigma, floral presentation similar to *P. hexandrum* in some populations, and the petals which eventually become reflexed like a *Cyclamen*. These traits suggest outbreeding behaviour, but the scent, pollinator and compatibility remain unknown.

The discovery that anther length varies between the northern *P. aurantiocaule* subsp. *aurantiocaule* and the southern subsp. *furfuraceum* invites speculation that there may be a corresponding change in the pollen to ovule ratio, and a consequent change in the breeding system.

SECTION *HEXANDRA*

The scent of *P. hexandrum* flowers is not strong but in the clones sampled rather unpleasant, being reminiscent of drains. Flies have been seen visiting flowers in the garden at Nottingham. Most clones are self compatible and self pollinate by a unique method that has only recently been discovered. Studies by a Chinese team (Xu et al., 1997) revealed that as the flower opens the gynoecium is upright and surrounded by a ring of six regularly spaced stamens. When the flower is fully expanded the gynoecium tilts from the base until the stigma makes contact with an adjacent anther, after which the gynoecium returns to a normal upright position at the centre of the flower and immediately enlarges. This tilting process is complete within 4 to 6 hours and results in almost 100 per cent fruit set. Consequently, out-crossing appears to be rare in this species; when artificially attempted it produces mixed results. Crosses between very different lines can result in no seed set or produce plants with aborted anthers (*KWA* 20 × 'Major', for example). On the other hand, of 10 emasculated flowers in a cultivated colony of seedlings at Nottingham, 4 produced fruits, presumably from cross pollination. This behaviour favours the formation within the species of a large number of inbreeding lines which appear different from one another, while the individual plants within each line are fairly uniform. Some cultivar names in use apply to such lines, such as 'Major', 'Chinese'. This also helps to explain the maintenance of many different morphs in the wild. The discovery of some collections with unusually short anthers suggests that out-breeding does occur to a limited extent in the wild. It may be limited by pollinator availability.

Podophyllum hexandrum has also attracted attention as the only species in the genus to release pollen in tetrads rather than monads. This is perhaps related to the significantly larger number of ovules available in the ovary of this species than the others in the genus with up to 127 seeds per fruit being produced.

David Roberts of Kew has suggested that the Madagascan orchid *Disperis oppositifolia* Sw. may also be self-pollinated by movement of the ovary or column in a manner reminiscent of *P. hexandrum*.

CYTOLOGY OF *PODOPHYLLUM*

A consistent chromosome number has emerged for those species investigated of n = 6, 2n = 12 (Table 1). Earlier anomalous counts of 2n = 6, 8, ?14 or 16 (Mottier, 1897, 1905; Overton, 1905; Richards, 1909) may be due to the presence of satellite chromosomes or fragmentation (Newman, 1959, 1966). The diploid number 2n = 12 is shared with *Diphylleia grayi*, *D. sinensis* (Shao-Bin & Zhi-Hao, 1996), and probably *D. cymosa* (Ying *et al.*, 1984), along with more remotely related genera of the *Berberidaceae*, including *Achlys*, *Epimedium*, *Jeffersonia* and *Vancouveria* (Darlington & Wylie, 1955).

Karyotype studies have been reported for *P. peltatum* (Kosenko, 1979), *P. hexandrum* (Kuroki, 1965; Kosenko, 1979; Heyenga, 1989; Siddique *et al.*, 1990) and some members of the *P. pleianthum* complex. In the absence of voucher specimens the following determinations must be regarded as tentative; *P. pleianthum* (Kuroki, 1965; Li, 1986; Zhang *et al.*, 1991 [as *Dysosma pleiantha*]) and *P. versipelle* subsp. *boreale* (Zhang *et al.*, 1991 [as *Dysosma versipellis*]).

The karyotype is basically similar in all those species so far investigated. There are two pairs of large metacentric chromosomes (length 11.6–23.7 microns), two pairs of medium meta- or submeta-centric chromosomes (length 9.0–19.3 microns), and two pairs of shorter telo- or subtelo-centric chromosomes (length 6.8–11 microns), with one pair being telocentric with an extremely small shorter arm.

Kosenko (1979) reports that in *P. peltatum* the shortest telocentric chromosome pair have a secondary constriction at the mid-point of the longer arm. Several studies report in detail various chromosomal aberrations in this species, which include the formation of bridges and fragments during meiosis (Newman, 1959, 1966; Muhling & Wilson, 1961).

Section *Paradysosma* is, unfortunately, cytologically unknown at present. Karyological information would be particularly helpful in assessing its relationship to section *Hexandra*. In some unrelated angiosperm species a change in anther length appears to be linked to changes in chromosome number. For example, within *Veronica hederifolia* s.l. (*Scrophulariaceae*) the increase in ploidy level from *V. lucorum* (n = 36) to *V. hederifolia* (n = 54), (Fischer, 1975) is associated with an average decrease in anther length of 63 per cent, (from 0.4–0.8 to 0.7–1.2 mm), whilst within the *Poa annua* complex (*Gramineae*) the doubling of ploidy from *P. annua* (2n = 14) to *P. infirma* (4n = 28), (Moore, 1982; Stace, 1991), is associated with an average decrease in anther length of 100 per cent, (from 0.6–0.8 to 0.2–0.5 mm). The variation in anther length observed in section *Paradysosma* (*P. aurantiocaule*, 4.4–9.0 mm; *P. furfuraceum*, 3.0–5.6 mm), an average difference of 58 per cent,

TABLE 1. CHROMOSOME COUNTS IN *PODOPHYLLUM*

Taxon	Count	Reference
P. peltatum		
	2n=6, 8	Mottier, 1897
	2n=8	Mottier, 1905
	n=8, 2n=16	Overton, 1905, 1909, 1922
	2n=?14	Richards, 1909
	2n=12	Litardière, 1921
	2n=12	Lublinerowna, 1925a
	n=6, 2n=12	Kaufmann, 1926
	2n=12	Langlet, 1928
	n=6, 2n=12	Newman, 1959, 1966
	n=6	Muhling & Wilson, 1961
	n=6	Swanson & Sohmer, 1976a
	2n=12	Kosenko, 1979
P. hexandrum		
(*P. emodi*)	2n=12	Litardiere, 1921
(*P. emodi*)	2n=12	Langlet, 1928
(*P. leichtlinii*)	2n=12	Langlet, 1928
(*P. emodi*)	2n=12	Kuroki, 1965
	2n=12	Fedorov, 1969
	2n=12	Hara, 1971
	2n=12+0-2f	Hara, 1971
(*P. emodi*)	2n=12	Kosenko, 1979
	2n=12	Malla *et al.*, 1981
	2n=12	Heyenga, 1989
	2n=12	Siddique *et al.*, 1990
P. pleianthum		
	2n=12	Miyaji, 1930
	2n=12	Kuroki, 1965
	2n=12	Chen *et al.*, 1977
	2n=12	Li, 1986
	2n=12	Zhang *et al.*, 1991
P. versipelle subsp. **boreale**		
(*P. versipelle*)	2n=12	Darlington, 1936
(*D. versipellis*)	2n=12	Zhang *et al.*, 1991
(*D. versipellis*)	2n=12	Shao-Bin & Zhi-Hao, 1996
P. × inexpectatum		
	2n=12	T. Hartman in Shaw, 1999
P. delavayi (*D. veitchii*)	2n=12	Shao-Bin & Zhi-Hao, 1996

may indicate the existence of cytological diversity that remains uninvestigated to date.

Although there appears to be a link between anther length and chromosome number and the change in length observed between related taxa is of the same order (50–100 per cent), the relationship is not of a simple nature. In *Poa* increase in polyploidy is apparently linked to a decrease in anther length, where as in *Veronica* the anther length increases with the ploidy level. It is therefore not possible to predict the nature of the cytological diversity that may be found in section *Paradysosma* from these examples. It is possible that other factors, such as pollen to ovule ratio or breeding system, are involved in moderating anther length in this section.

Polyploidy is unknown in *Podophyllum* and *Diphylleia*, but is recorded from another berberidaceous genus, *Achlys* (Fukuda, 1967; Fukuda & Baker, 1970).

In the case of *P. hexandrum*, according to Hara (1971), the karyotype seems to vary considerably from one population to another. He also reports that in material examined from Zojila, Kashmir, some chromosome pairs were heteromorphous. In some accessions a pair of large submedian chromosomes had submedian constrictions and in a collection from Bhutan, a pair of smaller subterminal chromosomes also frequently manifest submedian or subterminal secondary constrictions (Hara, 1971). Some material from Kashmir displays one or two very small extra metacentric chromosomes or B-chromosomes (Hara, 1971). This is of particular interest in view of the wide foliar variation observed in *P. hexandrum*. Work on *Datura* (*Solanaceae*) by Blakeslee (1922, 1931) appeared to link the occurrence of specific foliar and other morphological variants to the presence of specific extra chromosomes. It seems possible that the two phenomena, and also the variable anther length, may be related in *P. hexandrum*. Siddique *et al.* (1990) reported that in the majority of complements examined, four chromosomes bore secondary constrictions. Kosenko (1979) reported only one secondary constriction, which occurred on the shorter arm of the second longest chromosome pair. Chromosome variation in tissue-cultured *P. hexandrum* is discussed by Arumugam & Bhojwanii (1994).

A secondary constriction occurs on the shorter arm of the longest metacentric pair in *P. pleianthum* in two reported counts (Kuroki, 1965; Zhang *et al.*, 1991), whereas Li (1986) reports this constriction on the longest arm of the longest chromosome and found an additional secondary constriction on the longest arm of one chromosome of the longer subterminal pair; *P. versipelle* differs with the secondary constriction being on the longest arm of the third longest chromosome (Zhang *et al.*, 1991). These two taxa also differ in the total length of the chromosomes, that being 102.34 microns in *P. pleianthum* and 112.99 microns in *P. versipelle* subsp. *boreale* (Zhang *et al.*, 1991).

Although the chromosome number seems to be stable in the genus, it appears that cytological diversification has occurred, involving structural change without altering the basic number. Evidence for this includes the differences between reported karyotypes of *P. pleianthum* (Li, 1986; Zhang *et al.*, 1991) and the occurrence of heteromorphic chromosome pairs within *P. hexandrum* (Hara, 1971). This heterozygosity may be the result of an unequal chromosome interchange. It seems likely that these phenomena may be related to hybridisation, which appears to be widespread in the genus.

Several authors comment on the similarity of karyotype between *Podophyllum* and *Diphylleia* (Kuroki, 1965; Kosenko, 1979) grouping the two genera together in their schemes. Kuroki (1965) goes so far as to state that from the viewpoint of the karyotype, "it seems unreasonable that both genera are separated from each other". Li (1986) sees evidence in the karyotypes that *P. hexandrum* and *P. peltatum* are derived from section *Dysosma*; while Zhang *et al.* (1991) comment on the close interspecific relationship between *P. versipelle* and *P. pleianthum*.

TAXONOMIC TREATMENT OF PODOPHYLLUM

Podophyllum L., Sp. Pl., ed. 1: 505 (1753) & Gen. Pl., ed. 5: 223 (1754). Type species: *Podophyllum peltatum* L.
Anapodophyllum Tourn., Inst. Rei Herb.: t.122 (1700).
Sinopodophyllum T.S.Ying in Acta Phytotax. Sin. 17(1): 15–16 (1979).

GENERIC DESCRIPTION. *Herbaceous perennials. Rhizome* scaly with thick fibrous contractile roots and large circular stem scars along upper surface; internodes short and thick in most Asiatic species, long and thin in the American; apical bud with cataphylls. *Stems* annual, erect, unbranched, smooth, up to 90 cm, though usually less. *Leaves* 1–3, alternate, above the middle of the scapes, exstipulate, simple peltate or deeply palmately lobed, with juvenile and adult forms, folded like an umbrella upon emerging above ground; lobes ranging from shallow, triangular, subentire to deep, lanceolate and deeply dissected; sinuses acute or rounded, penetrating from about half to almost the entire radius; margins with small teeth, hirsute or glabrous. *Inflorescence* of 1–18 flowers, terminal, nodal or borne on the petiole of the upper leaf. *Flowers* actinomorphic, bisexual, erect or pendulous, white, cream, rose-pink or dark reddish purple, variously scented usually malodorous. *Sepals* 3–6, greenish or inner ones coloured and petaloid, fugacious, imbricate. *Petals* (4–)6(–9), imbricate, ovate, lanceolate, obovate or spathulate with a rounded or irregular apex, occasionally deeply notched, showy. *Stamens* 6–12(–18), usually as many or twice as many as the petals, typically with a distinct filament; anthers oblong, opening by longitudinal slits, extrorse, often with the connective produced beyond the thecae to form a small mucro of variable length. *Pollen* 3-colpate, usually released in monads or as tetrads in *P. hexandrum*. *Ovary* superior, unilocular, subspherical or ovate, multiovulate; placentation marginal, multiseriate. Style short to elongated in the *P. delavayi* group; stigma large, peltate. *Fruit* a fleshy berry, rupturing irregularly with many seeds embedded in pulp, yellow or red, spherical or elliptic, often emitting a fruity odour, edible when ripe. *Seeds* obovoid maroon or subreniform, yellowish, rarely polyembryonic, endosperm copious, embryo small; germination epigeal; cotyledons rounded with entire or toothed margins, fused along adaxial margin and produced downwards to form a thin tube at the centre of the join; shoot emerging by rupturing this tube near the base. $2n = 12$. Self-incompatible or rarely self-compatible, some species obligate outbreeders.

Key to the species of *Podophyllum*

1. Stamens 12–18, rhizome with long, thin internodes 3–20 cm long (Section *Podophyllum*) 2
1. Stamens 6(–8), rhizome with short, thick internodes 1–2 cm long 3

2. Pedicel not jointed, ebracteate .... 1. ***P. peltatum***
2. Pedicel jointed, subtended by small bract .... 1. × 2. ***P. × inexpectatum***

3. Mature leaf lamina with one sinus penetrating to petiole apex; pollen in tetrads (Section *Hexandra*) .... 14. ***P. hexandrum***
3. Mature leaf lamina with undivided ring around petiole apex; pollen in monads 4

4. Leaf margins with teeth of several sizes; anther mucro absent or < 0.4 mm; petals reflexed with age (Section *Paradysosma*) .... 13. ***P. aurantiocaule***
4. Leaf margins with uniform teeth; anther mucro present 1–13 mm; petals at most patent or spreading, never reflexed (Section *Dysosma*) 5

5. Petals oblong to ovate, or obovate 0.5–6 cm; anther mucro < 3 mm (Series *Pleianthae*) 6
5. Petals lanceolate to ligulate, (3–)4–10 cm; anther mucro > 4 mm (Series *Delavayae*) 16

6. Pedicels hairy, leaf margin usually with irregularly spaced teeth, 1–3 per cm, rarely with regular closely spaced teeth 7
6. Pedicels glabrous, leaf margin with regularly closely spaced teeth, 5–11 per cm .. 12

7. Leaf lobes with parallel or usually concave margins, often trilobulate 8
7. Leaf lobes with convex margins or rarely straight sided triangular, without lobules 10

8. Leaf lobes parallel-sided, sinuses acute .... 9. ***P. guangxiensis***
8. Leaf lobes convex or concave sided, narrowed at base, sinuses rounded 9

9. Leaf lobes with narrow *c.* 2 cm wide base, deeply trilobulate; petals < 2 cm long .... 8. ***P. majoense***
9. Leaf lobes with broad *c.* 5 cm wide base; shallowly trilobulate or elobulate; petals > 2.5 cm .... 3. ***P. versipelle***

10. Petals < 1.2 cm long; leaf undersides glaucous .... 6. ***P. glaucescens***
10. Petals > 1.5 cm long; leaf undersides pale green 11

11. Upper leaf irregularly and indistinctly lobed; upper surface with velvety sheen from papillate trichomes; inflorescence 1–3 flowered .... 7. ***P. difforme***
11. Upper leaf with 4–7 lobes on one side, upper surface not papillate; inflorescence 4–9 flowered .... 3. ***P. versipelle***

12. Leaf lobes wider than long, sinuses penetrating less than $^1/_3$ of radius, flowers often in or near fork of petioles, rarely about midpoint 13

12. Leaf lobes longer than wide, sinuses penetrating $1/2$ to $4/5$ of radius, flowers always on petiole near lamina 14
13. Inflorescence at or below midpoint of petiole on wild collected specimens; pedicel glabrous or hairy .... 2. × 3. ***P. pleianthum*** × ***P. versipelle***
13. Inflorescence in fork of petioles; pedicel glabrous .... 2. ***P. pleianthum***
14. Leaf lobes very long to 15 cm, sinuses penetrating to $4/5$ of radius .... 4. ***P. hemsleyi***
14. Leaf lobes short to medium to *c*.10 cm, sinuses penetrating to $1/3$ of radius 15
15. Petals short, 15–20 mm, ovate, spreading, forming open bowl-shaped flower .... 5. ***P. mairei***
15. Petals > 2 cm, oblong, curved forming closed spherical or funnel-shaped chamber .... 3. ***P. versipelle***
16. Inflorescence in fork of petioles .... 10. ***P. delavayi***
16. Inflorescence along petiole just below leaf or about midpoint or held above leaves on elongated peduncle 17
17. Inflorescence simple, 5–9-flowered positioned immediately below leaf or held above leaves on elongated peduncle .... 10. × 7. ***P. delavayi*** × ***P. difforme***
17. Inflorescence simple 3–5-flowered or compound 10–12-flowered, positioned about midpoint on petiole 18
18. Plants 20–25 cm tall, inflorescence with 3–5 flowers, petals 4–5 cm long .... 11. ***P. trilobulus***
18. Plants 40–50 cm tall, inflorescence compound with 10–12 flowers, petals 6.5–7.5(–8.5) cm long .... 12. ***P.* sp. A**

Section i. PODOPHYLLUM

(Species number 1)

This section comprises but a single species *P. peltatum*, distinguished by the following characters: rhizomes with long slender stolon-derived internodes; leaves peltate with deep sinuses and long bilobulate lobes, or subpeltate-palmate with one sinus penetrating to the petiole apex. Flower solitary, in the fork of the petioles; pedicel reflexed; petals fleshy, white, cream, rarely pink. Stamens 12–18. Fruit subglobose-spherical, yellow or rarely red, with fruity aroma when ripe. Seed yellowish or rarely maroon, subreniform. 1 sp. Eastern North America. Type species: *Podophyllum peltatum* L.

1. PODOPHYLLUM PELTATUM

This *Podophyllum* was the first to be encountered by a European and was briefly described by the French explorer of Canada Samuel de Champlain (1567–1635) in 1615 when relating plants used by Huron Indians — a tribe noted for their horticultural activity. "One

Podophyllum peltatum

SYDENHAM EDWARDS

of their berries was new to us. It looked rather like a small lemon but tasted more like a fig... in some parts of the country these berries are plentiful and they are extremely good to eat" (Erichsen-Brown, 1979). Interestingly, some of the isolated populations of *P. peltatum* to be found in Canada today are thought to be relics of Huron cultivation. By 1664 it had been introduced into cultivation in Britain and at about the same time into France.

The numerous varieties described by Rafinesque-Schmaltz (1828) were based mainly on the continuous variation in leaf lobe shape and pigmentation. They indicate the considerable degree of variation found in foliar characters throughout the range of this species. This variation has been studied in great detail by Martin (1958), who concluded that it had arisen from introgression with a species similar to those from Asia. He developed a way of measuring characters of the middle lobe of the leaves so as to assign a numerical index value to each plant. Plants with a high index value displayed characters that resembled those of section *Dysosma* from Asia. Interestingly the areas of the range of *P. peltatum* which display the greatest extremes in foliage variation and most closely approach section *Dysosma* are the northern areas of distribution, thought to have once been glaciated, including Pennsylvania. These are also the regions in which there have been discoveries of *P. peltatum* plants with Asiatic characteristics such as red fruits, pink flowers, inflorescences borne along the petiole and shoots bearing three leaves (Porter, 1877; Foerste, 1884; Harris, 1909). Plants thought to be most similar to the original ancestral American form appear to be concentrated in the Ouachita region and parts of the Appalachian mountains. The forms with red-purple pigmentation all occur at the limits of the natural range of *P. peltatum*, and are especially frequent in Pennsylvania (Wherry *et al.*, 1979). It is not yet clear whether these forms represent stress induced changes triggered by environmental factors or if they are indeed relics of past introgression from plants of section *Dysosma*. Certainly it is interesting that the appearance of the stems of the hybrid *P.* × *inexpectatum* (*P. peltatum* × *P. pleianthum*) which are flecked with red, matches the description of the lavender flecks observed in f. *deamii*.

The Mayapple is a very hardy and reliable woodland garden plant, almost a weed in some American gardens, but it does tend to die back early in a hot summer. It is very easy to propagate by division of the rhizome, although obtaining seed can be a problem in cultivation since many clones are self incompatible.

Podophyllum peltatum is readily distinguished from all other *Podophyllum* by its flowers with 12 to 18 stamens and the long thin internodes of the rhizome retained at maturity. The name *peltatum*, Latin *peltatus* (shield-shaped), refers to the peltate shape of the leaves with the stalk attached towards the centre of the blade rather than at the edge. Although this feature is common to most species of the genus, at the time of Linnaeus it was the only species known.

1. Podophyllum peltatum L., Sp. Pl., ed.1: 505 (1753); Gronov., Fl. Virgin.: 80 (1762); Michx., Fl. Bor.-Amer. 1: 309 (1803); Pursh, Fl. Amer. Sept.: 366 (1814); Curtis's Bot. Mag. 43: t.1819 (1816); Torr., Comp. Fl. N. Middle Stat.: 217 (1826); Raf., Med. Fl. 1: 59 (1828); Torr. & Gray, Fl. N. Amer. 1: 54 (1838); Bigelow, Fl. Boston.: 229 (1840); Torr., Fl. New York 1: 35 (1843); Gray, Gen. Amer. Bor. 1: 87, t.35, 36 (1848); Dawe &

Collins, Fl. Middlesex Co., Mass.: 3 (1888); Britton, Cat. Pl. New Jersey: 42 (1889); Chapman, Fl. South. U.S.: 16 (1897); Cole, Grand Rapids Fl.: 73 (1901); Small, Fl. Pennsylvania: 143 (1903); Stone, Pl. S. New Jersey 1: 459 (1911); Britton & Brown, Illust. Fl. N. U.S. 2: 130, t.(1913); Small, Fl. S.E. U.S.: 458 (1913); House, Wild Fl. New York: 114, t.76 (1918); Wiegand & Eames, Fl. Cayuga Lake Bas.: 218 (1926); Pepoon, Annot. Fl. Chicago: 316 (1927); Deam, Fl. Indiana: 475 (1940); Steyerm., Spring Fl. Missouri: 211, t.55 (1940); Jones & Fuller, Vasc. Pl. Illinois: 220 (1955); Martin, Variation & Morphology of *P. peltatum*. Thesis (1958); Jones, Fl. Illinois: 59 (1963); Steyerm., Fl. Missouri: 710 (1963); Morley, Spring Fl. Minnesota: 124 (1969); Seymour, Fl. New England: 276 (1969); Correll & Johnston, Man. Vasc. Pl. Texas: 654 (1970); Weishaupt, Vasc. Pl. Ohio: 108 (1971); Hiroe in Pl. Basho's & Buson's Hokku Lit. 8(3): 327 (1973); Krochmal *et al.*, USDA For. Ser. Res. Paper NE-296 (1974); Meijer in Econ. Bot. 28: 68–72 (1974); Fassett, Spring Fl. Wisconsin: 162 (1976); Scogan, Fl. Canada 3:762 (1978); Gupton & Swope, Wild Fls. Shenandoah & Blue Ridge: 12 (1979); Rickett, Wild Fls. Amer.: 31, t.141 (*c.* 1979); Strausbaugh & Core, Fl. W. Virginia: 402 (1979); Swink & Wilhelm, Pl. Chicago region: 536 (1979); Duncan & Kartesz, Vasc. Fl. Georgia: 60 (1981); Steffey in Amer. Hort. 64(4): 8 (1985); Voss, Michigan Fl. 2: 232 (1985); Cox, Common Fl. Pl. Northeast: 45, t.21 (1985); Stearn in Walters (ed.), Eur. Gard. Fl. 3: 396 (1989); Wofford, Guide Vasc. Pl. Blue Ridge: 181 (1989); Morton & Venn, Checklist Fl. Ontario: 71 (1990); Ownbey & Morley, Vasc. Pl. Minnesota, 10 (1991); Phillips & Rix, Perennials 1:39 (1991); Stones, Fl. Louisiana: 35 (1991). Type: "Habitat in America septentrionali" (Lectotype, designated in Jarvis *et al.*, (1993), Herb. Linn. no. 667.1 LINN).

Aconitifolia humilis flore albo unico campanulato fructu cynosbati Mentzel, Ind. Nom. Pl.: t.11 (1682).

Anapodophyllum canadense Catesby, Nat. Hist. Carolina 1: 24, t.24 (1731).

P. peltatum var. *pumilum* Raf., Med. Fl. 1: 59 (1828).

P. peltatum var. *elatior* Raf., Med. Fl. 1: 59 (1828).

P. peltatum var. *grandiflorum* Raf., Med. Fl. 1: 59 (1828).

P. peltatum var. *odoratum* Raf., Med. Fl. 1: 59 (1828).

P. peltatum var. *heterophyllum* Raf., [as hetorophyllum] Med. Fl. 1: 59 (1828).

P. peltatum var. *oligodon* Raf., Med. Fl. 1: 59 (1828).

P. peltatum var. *triphyllum* Raf., Med. Fl. 1: 59 (1828).

P. peltatum var. *extraxillare* Raf., Med. Fl. 1: 59 (1828).

P. montanum Raf., Med. Fl. 1: 59, t.73 (1828).

P. montanum var. *acuminatum* Raf., Med. Fl. 1: 60 (1828).

P. montanum var. *parviflorum* Raf., Med. Fl. 1: 60 (1828).

P. peltatum f. *polycarpum* Clute in Amer. Bot. 21: 93, t.(1915) & Amer. Bot. 24: 113 (1918); Pepoon, Annot. Fl. Chicago: 316 (1927); Plitt in Rhodora 33: 229 (1931); Rix in Plantsman 4(1): 9 (1982). Type not cited.

P. peltatum f. *aphyllum* Plitt in Rhodora 33: 228–229 (1931); Deam, Fl. Indiana: 476 (1940). Type not cited.

P. podophyllum Steffey in Amer. Hort. (Alexandria) 64(4): 8 (1985) *nom. illegit.*

ILLUSTRATIONS. Bot. Mag. 43: t.1819 (1816); Britton & Brown, Illus. Fl. N. U.S. 3: 130 (1913); Mathews, Field Book Amer. Wild Fl.: 155 (1927); Phillips & Rix, Perennials 1: 39 (1991); Niering, National Audubon Soc. Field Guide to N. American Wildflowers: pl. 87 (1995).

DESCRIPTION. *Perennial rhizomatous herb* with annual herbaceous shoots. *Rhizome* with long, thin stolon-derived internodes *c.* 5 mm thick and 3–20 cm long, between short, thicker rounded nodal regions, branching at the nodes to form complex, extensive networks with one clone often covering a large area; internodal regions formed from subterranean stolons with regularly spaced paired scales. Terminal buds enclosed in several thick white scales and prophylls. Aerial stems 15 to 40 cm, smooth, glabrous, green or tinged reddish when growing in strong light, usually with 1–3 leaves. *Juvenile leaves* borne singly on stems from the rhizome, peltate, radially symmetrical, with 4 to 8 lobes; lobes bilobulate or rarely entire, margins entire or with a few coarse teeth towards the apex; sinuses deep, slim, penetrating up to $^{11}/_{12}$ of the radius, leaving a narrow disk of undivided lamina around the petiole apex; leaf margin with fine white hairs when young. *Adult leaves* either peltate, with a ring of undivided lamina around the petiole apex or more frequently palmate, with the lamina retarded on the adaxial edge so that the resulting cleft reaches to the petiole apex, bilaterally symmetrical with (4–)5–8 lobes; lobes narrow at the base, narrowly to broadly ovate-lanceolate or obovate, usually bilobulate at the apex, or rarely apex entire, acute; lateral lobes often partly fused; margins with a fringe of hairs when young, entire or with irregular coarse teeth towards the apex, one or two apical teeth sometimes enlarged to form lobules; sinuses long and narrow, frequently obscured by overlapping lobes, penetrating to near the petiole, $^{11}/_{12}$ of the radius or more. First leaf larger than second with 7–8 lobes, second leaf with 4–5 lobes. *Inflorescence* a solitary flower, 3.5–6.5 cm in diameter, in the fork between the two petioles, very rarely with 2 flowers or positioned along the petiole of the second leaf. Pedicel glabrous, erect in bud, deflexed at anthesis and in fruit. *Sepals* (3–)6, ovate-lanceolate, the outer very narrow, pale green, fugaceous. *Petals* (5–)6(–8), white, cream or rarely pink, fleshy, obovate to ovate or oblong, rounded or irregularly toothed at the apex, up to 30 mm long × 16 mm wide. Stamens (7–)12(–18), *c.* 10 mm long with a short filament; apical mucro occasionally present, small *c.* 0.2 mm long. *Ovary* elliptic, with a large peltate, corrugated stigma. *Fruit* a sub-globose fleshy berry, 2–5 cm in diameter, yellow or rarely red, with a strong fruity aroma when ripe. *Seeds* elongated, slightly curved, 7–9 mm long, pale yellow; testa smooth; cotyledons ovate, entire. 2n = 12. Typically self-incompatible, but varying from clone to clone.

DISTRIBUTION. Eastern North America, east of longitude 97°W. Absent from Florida, except the extreme NW. Possibly introduced in much of Canada. An excellent map is provided in Meijer (1974). A record from Japan (Ito, 1887) appears to be based on cultivated material.

ECOLOGY. Common as an understorey component in deciduous woodland, at elevations from sea level up to 1400 m in the Appalachians. One of the first spring flowers to appear, forming large colonies of 1000 or more shoots and often persisting after woodland has been removed. Populations vary in breeding system, most being self-incompatible, others self-compatible to some extent (Policansky, 1983; Whisler & Snow, 1992). There is an unconfirmed report of an apomictic population (Bernhardt, 1975). The following report ecological studies; Krochmal *et al.*, 1974; Meijer, 1974; Swanson & Sohmer, 1976a, 1976b; Frye, 1977; Sohn & Policansky, 1977; Rust & Roth, 1981; Laverty & Plowright, 1988; Benner & Watson, 1989).

In North America this species is host to two parasitic fungi. *Septotinia podophyllina* infects the newly emerged leaves causing blotches and eventual necrosis (Whetzel, 1937; Gremmen, 1987). Strausbaugh and Core (1979) report that in west Virginia the leaves of *P. peltatum* are often infected with a rust fungus, *Puccinia podophylli*. An analysis of the host pathogen relationship is provided by Parker (1989).

Key to the varieties and forms of *P. peltatum*

1. Petals white or cream; seeds pale yellow; berry yellow or orange 2
1. Petals pink; berry pink or maroon 4
2. Mature leaves with one sinus penetrating to petiole apex (var. **peltatum**).............. 3
2. Mature leaves with an undivided zone of lamina around the petiole apex .... var. **annulare**
3. Fruit yellow .... f. **peltatum**
3. Fruit apricot orange .... f. **biltmoreanum**
4. Berry pink .... f. **callicarpum**
4. Berry maroon; seeds maroon .... f. **deamii**

var. **peltatum**
DESCRIPTION. The mature leaves always possess one deep sinus on the adaxial side that penetrates to the petiole apex, so that the leaf is not strictly peltate but palmate.

Forms of var. *peltatum* with white flowers

Plants with white or cream petals and yellow fruits with straw-coloured seeds. These are the clones commonly encountered both in cultivation and the wild. Almost every clone appears to be slightly different in some minute detail of leaf shape, and many differ in degree of self-compatability. There is only one known occurrence of a plant with twin flowers (*Pitch s.n.,* US). A clone that produces lobed fruits which appear to be clusters of berries (probably mutiple carpels from a single flower) has been found several times and named f. *polycarpum*. It is not thought to be in cultivation.

Few cultivars have been named in this white-flowered group as yet, although there is potential, for example, plants with extra petals producing 'double' flowers have been recorded on several occasions. Recently one such collection has been taken into cultivation by Jim McClements to assess its stability.

'Bulwood Bronze' is a cultivar distinguished by bronze foliage and white flowers with narrow spathulate petals that do not overlap along the margins; this clone has been distributed in Britain as 'bronze leaved form' by Bulwood Nursery, which has since ceased trading. Fortunately, Gary Dunlop maintains this clone, which has a tendency to die back early in the season, in his collection at Ballyrogan Nurseries, Northern Ireland.

forma **biltmoreanum** Steyerm. in Rhodora 54: 134 (1952) & Fl. Missouri: 711 (1963). Type: USA, Illinois, Lake Co., North-facing open oak-hickory wooded slope on N.E. side of creek, S. of Eton Drive and W. of Kimberley Road, on property of George Foster, S.W. of Bordener's house, Biltmore Estates subdivision, 5 miles N. of Barrington, 16 August 1950, *Bordener & Steyermark s.n.* (syntypes: MO, F).
This clone has only been found once in the wild, in Lake county, Illinois. In this plant the fruit colour, rich apricot or mango orange, is the only distinguishing feature. It may still be grown at Missouri Botanic Garden.

The name *biltmoreanum*, is derived from the type locality, Biltmore Estates.

Forms of var. *peltatum* with pink to red flowers

In 1927, C.C.Deam found a plant with maroon fruit in woods at Arthur Miller farm, near Mauckport, which he took and grew in his garden. In 1937 he sent four large fruits that had formed to Dr. Edgar Anderson of Missouri Botanic Garden for investigation. Some years later Anderson encouraged one of his students, F.W.Martin to carry out an in depth study of variation within *P. peltatum*, the resulting thesis was completed in 1958. This clone is thought still to be in cultivation at Missouri Botanic Garden and Montreal Botanic Garden, Canada where it is given cultivar status as 'Deamii'. Recently, Jim McClements and colleagues have succeded in finding several new colonies of red flowered forms in Pennsylvania. These include plants with pale pink petals and 'picotee' (with a pink fringe on the petals), usually with a red ovary. Amongst these there appear to be several worthwhile garden plants.

The epithet *deamii*, recalls the plant's discoverer Dr Charles Clemon Deam (1865–1953).

forma **deamii** Raymond in Rhodora 50: 18 (1948); *vide* Deam, Fl. Indiana: 476 (1940); Steyerm., Fl. Missouri: 710 (1963); Rix in Plantsman 4(1): 9 (1982). Type: USA, Indiana, Harrison Co., Arthur Miller farm near Mauckport, cultivated at Montreal B.G., 755–43. *Deam s.n.* (syntypes: MT, MTJB).
P. peltatum f. *roseum* Bush, *nom. nud.*, in Steyermark, Fl. Missouri: 710 (1963).

DESCRIPTION. *Apical bud of rhizome* dark purple-wine coloured. *Aerial stems* and petiole flecked with lavender. *Pedicel* purplish, petals pink or pinkish. *Ovary* deep wine-purple; berry maroon-purple; seeds dark purple-brown.
DISTRIBUTION. Isolated colonies known from Illinois; Indiana; Cole Co., Missouri; several locations in Pennsylvania (Wherry *et al.*, 1979); and possibly Texas (but see below under f. *callicarpum*).

forma **callicarpum** (Raf.) J.M.H.Shaw **comb. nov.**
P. callicarpum Raf., Fl. Ludov: 14 (1817) & Med. Fl. 1: 60 (1828). Type material may exist amongst the Rafinesque collections at DWC, PH and WIS.

This colour form does not appear to have been recollected since and has been ignored or treated as a synonym of f. *deamii*. Rafinesque distinguished this taxon on the following characters; flowers large, smelling like orange flowers, 'berry oblong, white coloured of rose [probably pale pink].' This fruit colour is quite different from maroon or orange; also the floral scent may be significant. It probably represents the earliest known record of a colour form of *P. peltatum*, but its relationship to the better known forms enumerated above remains obscure. As it was described from Louisiana and Texas, it may relate to the Texan record listed above under f. *deamii* (See Correll & Johnston, *Man. Vasc. Pl. Texas*, 654 (1970)). Live material should be sought in Texas and should be compared with those from Indiana and Pennsylvania.

The epithet *callicarpum*, Greek *kalli-* (beautiful-) *karpos* (fruit), refers to the pale pink pigmentation of the berry.

var. **annulare** J.M.H.Shaw in New Plantsman 6(3): 158 (1999). Type: North America, Eastern Virginia, 1771, Herb. *P.D.Giseke* 1842 (E).
DESCRIPTION. *Plants* with white petals and yellow fruit; leaves with an undivided ring of lamina about 1 cm wide around the petiole apex on mature leaves.
DISTRIBUTION. Known from the vicinity of Washington D.C., E. Virginia and Philadelphia.

During the spring and summer of 1771, Giseke stayed with Linnaeus at Uppsala, from whom he acquired herbarium specimens as well as from the gardens at Uppsala and Hammarby, (Hedge, 1967).

This interesting variant provides yet another character associated with section *Dysosma*, this plant may represent an ancient form of *P. peltatum*.

The name *annulare*, Latin *annularis* (ring-shaped) refers to the undivided zone of the lamina towards the centre of the leaf.

1. × 2. PODOPHYLLUM × INEXPECTATUM

A garden hybrid known only from cultivation, derived from *P. peltatum* × *P. pleianthum*. The cross was first made at Nottingham, UK, in 1993 and this flowered in 1997. Seedlings may be identified at the cotyledon stage by the presence of a few small teeth on the margins. This feature is intermediate between *P. peltatum* with entire margins and *P. pleianthum* with many prominent teeth on the margins. The hybrid has since been synthesised by Jim McClements (Delaware, USA) using different parental clones. He obtained a batch of seedlings, many of which had intermediate, slightly toothed cotyledons. This hybrid is placed in section *Podophyllum* rather than section *Dysosma* on account of its twelve stamens.

The epithet *inexpectatum*, Latin *inexpectatus* (unexpected), refers to the perceived difficulty in producing a hybrid plant of this parentage in view of the large geographical distance between the origins of the parent species in North America and China. However viable hybrids have been produced between other widely separated species such as *Buddleja* × *weyeriana* (i.e. *B. davidii* from China × *B. globosa* from Chile), providing an interesting approach to the study of disjunctions.

An initial cytological investigation by Dr. Tom Hartman of somatic chromosomes from root tips has revealed a count of $2n = 12$. The chromosomes are undisturbed at anaphase. There appears to be good pairing with no fragments, indicating a close relationship between the parent species.

Observations made from a living plant were that the flower was scentless when it first opened, at which stage a bumble bee was seen to visit. After two to three days a scent of over-ripe cheese developed and persisted for about one week. A fruit began to develop from self-pollination but due to the hot weather the whole plant went dormant early and no seed developed.

As only a single seedling resulted from the Nottingham cross the cultivar name 'East meets West' has been given to distinguish it from other clones resulting from the same hybrid cross made elsewhere, which are likely to differ slightly.

Podophyllum × inexpectatum J.M.H.Shaw in New Plantsman 6(3): 159 (1999). Type: Cultivated at Nottingham, *J.M.H.Shaw s.n.*, May 1997 (Herb. J.M.H.Shaw).

ILLUSTRATION. New Plantsman 7(2): 111 (2000).

DESCRIPTION. *Perennial* to 35 cm; rhizome thick, internodes short. *Aerial stem* to 20 cm with cataphylls at the base; petioles to 12 cm, light green, glabrous with numerous red flecks. *Juvenile leaves* solitary, 4–5 lobed; lobes weakly trilobulate or shortly bifid with each lobule asymmetrically bifid; lobes asymmetrically arranged so that the leaf is almost divided in two by a pair of very deep sinuses penetrating almost to the petiole like a *Diphylleia*. *Mature leaves* 2; first leaf 6-lobed; lamina 27 cm across, basal sinus pentrating to 3.5 cm from centre; lobed to $^2/_3$ of the radius; lobes oblong to slightly obovate, weakly trilobulate; leaf margin with fine simple hairs, toothed; larger teeth *c.* 5 mm apart with smaller teeth in between, tooth apex 1.5 mm, hard; second leaf with 6 lobes, lamina 23 cm across. *Inflorescence* of a solitary flower, 6 cm across when fully open, bowl-shaped, inserted at fork of petioles, base of pedicel with membranous sheath; pedicel jointed *c.* 1 cm from base, with a minute membranous bract at joint; pedicel continuing above joint for 2.7 cm, dilated above towards apex, 4 mm wide. *Bud* ovoid. *Sepals* 6, green, ovate-lanceolate with membranous transparent margins, several inner sepals fused together; outer sepals free, 15–22 × 6–8 mm. *Petals* 6, rose pink, streaked, white at base, 4 cm × 15–19 mm, waxy, oblong-lanceolate with apex irregularly toothed. *Stamens* 12; filaments white, 10 mm; anthers curved, 9–10 mm, yellow with mucronate apex, 1–1.5 mm. *Ovary* ovoid, cream to pale green, 18 mm long with large capitate, sculptured stigma. *Immature fruit* dark green, ellipsoidal. *Mature fruit* and seeds not known.

DISTRIBUTION. Known only from cultivation.

Fig. 105. **Podophyllum × inexpectatum** (*P. peltatum* × *P. pleianthum*), an artificial hybrid. Photograph: Julian Shaw.

Section ii. DYSOSMA

(Species number 2–12)

Section *Dysosma* (Woodson) J.M.H.Shaw in New Plantsman 6(3): 159 (1999). Type species: *P. pleianthum* Hance; for note on typification see New Plantsman 6(3): 160 (1999).

Dysosma Woodson in Ann. Missouri Bot. Gard. 15: 338–339 (1928); Loconte in Kubitzki *et al.*, Fam. & Gen. Vasc. Pl. 2: 151 (1993).

Multi-flowered inflorescences of typically dark purple-red (rarely white), malodorous flowers and well developed thick anther connective and mucro are the characters that most readily separate this section from sections *Peltatum* and *Hexandra*. The fine, regular foliar marginal teeth, together with an absence of hairs on the upper leaf lamina also distinguish this section from *Paradysosma*. The ten or so species included here are all fly pollinated and probably all self-incompatible, necessitating cross pollination between different clones to produce seed.

Some species from this section are particularly garden worthy on account of the astonishing foliar patterns. They were described by Dan Hinkley of Heronswood nursery as "among the most dramatic foliage plants that I currently have in my garden." Recently, there have been several introductions new to cultivation by collectors visiting China, from Japanese nurseries and the Kaichen nursery in China.

The plants in this section tend to cluster around two groups of species. Extending from Taiwan and eastern China through to Yunnan in western China is the *P. pleianthum-P. versipelle* complex with often large peltate leaves and clusters of dark red-purple globose flowers. The other group centres on *P. delavayi* and predominates in Yunnan and western China. These plants are characterised by smaller richly patterned leaves with trilobulate lobes and longer petals. Then, as always, there are the individualists — a small assemblage of taxa, presumably of hybrid origin, that combine characters from both these groups with a few of their own. They are scattered in western to central China, are usually rarely collected and even more rarely grown. The best known of these is the beautiful *P. difforme* which is now well established in cultivation.

Woodson (1928) made several errors in his description of *Dysosma*, including some of the characters on which the genus was based. For example, he describes the subterranean organ as a tuber, when it is a sympodial rhizome, usually with short internodes. The anthers were said to exhibit introrse dehiscence, whereas subsequent observation by Kumazawa (1936) and Terabayashi (1983) has shown dehiscence to be extrorse. Unfortunately Woodson has been followed uncritically by later workers as recently as Loconte (1993).

Series A. PLEIANTHAE

(Species 2–9)
Series *Pleianthae* J.M.H.Shaw **ser. nov.** Petala ovatia vel oblongus ad 6 cm longa; mucrones antherae parvuli ad 3 mm longi. Type species: *Podophyllum pleianthum* Hance.

This series contains the *Podophyllum pleianthum-versipelle* complex (species 2–6) which comprises a group of closely related critical taxa, the members of which tend to intergrade, probably through hybridisation. The three other species (7–9) appear to represent hybrids between the *P. pleianthum-versipelle* complex and *P. delavayi*. Some incomplete specimens cannot be definitely determined. Most floristic treatments group all these under two species *P. pleianthum* Hance and *P. versipelle sensu auctt.*, usually under the synonym *Dysosma versipellis*. This latter taxon seems to have been used in the past as a general repository for anything which was not readily identifiable with other known taxa. It is here subdivided in an attempt to indicate the wide range of variation encountered within this important group. This has significant implications for conservation which must be borne in mind when reading the distribution data given for "*Dysosma versipellis*" in Red Data books such as Fu & Jin (1992) and Walter & Gillett (1998). Many of these taxa are very local and rare and their conservation status needs to be reviewed urgently as they are a focus of collecting activity by medicinal plant gatherers and more recently for horticultural export.

2. PODOPHYLLUM PLEIANTHUM

The British Consul in Taiwan, T.Watters, made the type collection from Tamsui in 1881. He passed his material on to fellow Consul, H.C.Hance on the Chinese mainland, who published a description in 1883 (Hance, 1883a). Hance pursued botanical activities to such an extent that by 1885 he had amassed a herbarium of over 22,000 specimens. In 1885 live plants were sent to Kew from the Hong Kong botanic garden along with *Eomecon chionantha* and were cultivated in pots in a cold frame until 1889 when, after planting in the border of a cool house, they flowered enabling production of the plate in *Curtis's Botanical Magazine*.

This species appears to intergrade, through hybridisation, with *P. versipelle* on the mainland and possibly in some parts of Taiwan, but the material seen is usually separable by leaf lobing, if not inflorescence position. In mainland China it is now restricted to the east coast and lower Yangtze valley and may have been replaced over much of its former range by *P. versipelle*. *Podophyllum pleianthum* has the inflorescence inserted usually in the fork of the upper two petioles, rarely is it produced about midway along the petiole of the upper leaf. This has been seen in cultivation in two-leaved plants that bore three-leaves the previous season. Rarely cultivated specimens have been seen in which the inflorescence is immediately below the leaf lamina, as it invariably is in *P. mairei* and *P. versipelle*; these specimens can be separated by leaf lobe length. This atypical positioning of the inflorescence may be an artifact of cultivation. However, recently Dan Hinkley observed populations in central and northern Taiwan in which there seemed to be no consistency from one individual to another, the inflorescence position varying from the junction to mid petiole. Plants with this presumably atypical form will be difficult to separate from the putative hybrid *P. pleianthum* × *P. versipelle*, or may even be derived from this cross.

One intriguing variation is the habit of the plant. Most plants have the aerial stem longer than the petioles, while a dwarf form has the petioles longer than the aerial stem. This appears to be genetically controlled rather than induced by different cultivation techniques. These growth forms have been known in cultivation since at least the work of Kumazawa (1936), who illustrates the dwarf form. A report by Chen *et al.* (1977) indicates the existence of plants from northern Taiwan with a heteromorphic karyotype and suggests that this may be linked with differences in 'plant habit' but does not elaborate this statement. Roy Lancaster has related observing batches of *P. pleianthum* seedlings at Heronswood nursery of different accessions which were readily distinguishable as dwarf or large plants. Currently, he has one of each type in his garden and they retain this character. The more robust accession originated from Taiwan, freely produces large amounts of viable seed and bears the flowers approximately mid way between the two petioles. The smaller plants from seed obtained from North Carolina State University consistently produce the flowers precisely at the juncture of the petioles. Additionally there is variation amongst other collections in the length of the foliar marginal teeth with some individuals producing relatively long narrow teeth.

Material in cultivation represents clones from both Taiwan and mainland China, additionally home grown seedlings are available. Cross pollination between different

Podophyllum pleianthum

MATILDA SMITH

clones is essential for seed set. This is easily performed with large numbers of viable seeds produced, which germinate readily if sown soon after harvest before they have had chance to dry out. Storage in a dry paper packet for more than a day or two is fatal, as the seeds are extremely sensitive to desiccation.

The Chinese name Pa-chio-lien (which may also be applied to *P. versipelle* subsp. *boreale*) describes the appearance of the eight-angled leaves, similar to those of a waterlily. The name K'uei-chiu, literally 'Devil's mortar', may refer to the slightly concave peltate leaf lamina or to the use of the rhizome, when combined with cannabis, in hallucinogenic pills which were taken in order to 'see spirits', as recorded by Meng Shen *c.* 670 AD (Li, 1977). *Podophyllum pleianthum* is still employed in Chinese herbal medicine under the name Baijiaolian root as a treatment for snake bite, lymphadenopathy and tumours, sometimes with severe adverse effects (Bracchi & Routledge, 1996).

The epithet *pleianthum,* Greek *pleio-* (more than usual-) *anthos* (flower) recalls the excitement expressed by Hance at receiving a *Podophyllum* which produced more than a solitary flower from each aerial shoot.

Podophyllum pleianthum Hance in J. Bot. (London) 21: 174–175 (1883); Forbes & Hemsl. in J. Linn. Soc., Bot. 23: 33 (1886); Hook. in Curtis's Bot. Mag. 116: t.7098 (1890); Henry in Trans. Asiat. Soc. Japan, 24 (suppl.): 17 (1898); Matsum. & Hayata, Enum. Pl. Formos. in J. Coll. Sci. Imp. Univ. Tokyo 22: 19 (1906); Kawak., List Pl. Formosa: 5 (1910); Hayata, Icon. Pl. Formos. 1: 41 (1915); Kudo & Masam., Gen. Pl. Formos. 1: 81 (1932); Kumaz. in Bot. Mag. (Tokyo) 50: 268–276 (1936); Liu in Li, Fl. Taiwan 2: 520–521, t.396 (1976); Rix in Plantsman 4(1): 9 (1982); Brickell & Sharman, The Vanishing Garden: 182–183, t.64 (1986); Stearn in Walters, Eur. Gard. Fl. 3: 396 (1989); Phillips & Rix, Perennials 1: 39 (1991). Type: Taiwan, 'In bambuseto impeditissimo, ad septentrionem oppidi Tam-sui', April 1881, *T.Watters s.n.* in herb. Hance 21697 (holotype BM, isotype K).

P. onzoi Hayata, Icon. Pl. Formos. 5: 2–4, t.1 (1915). Type: Taiwan, 'Mt. Arisan, senninbora, secus fossas crescens, rarissima', April 1914, *Onzo s.n.* (holotype not traced). The name also appears as 'ontzoi', auctt. in syn.

Dysosma pleiantha (Hance) Woodson in Ann. Missouri Bot. Gard. 15: 339 (1928) *pro parte*, *quoad P. pleianthum, P. onzoi* in syn; Hiroe in Pl. Basho's & Buson's Hokku Lit. 8(3): 327–328 (1973) *pro parte*, excl. *P. versipelle*; Yang, List Pl. Taiwan: 613 (1982); Chen Ji-Sheng Zheng (ed.), Chinese Toxic Pl.: 131, t.21 (1987); Qi Chengjin *et al.*, List Hunan Fl.: 66 (1987); Chen & Cheng, China Floral Encyclopaedia: 283 (1990); Su & Liu in J. Wuhan Bot. Res., 10(4): 385 (1992) & 12(3): 217–219 (1994).

P. chengii S.S.Chien in Contr. Biol. Lab. Chin. Assoc. Advancem. Sci., Sect. Bot. 10(2): 108, t.16 (1936); Rix in Plantsman 4(1):10 (1982). Syntypes: China, Chekiang, Yutsien, W. Tienmushan, Sientin, 1500 m, in open places, 23 April 1931, *W.C.Cheng* 2413; Suichang, Peimashan in woods, 30 April 1933, *S.Chen* 1254 (syntypes ?PE, not traced).

Dysosma chengii (S.S.Chien) Hiroe in Pl. Basho's & Buson's Hokku Lit. 8(3): 328 (1973).

Diphylleia pleiantha Griffiths, R.H.S. Ind. Gard. Pl.: 913 (1994) *comb. invalid., sine basionym.*

Fig. 106. ***Podophyllum pleianthum*** in fruit; cultivated plant originating from Taiwan. Photograph: Julian Shaw.

ILLUSTRATIONS. Bot. Mag. 116: t.7099 (1890); The Garden 2: 299 (1899); Hui-Lin, Flora of Taiwan 2: 520, t.396 (1976); Brickell & Sharman, The Vanishing Garden: 182, pl. 64 (1986); Phillips & Rix, Perennials 1: 39 (1991); Hinkley, The Explorer's Garden: 116 (1999); New Plantsman 7(1): 39; 7(3): 145, pl.1A, (2000).

DESCRIPTION. *Perennial herb* with annual stems up to 60 cm tall. *Rhizome* thick with short internodes, with circular, annual stem scars along upper surface, and numerous adventitious roots along underside. *Terminal bud* enclosed in large scales often with minute undeveloped lamina at their apex. *Stem and petiole* covered with waxy bloom, petioles grooved along upper side. *Leaves* 1–3, peltate, up to 40 cm diameter, with about 8 shallow lobes. *Juvenile leaves* usually borne singly on stems from the rhizome, radially symmetrical with 4–8 shallow, triangular or rounded lobes, which are broader than long; sinuses penetrating $1/4$ to $1/3$ of the lamina radius, acute or rounded, keyhole shaped. *Adult leaves* 2 or 3 on a stem, lobed, with the lobing attenuated on the adaxial side or reduced to slight curves, with the greatest reduction on the second and third leaves; lobes on the abaxial edge $1/4$ to $1/3$ of the radius, broader than long. *Lamina* usually glabrous on both sides or with a few *c.*1 mm long hairs erratically distributed on the underside; dark green above, pale green below, main veins equalling the number of lobes, radiating from the petiole like the ribs of an umbrella, prominent on underside, secondary veins slightly so, veinlets smooth; leaf margin with small green teeth 0.6 mm long, about 10 per cm. *Inflorescence* inserted at the junction of either first and second, or second and third leaves if present, or very occasionally in cultivation emerging from about midway along the petiole of the upper leaf. *Pedicels* 5–10, each bearing one flower, *c.* 1 mm thick and 3–4 cm long, dilated to 2.5 mm below the receptacle. *Sepals* 6(–9), 3 outer, 3 inner, fugaceous, pale glaucous green, 2–2.5 cm long, elongating slightly as flower opens. *Petals* 6(–9), spathulate to lanceolate with a rounded or apiculate apex, dark reddish purple, up to 5 cm long. *Flowers* like small balloons when open, *c.* 2 cm diameter, the petals arching convexly to form a cage around the anthers and stigma, smelling somewhat unpleasantly. *Stamens* 6, up to 16 mm long, with a broad enlarged connective, produced apically to form a mucro 0.6–2 mm long; filament 6.5 mm; anthers 8.5 mm, extrorse, opening longitudinally by slits. *Pollen* pale yellow, in monads, 3-colpate. *Ovary* elliptic, style rather short, stigma peltate with very convoluted surface. *Fruit* a fleshy berry, ovoid, yellowish green, the colour sometimes obscured by silver-white waxy bloom, *c.* 3 cm long. *Seeds* numerous, semi-circular to subreniform, yellow-straw coloured; cotyledons orbicular, united along adaxial edge, with mucronately toothed margins. $2n = 12$, (Chen *et al.*, 1977; Li, 1986). Self-incompatible, obligate outbreeder.

DISTRIBUTION. China, Taiwan. Formerly widespread on Taiwan, on the mainland confined to the southeast. Recorded from Shanghai municipality, Anhui, Zhejiang, Fujian, Jianxi and Guangxi, (Icon. Cormoph. Sinic., 1980), Hubei (Fu, 1976).

ECOLOGY. Valleys and hillsides under shrubs, on wet slopes and moist places below 1500 m, (Fu, 1976). On Taiwan, central ranges between 1000 and 2500 m (Liu, 1976). Formerly common but now rare and endangered on Taiwan by over collecting (Hu, pers. comm. 1986).

Key to varieties of *P. pleianthum*

Petals dark maroon red .... var. **pleianthum**
Petals white.. var. **album**

var. **pleianthum**
Flowers with dark red-purple wine coloured petals and a strong unpleasant odour. Occurs throughout the range of the species.

var. **album** Masam. in Kudo & Masamune, Gen. Pl. Formos. 1: 81 (1932).
This is the only record of a white flowered form of this species, from the mountains of Taiwan. It would be interesting to know what pollination syndrome this white flowered form exhibited. There are no recent records and unfortunately it is probably extinct.

2. × 3. PODOPHYLLUM PLEIANTHUM × P. VERSIPELLE

A putative hybrid of which no material has been seen; it is probably not easy to distinguish from cultivated material of *P. pleianthum* which sometimes produces the inflorescence along the petiole. One would expect to find this hybrid along the western limit of *P. pleianthum* on the Chinese mainland and it appears to be represented in the *Flora of Zhejiang*. Cytological investigation of suspected hybrid populations may yield helpful results. A useful preliminary to this is the comparison of karyotypes of the two putative parents by Zhang *et al.* (1991).

Podophyllum pleianthum × Podophyllum versipelle
[*Dysosma versipellis sensu* Wang, Flora of Zhejiang 2: 319, f.2–422 (1986) *pro parte*]

ILLUSTRATION. Wang, Flora of Zhejiang 2: 319, f.2–422 (1986).
DESCRIPTION. *Leaves* with shallow lobes that are broader than long. *Inflorescence* inserted on the petiole near to the fork. Plants recorded from around 500–800 m in Zhejiang province, China, appear to represent this hybrid, with the features of *P. pleianthum* dominant.

3. PODOPHYLLUM VERSIPELLE

This was the second dark reddish-purple flowered species of the genus to come to the attention of western botanists, when B.C.Henry collected plants on a mountain in Guangdong province in 1883. B.C.Henry, a missionary who collected the type specimen of subspecies *versipelle*, is not to be confused with the better known Irish doctor Augustine Henry, who worked for the Imperial Customs. Augustine Henry was based further north in Hubei province from 1881 until his transfer to the island of Hainan in 1889 and, as a consequence of his location, collected specimens of subspecies *boreale*.

Podophyllum versipelle subsp. *boreale*

MATILDA SMITH

3. PODOPHYLLUM VERSIPELLE

The name *versipelle*, Latin *versi-* (changeable, different), *pellis* (skin), evidently makes reference to the leaf shape, since Hance wrote "The leaves vary marvellously in outline from a square, parallelogram, triangle or pentagon to a circle, and are either with or without lobes" (Hance, 1883b). Although Hance's type specimen has upper and lower mature foliage present, this apparent degree of variation was in part due to the poor state of the collection with many of the leaf margins partly decayed. Hance earlier revealed the probable cause of this, "His [B.C.Henry's] specimens arrived in a very unsatisfactory condition, owing to the torrential rains to which he was exposed" (Hance, 1883b). The leaves on the type are rather small for the species and may be influenced by growth conditions; specimens in herbaria display leaves of widely differing dimensions.

Subsp. *versipelle* consists of isolated scattered populations, now confined to mountains, some of which have been given specific names. It has recently been collected and photographed in habitat by Guy Gusman of Brussels from Mt. Bavi, north of Hanoi, growing in wet broad-leaved evergreen forest along with *Paris (Daiswa) vietnamensis*, *Arisaema balansae* and *A. petelotii*. Plants which probably belong to this subspecies have only recently appeared in cultivation.

Subspecies *boreale* is in cultivation at Mt. Usher gardens, near Dublin, from an original collection by Augustine Henry (Morley, 1980, 1986). It was also cultivated during the early twentieth century at Kew from a 1903 Veitch introduction, possibly from an E.H.Wilson collection, and figured in Hooker's *Icones Plantarum*. At some point after 1936 it died out at Kew and Edinburgh; plants being cultivated under this name in the 1970s were *P. hexandrum*. More recently it has been reintroduced from China via the Kaichen nursery (sometimes listed as *D. veitchii*) and is in cultivation at Heronswood Nursery, Washington State, and Crug Farm Plants, Wales (where it has been listed as *P. mairei*). Some juvenile plants can be difficult to separate from immature *P. pleianthum*, however *P. versipelle* often has red-tinted lower stems and a concave leaf lamina with deeper lobes, while stems of *P. pleianthum* are invariably green with a flat leaf lamina.

The epithet *boreale*, Latin *borealis* (northern), is with reference to the geographical distribution of the subspecies to the north of subspecies *versipelle*.

Podophyllum versipelle Hance in J. Bot. (London) 21: 362 (1883). Type: China, Guangdong province, 'in fissuris montium Lo-fau-shan, prov. Cantonensis, juxta coenobium buddhicum Put-wan-mun, alt. 2000 ft. et ultra', May 1883, *B.C.Henry*, in herb. *Hance* 22200 (holotype: BM, K drawing).

DESCRIPTION. *Rhizomatous perennial* with annual aerial stems 15–80 cm, often grooved in herbarium sheets, bearing 1–3 peltate leaves. *Rhizome* 1.5–2 cm thick with short internodes; cataphyllary scales present at base of stem, up to 5 cm long, dark red-brown. *Juvenile foliage* borne singly on stems from the rhizome, radially symmetrical with 5–7 shallow, triangular-ovate lobes and shallow angular sinuses. *Adult leaves* 15–40 cm diameter, lower leaf larger with 6–8 lobes; margins with small mucronate teeth, 1–15 mm apart; upper surface dark green, glabrous; lower surface pale green with sparse hairs to tomentose, main veins slightly prominent more densely tomentose; lobes shallow or deep, elobulate or weakly trilobulate, broadly triangular-ovate, with an acute or mucronate apex; sinuses shallow or deep, penetrating $1/5-2/3$ the radius, gently curved or angular, sometimes keyhole-shaped, occasionally obscured by overlapping obovate lobes; lobules when present consisting of an enlarged pair of teeth below the lobe apex; upper leaf above the inflorescence with 4–5 lobes present only on the abaxial side, which are deeper than on lower leaves, with the adaxial edge often reduced to a smooth semi-circular arc. *Inflorescence* with 4–19 flowers, borne on the petiole just below the upper leaf. *Pedicels* 2.5–10 cm long, finely hirsute or glabrous, hairs white to brown. *Sepals* 6, elliptic to ovate, fugaceous, 1.5–2 cm long, 0.5–1 cm wide, papery when dry, the outer 3 shorter and narrower, sometimes sparsely hairy on the exterior. *Petals* 6, oblong with a rounded or slightly irregular apex, dark red-purple, 2.5–3.5 cm long, *c.* 1 cm wide. *Stamens* 6, slightly curved, 15–20 mm long; filament 5–7 mm, thin, flattened at the base; anther 7–12 mm; connective thickened and pigmented, extended to form a mucro 1–3 mm in length. *Ovary* subglobose to elliptic, with a short style 2–3 mm long; stigma peltate, large, corrugated. *Fruit* a fleshy ellipsoidal berry *c.* 3 cm across; seeds numerous, reniform, 5 mm long, cotyledons unknown. Self-incompatible, obligate outbreeder.

Leaf lobes elobulate, upper leaf with 4–7 lobes on one side, lobes shallow $1/5-1/3$ of radius; inflorescence 4–9 flowered; pedicels with hairs .... subsp. ***versipelle***

Leaf lobes often with lobules, upper leaf with 4–5 lobes on one side, lobes deep, $1/2-2/3$ of radius; inflorescence 4–10(–19) flowered; pedicels glabrous or rarely hairy.. subsp. ***boreale***

subsp. **versipelle**

P. versipelle Hance in J. Bot. (London) 21: 362 (1883); *sensu* Forbes & Hemsl. in J. Linn. Soc., Bot. 23: 33 (1886) *pro parte, quoad B.C.Henry s.n., Ford s.n.*; *sensu* Oliver in Hooker's Icon. Pl. 20: t.1996 (1891) *pro parte, quoad B.C.Henry s.n*; Dunn & Tutcher, Fl. Kwangtung & Hongkong in Bull. Misc. Inform. Kew, Addit. Ser. 10: 33 (1912).

P. esquirolii Lév. in Repert. Spec. Nov. Regni Veg. 11: 298 (1912) & Fl. Kouy-Tcheou: 50 (1914); Rix, Plantsman 4(1): 10 (1982). Type: China, Kweichow, 'bois humide de Na-Laou, 800 m, et enfoncement de Pfa-Y-To, 1250 m', 19 April 1902, *Esquirol* 2800 (holotype E).

Dysosma pleiantha (Hance) Woodson in Ann. Missouri Bot. Gard. 15: 339 (1928) *pro parte, quoad P. versipelle, P. esquirollii* in syn.; Hu in Bull. Fan Mem. Inst., Bot. Ser. 8: 36–37 (1937) *pro parte, quoad P. hispidum*; Lauener in Notes Roy. Bot. Gard. Edinburgh 24(1): 74 (1962) *pro parte, excl. P. pleianthum*; Hiroe in Pl. Basho's & Buson's Hokku Lit. 8(3): 327 (1973) *pro parte, quoad P. versipelle* in syn.; Lauener in Plantsman 4(2): 127 (1982) *pro parte, excl. P. pleianthum*.

P. hispidum Hao in Repert. Spec. Nov. Regni Veg. 36: 223 (1934) & Contr. Nat. Acad. Peiping 3: 2 (1935); Rix in Plantsman 4(1):10 (1982). Syntypes: China, Kwangsi, 'in den waldern des Yao-schan-Gebirges', 1800 m, May 1931, *S.S.Sin* 21089; 1700 m, 25 May

1931, *S.S.Sin* 21418; May 1931, *S.S. Sin* 20172, 9018 (syntypes ?PE, B not found, BM drawings) [The syntypes of *P. hispidum* are not at Berlin. Collections of S.S.Sin were sent to Berlin between 1927 and 1929. If the types (collected in 1931) were also deposited in Berlin, they would have been destroyed in 1943 (Hiepko, pers. com. 1993)].

P. tonkinense Gagnep. in Bull. Soc. Bot. France 85: 167–168 (1938) & in Humbert, Suppl. Fl. Gen. Indo-Chine 1:146–147, t.13 (1939); Rix in Plantsman 4(1): 10 (1982) [as *tonkinse*]. Type: Vietnam, Tonkin, Chobo, plateau, 400 m, *Mlle. Colani* 3008 in herb. *Pételot* (holotype P).

Dysosma tonkinense (Gagnep.) Hiroe in Pl. Basho's & Buson's Hokku Lit. 8(3): 328 (1973).

Dysosma hispida (Hao) Hiroe in Pl. Basho's & Buson's Hokku Lit. 8(3): 329 (1973).

Dysosma versipellis (Hance) M.Cheng [in Icon. Cormophyt. Sinic. 1: 760 (1972) *pro parte*, excl. t.1519; Fu, Fl. Hupehensis 1: 401 (1976) excl. t.567, *comb. inval. sine basionym*], validated by Ying in Acta Phytotax. Sin. 17(1): 18 (1979), *pro parte, quoad P. versipelle* Hance in syn.

ILLUSTRATIONS. Humbert, Suppl. Fl. Gen. Indo-Chine 1: t.13 (1939); New Plantsman 7(3): 145, pl.1C (2000).

DESCRIPTION. *Adult leaves* with small mucronate teeth 1–15 mm apart, lobes shallow, broadly triangular to ovate, sinuses penetrating $1/5$–$1/3$ the radius, gently curved or angular. *Upper leaf* with lobes only on the abaxial side, with the adaxial margin often reduced to a smooth semi-circular arc. *Flowers* 4–9, pedicels hirsute.

DISTRIBUTION. Known from southern China, Guangdong, Guangxi and Guizhou provinces, and adjacent Vietnam. The distribution reported under *Dysosma versipellis* in the IUCN Red list of Threatened Plants (Walter & Gillett, 1998) represents multiple species, as interpreted here.

ECOLOGY. Woodland areas on mountains.

subsp. **boreale** J.M.H.Shaw in New Plantsman 6(3): 160 (1999). Type: China, Hubei province, So Patung, received 1888, *A.Henry* 5372 (holotype K, isotypes K, E, BM). The type collection is numbered 5372 without any suffix. There are many other sheets with the same number but distinguished by a letter: 5372A, 5372B etc. these do not represent the type collection. At some point Augustine Henry began to group collections from different areas and times which he believed to be conspecific under one number and add a suffix letter (Morley, 1979).

[*P. versipelle sensu auctt., non* Hance, Forbes & Hemsl. in J. Linn. Soc., Bot. 23: 33 (1886) *pro parte*, excl. *B.C.Henry* s.n .; Oliver in Hooker's Icon. Pl. 20: t.1996 (1891) *pro parte*, excl. *B.C.Henry s.n.* & *F.Faber* 33; Diels, Fl. Central-China in Bot. Jahrb. Syst. 29(2–5): 336 (1901); Hutchinson in Bot. Mag. 133: t.8154 (1907) *pro parte*, excl. *B.C.Henry s.n.*; Rix in Plantsman 4(1): 9 (1982) *pro parte*, excl. ref. to white flowered plant in Gard. Chron., (1938); Morley in Garden 111(8): 395 (1986); Brickell & Sharman, The Vanishing Garden: 183 (1986) *pro parte*; Stearn in Walters, Eur. Gard. Fl. 3: 396 (1989), as *P. versipelle* misapplied.]

[*P. pleianthum sensu* Diels, *non* Hance, Fl. Central-China in Bot. Jahrb. Syst. 29(2–5): 336 (1901)]

[*Dysosma versipellis sensu auctt. non* (Hance) M.Cheng [in Icon. Cormophyt. Sinic. 1: 760, t.1519 (1972), *comb. inval. sine basionym*; Fu, Fl. Hupehensis 1: 401, t.567 (1976)]; validated by Ying in Acta Phytotax. Sin. 17(1): 18 (1979); Icon. Cormophyt. Sinic. ed. 2, 1: 760, t.1519 (1980); Chen Ji-Sheng Zheng, Chinese Toxic Plants: 135 (1987) [as *versipoellis*]; Qi Chengjin *et al.*, List Hunan Fl.: 66 (1987); Su & Liu in J. Wuhan Bot. Res. 10(4): 385 (1992)].

[*P. pleianthum sensu* Morley, *non* Hance, in Garden 105(7): 287 (1980), *pro parte, quoad* ref. to plant cultivated at Mt. Usher Gardens.]

ILLUSTRATIONS. Hook. Icon. Pl. 20: t.1996 (1891); Bot. Mag. 133: t.8154 (1907); Garden 111(8): 395 (1986); Hinkley, The Explorer's Garden: 116 (1999); New Plantsman 7(3): 145, pl.1B (2000).

DESCRIPTION. *Adult foliage* of 2–3 leaves borne on unequal petioles with marginal teeth 1–2 mm apart, the lower leaf larger with 6–8 lobes, the lobes somewhat attenuated on the adaxial side; second leaf smaller with 4–5 lobes and very reduced lobes along the adaxial margin, lobes deep, obovate to oblong, $^1/_2$–$^2/_3$ the radius, entire or weakly trilobulate, with an acute or apiculate apex; lobules when present consisting of an enlarged pair of teeth below the lobe apex; sinuses deep, rounded sometimes keyhole shaped, occasionally obscured by overlapping obovate lobes. *Inflorescence* with 4–10(–19) flowers, pedicels glabrous (or hirsute var. *sichuanense*). 2n = 12 (Darlington, 1936).

DISTRIBUTION. China, provinces along the Chanjiang [Yangtze] river (Fu, 1976); Hubei, Jiangxi, Sichuan, Yunnan. Ying (1979) also records this taxon from Henan. The reported occurrence of this taxon in Tibet (Griffiths, 1994) is apparently based on the suggestion by W.T.Stearn in *The European Garden Flora* 3: 396 (under *P. versipelle* misapplied) that it may be the same taxon as *D. tsayuensis* Ying (here treated as a synonym of *P. aurantiocaule*) described from S.E.Tibet. While it is not impossible that *P. versipelle* may occur in Tibet, this is unlikely; either way, this requires confirmation.

ECOLOGY. Common in woods and shaded situations in mountains, occurs in colonies of sometimes over 100 plants, (Henry in Hooker, 1891). These colonies may be clonal from adventitious root buds. Undisturbed forest on hills in humus rich soils, 500–2400 m, (Fu, 1976).

Two varieties of subsp. *boreale* may be distinguished as follows:

Leaf lobes with convex margins, entire or trilobulate; leaf underside and
    pedicels usually glabrous .... var. **boreale**
Leaf lobes with concave margins, usually trilobulate; leaf underside and
    pedicels usually hairy .... var. **sichuanense**

var. **boreale**

Throughout the range of the subspecies.

var. **sichuanense** J.M.H.Shaw in New Plantsman 6(3): 161 (1999). Type: China, Sichuan province, Kuan-hsien [Guan Xian, 31.01 N, 103.40 E], 25 May 1936, *S.S.Chien* 5710 (holotype A, isotype A, E). Other collections seen: *Faber* 33 (BM); *Sun & Chang* 1773 (A).

The name *sichuanense* refers to the type collection originating from Sichuan province.

ILLUSTRATION. New Plantsman 6(3): 161 (1999) & 7(3): 145, pl.1E, (2000).
DESCRIPTION. Resembles subsp. *versipelle* in its pubescent pedicels and leaf undersides, although the indumentum is usually denser, but easily distinguished in typical collections by the leaf lobe shape and length. Some individuals from further east have lobes with less concave or slightly convex sides. Differing from typical subsp. *boreale* in the dense indumentum on the leaf underside, pubescent pedicels and concave sided leaf lobes.
DISTRIBUTION. Sichuan, often collected on Emei Shan (Mt. Omei).

Informal list of other variants of *P. versipelle*:

var. *tomentosum* Yu, Wang & Chen *nom. nud.* [Syn. *Dysosma versipellis* var. *tomentosa* Yu, Wang & Chen, J. Nat. Prod. (Lloydia) 54(5): 1422 (1991)]. The identity of this taxon is uncertain. It has not been possible to trace any validating publication or examine any specimens, although a voucher specimen is deposited at Dept. of Chemistry, Shanghai Second Medical University. It relates to material collected in Anhui province, and determined by Professor Mao-Bing Deng, which evidently has tomentose leaves (undersides ?). Its identity and relationship to the *P. versipelle* group requires investigation. The epithet *tomentosum*, Latin *tomentosus* (dense interwoven haircovering) describes the leaf underside.

Variants with trilobulate-lobed leaves
Throughout the range of *P. versipelle* individuals occur with varying degrees of foliar lobulation. They probably represent introgression from the *P. delavayi* group. Some of these collections also display hairs along the back of the outer sepals, elongated ovaries, a greater degree of foliar asymmetry and a slightly longer anther mucro, which is further evidence in favour of such introgression.

Obovate-lobed leaves
Occasionally encountered are plants from Hubei with strongly obovate leaf lobes which overlap at the proximal margins (*Henry* 5372E, BM). See illustration in *The New Plantsman* 7(3): 145, pl. 1D, (2000).

Long-acuminate lobed leaves
A collection from Sichuan (*Farges* s.n, P) displays pronounced acuminate apices to the leaf lobes. It may represent a distinct variety. See illustration in *The New Plantsman* 7(3): 145, pl. 1F, (2000).

4. PODOPHYLLUM HEMSLEYI

This species is remarkable for its very large deeply lobed leaves, giving the impression of a cart wheel. It is most similar to *P. versipelle* subsp. *boreale* and can be distinguished as generally a larger plant with very deeply divided leaves, the spathulate-oblong leaf lobes to 22 cm, penetrate to $^4/_5$ of the radius.

There are two sheets of *Wilson* 432a at Kew, the holotype, consisting of an upper leaf with inflorescence and a second sheet consisting of a single larger leaf, which appears to be the lower leaf from the same individual. A further collection, *Wilson* 3203 [Arnold Arboretum exped., 1907–1909] from Western Hubeh (K), which also seems to be one plant divided between two sheets, may also represent this taxon, but differs in the lobes which are slightly trilobulate. It may be that Wilson revisited sites known to him from earlier expeditions and collected this specimen from near the type locality (Howard, 1980; Clausen & Hu, 1980). A further sheet of *Wilson* 3202 at B has lobes with convex margins which are only slightly longer than some collections attributed to *P. versipelle* subsp. *boreale*; intermediate collections may yet come to light. This taxon probably consists of a very localized population, which has not been recollected since Wilson's expedition, unless specimens exist in Chinese herbaria.

The epithet *hemsleyi* commemorates William Botting Hemsley (1843–1924), from 1899 to 1908 Keeper of the Kew Herbarium. During his period as keeper, Kew received for identification the herbarium specimens collected by E.H.Wilson on his plant collecting expeditions to China from 1899 to 1901 and 1902 to 1903 for Messers. James Veitch and Sons. In this unexpected task, Hemsley, who had done most of the compilation of Forbes and Hemsley 'Enumeration of all plants known from China' (*J. Linn. Soc., Bot.* vols. 23, 26, 36; 1886–1903), took a major part. He and Wilson published two new species of *Podophyllum*; *P. difforme* and *P. veitchii* (conspecific with *P. delavayi* Franch.) and left unpublished a manuscript Latin description of this species in the Kew Herbarium but provided no name. It was accordingly named *P. hemsleyi*.

Podophyllum hemsleyi J.M.H.Shaw & Stearn in New Plantsman 6(3): 161–162 (1999). Type: China, Hupeh province, in woods 1500–1800 m, fl. late May, *E.H.Wilson* 432a (Veitch Exped., 1899–1901) (holotype K).

ILLUSTRATION. New Plantsman 7(3): 153, pl. 3A (2000).
DESCRIPTION. *Perennial herb* to 40 cm; stem grooved on herbarium sheets. *Rhizome* 1–1.4 cm wide at nodes with short internodes, upper surface with circular annual stem scars, and numerous concentric scars from cataphylls. *Juvenile leaves* unknown. *Mature leaves* 2, large, to 35 cm diameter, with deep spathulate-oblong lobes with rounded sinuses penetrating to $^4/_5$ of radius; lobe apex rounded with acute to acuminate tip; margin with fine teeth *c.* 1 mm long, 5–6 per cm; upper surface dark green, lower surface glaucous green with prominent main veins bearing a few scattered simple 0.4 mm hairs. *Lower leaf* 35 cm across, with radius 20–22 cm at longest lobes. *Upper leaf* 28 × 22 cm with 5–6 lobes; lobes spathulate-oblong,

to 16 cm long, 2.5–3(–4) cm wide at base, 4–5.5 cm across at widest point some 6–7 cm below apex; lobing strongly retarded on adaxial side. *Inflorescence* with 4 flowers inserted on petiole 2 cm below lamina. *Pedicels* glabrous, 3–4 cm long, dilated to 2 mm wide below receptacle. *Bud* ovoid, glabrous. *Sepals* ovate-lanceolate, 14–16 mm long; outer sepals 6–8 mm wide, inner sepals 3–4 mm, glabrous, apex acute. *Petals* dark red, 3–3.5 cm long, oblong-spathulate, widest below apex; apex rounded, entire. *Stamens* 6; filaments flattened 7–8 mm long, anther loculi 9–10 mm, connective visible on both sides with small apical mucro *c*. 1 mm in dried state. *Ovary* globose to pear-shaped, 6–7 × 4 mm; style 2–3 × 1 mm; stigma globose, corrugated, 3–5 mm across. *Berry* and seeds unknown.

DISTRIBUTION. China, Western Hubeh.

5. PODOPHYLLUM MAIREI

Initially I considered this to be part of the taxon here named *P. versipelle* subsp. *boreale*, however the petal length and flower shape are different. It appeared that these characters could have been artifacts of drying an incompletely expanded flower (although the petals on the type in Paris do not show the wrinkling characteristic of partly expanded petals) and I was uncertain until seeing a photograph of live material, which shows that the flower differs significantly from that of live material of *P. versipelle* subsp. *boreale*. It differs from *P. versipelle*, with which it has been confused, in the smaller ovate petals which form a open bowl-shaped flower rather than a funnel-shaped or spherical chamber [This difference is well illustrated by the colour image currently (2001) displayed on the Kaichen nursery web page at http://www.bjkaichen-hgd.com.cn under the name '*Dysosma pysosma*']. This open flower is unusual in section *Dysosma* and may indicate a subtle change in pollination system, perhaps associated with introgression from the open-flowered *P. aurantiocaule* subsp. *furfuraceum*. These floral characters possibly induced T.S.Ying (1979) to reduce this taxon to synonymy under *P. aurantiocaule*.

Plants that appear to belong to this taxon have occasionally appeared in batches of rhizomes sent as *Dysoma versipelle* from the Kaichen nursery; one such was kindly forwarded to me by Jim McClements. Unfortunately, the presence of such plants and the size of the rhizomes suggests that these plants have been collected from the wild for export rather than being nursery grown stock. It was self-incompatible and did not produce seed when crossed with *P. pleianthum*, although the *P. pleianthum* ovaries did expand and produce small fruit which aborted before maturity. This Kaichen material differs from the type collection on several points: It is a smaller plant to 20 cm high with leaves to 14 cm across; marginal teeth fine, 0.5–1 mm long, *c*. 16 per cm. Lower lamina surface with scattered linear hairs 0.2–0.4 mm long, more numerous on the main veins. Sepals 11–12 × 3–5 mm, with hyaline margins. Petals dark red, 10–13 × 4–5 mm. Stamens 7–9 mm long; filament 1.5–2.5 mm, flattened; anther 6–7 mm with 1–1.5 mm mucro. Ovary 5 × 3 mm with 3 mm style.

The name of this species was chosen by François Gagnepain of the Muséum National d'Histoire Naturelle to commemorate Edouard Ernest Maire (1848–1932), a French missionary who collected in Yunnan, China between 1905 and 1914. Printed labels on his herbarium specimens at Paris bear the name R.-P.Maire, in deference to his ecclesiastical office (Fournier, 1932). A label on one of the type sheets is initialled "E.E.M." in Maire's own hand.

Podophyllum mairei Gagnep. in Bull. Soc. Bot. France 85:167 (1938). Type: China, Yunnan, 'montagne de Tchen-fong-chan', 750 m, *E.E.Maire s.n.* (holotype P, isotype P).
Dysosma mairei (Gagnep.) Hiroe in Pl. Basho's & Buson's Hokku Lit. 8(3): 328 (1973).
[*Dysosma aurantiocaulis sensu* Ying in Acta Phytotax. Sin. 17(1): 19 (1979) *pro parte*, excl.
  P. aurantiocaule Hand.-Mazz.].
Dysosma pysosma hort. A name used by the Kaichen nursery, China.

ILLUSTRATION. New Plantsman 7(3): 153, pl.3B, (2000).
DESCRIPTION. *Perennial herb* with annual aerial stems to 60 cm, usually bearing two leaves; rhizome with short internodes. *Adult leaves* peltate, 20–30 cm across, glabrous on both sides; pale on underside; lobes oblong to ovate, 4–7 cm long, acuminate, rarely with a slightly enlarged lateral tooth near the apex, but not becoming trilobulate; margin with small, 1 mm teeth, 3–4 mm apart; sinuses rounded, obtuse, penetrating to less than or just over half the radius; main nerves equalling number of lobes. *Lower leaf* with 7–8 lobes with widest margin on adaxial side; upper leaf, smaller with 5–7 lobes, lobing severely retarded on the adaxial side. *Inflorescence* with 5–7 flowers, inserted on petiole of upper leaf a few cm below lamina; pedicels 4–6 cm, glabrous, thickened towards apex. *Sepals* 4–5, ovate-oblong, pale green, 15 × 5–7 mm. *Petals* 5–6, orbicular to ovate, 15–17 × 11–13 mm, base abruptly narrowed, apex rounded, patent, held at right angles to floral axis forming a flat to shallow bowl-shaped flower. *Stamens* 5–6 with narrow, 5–6 mm long filaments; anthers 13 mm, mucro almost absent. *Ovary* oblong to ovoid, abruptly constricted at base; style short; stigma peltate, 3 mm across.
DISTRIBUTION. China, Yunnan province, possibly also Sichuan. Rather local.

6. PODOPHYLLUM GLAUCESCENS

A plant with square leaves balanced on a central stalk might be a fiction of a cubist designer. In this case it is the juvenile foliage of a *Podophyllum*, which had for many years excited comment at Harvard, although remaining unidentified.

This plant is distinguished by thin papery leaves with a glaucous underside, very shallow leaf lobes, pilose pedicels and flowers with short 10–12 mm pale reddish petals. It is most likely to be confused with *P. mairei* on account of its small flowers, but differs in the glaucous, papery leaves (*P. mairei* foliage has the thicker texture of *P. pleianthum*), leaf margins with wider spaces between teeth, long hairs on the pedicels, and shorter pale petals. It is easily distinguished from the small-flowered *P. majoense* by the leaf lobes; *P. majoense* leaves are strongly trilobulate.

It appears that *S.P.Ko 56055* from Guangxi, Chien Pien district, collected in moist woodland on 16 November 1935 (A, SYS), also represents this taxon. The herbarium sheet at Harvard consists of two juvenile leaves that are almost perfectly square, however the attached notes describe an immature green berry indicating that duplicate sheets with mature foliage exist elsewhere. This taxon may represent introgressed or otherwise hybridised individuals involving *P. difforme* and *P. versipelle* subsp. *versipelle*. The type specimen has been the source of pollen for studies by C.A. Meacham (1980) and Nowicke & Skvarla (1981).

The specific epithet *glaucescens*, from Latin *glaucus* (bluish-green) recalls the glaucous lower leaf surface.

Podophyllum glaucescens J.M.H.Shaw in New Plantsman 6(3): 162 (1999). Type: China, Guangxi, Loh hoh tsuen, Ling yun hsien, hillside, 1200 m, 1933, *A.N.Steward & H.C. Cheo* 210 (holotype A!, isotype ?NAS).

ILLUSTRATION. New Plantsman 6(3): 164 (1999) & 7(3): 149, pl.2A1–6 (2000).

DESCRIPTION. *Perennial herb* with aerial stem to 40 cm; rhizome with short internodes and circular stem scars; apical bud with rust brown cataphylls 4–6 cm long. *Juvenile leaves* dark green above, glaucescent below, with 4 shallow lobes, appearing 12 cm square; margins with small, 1 mm teeth, 8–15 mm apart. *Mature leaves* 2, glabrous, membranous, upper surface shiny, dark green, lower surface glaucescent; lower leaf trapezoid, oriented with adaxial edge longest, 23 × 15 cm with 4 lobes separated by very shallow undulating sinuses, lobing strongly retarded on adaxial margin; lobes with shortly acuminate apices; margin with small, 1 mm teeth 4–12 mm apart; upper leaf roughly pentagonal with 4 obvious shallow lobes and 2 obscure lobes on adaxial margin, *c.* 20 cm across. *Inflorescence* of 7 flowers inserted on petiole of upper leaf 2 cm below lamina. *Pedicels* thin, 4–6 cm, swollen near apex, densely covered with simple 2–3 mm hairs. *Sepals* 14 × 6 mm narrowly ovate, pale green with membranous margins. *Petals* 6, reddish-purple, ovate-lanceolate, 10–12 × 6–5 mm, with acute or slightly toothed apex. *Stamens* 6, 1 cm long; filament flattened 2–3 mm; anthers 6–7 mm with apical mucro 1.5–2 mm. *Ovary* ovoid 5 × 3 mm, style 2–3 mm with large peltate corrugated stigma 3 mm across. *Immature berry* green. *Mature fruit* and seeds unknown.

DISTRIBUTION. China, Guangxi province. Confined to moist woodland around Chien Pien district. Rare.

7. PODOPHYLLUM DIFFORME

Closely related to *P. delavayi* with which it shares the pilose pedicels and elongated mucronate stamens and probably derived from it by hybridization, the distribution of *P. difforme* follows an arc along the eastern limit of *P. delavayi*. *Podophyllum difforme* is probably the contributor to *P. versipelle* subsp. *versipelle* of the rather irregular leaf lobing seen in some collections. In the type collection and most others seen the inflorescence is inserted on the pedicel of the upper leaf, just below the lamina, but in cultivated material it is usually produced at the fork of the petioles. The significance of this variation has yet to be investigated.

There are some good foliage forms of this species in cultivation, particularly the pigmented juvenile leaf forms. The leaf surface is covered with papillate trichomes that scatter the light imparting a velvety sheen to the foliage. Experience has shown that this species prefers a little warmth; plants kept in a shaded cold greenhouse produce more leaves and seem to grow better. Some clones start into growth in the autumn. Seeds have been produced in cultivation, but compatibility was not noted.

The epithet *difforme*, from the Latin *difformis* (irregular), refers to the irregular leaf outline.

Podophyllum difforme Hemsl. & E.H.Wilson in Bull. Misc. Inform., Kew 1906: 152–153 (1906); Dunn, Suppl. List Chinese Fl. Pl. 1904–1910, in J. Linn. Soc., Bot. 39: 477 (1911); Rix in Plantsman 4(1): 10 (1982). Type: China, W.Hupeh, woods at elevations between 1200 and 1800 m, *Wilson* 966 (Veitch Exped., 1903–1905) (holotype K!, isotype K!).

P. triangulum Hand.-Mazz. in Anz. Akad. Wiss. Wien Math.-Naturwiss. Kl. 61: 163 (1924) & Symb. Sin. 7(2): 322 (1931); Rix in Plantsman 4(1): 10 (1982). Type: China, S.W.Hunan, 'in monte Yun-schan prope urbem Wukiang, inter 400 et 1420m, s. schisto argilloso' April 1919, *Wang-Te-Hui* in *Handel-Mazzetti* 47 (holotype W, not found; not at GZU, (Drescher, pers. comm. 1993)).

[*Dysosma pleiantha* (Hance) Woodson in Ann. Missouri Bot. Gard. 15: 339 (1928) *pro parte, quoad P. difforme* in syn.]

Dysosma difformis (Hemsl. & E.H.Wilson) T.H.Wang [in Icon. Cormophyt. Sinic. 1: 760, t.1520 (1972) *comb. inval. sine basionym*; Fu, Fl. Hupehensis 1: 402, t.569 (1976)]; validated by Ying, in Acta Phytotax. Sin. 17(1): 19 (1979); Icon. Cormophyt. Sinic. ed. 2, 1: 760, t.1520 (1980); Qi, C., Sun, X., Lin, S., List Hunan Fl.: 66 (1987); Su, Y. & Liu, Q. in J. Wuhan Bot. Res. 10(4): 385 (1992).

ILLUSTRATIONS. Icon. Cormoph. Sinic. 1: t.1520 (1980); Fu, Fl. Hupeh. 1: t.569 (1976); Hinkley, The Explorer's Garden 117, 118 (1999); New Plantsman 7(3): 148, pl.2D, (2000).

DESCRIPTION. *Perennial rhizomatous herb*; rhizome with short, thin internodes, *c.* 5 mm diameter, described as moniliforme by Handel-Mazzetti (1924). *Stems* 10–30(–40) cm high, glabrous, with 1–3 leaves. *Leaves* peltate, irregular and subtriangular or almost bilobed, up to 18 cm × 10 cm at the longest and broadest points. *Juvenile foliage* square or rectangular in outline, borne singly on stems from the rhizome. *Adult leaves* with the first leaf irregularly and shallowly obscurely lobed, 4–5 lobes present as alternating parts of the lamina edge slightly concave and convex, or 2–3 acute parts around the circumference, giving the leaf a triangular or bilobed appearance; second and third leaves progressively smaller, tending to be broadly triangular or obscurely bilobed and rounded along the adaxial edge; all adult leaves with sparse mucronate teeth, (4–)7–12 mm apart. *Inflorescence* of 1–3 flowers, inserted along the petiole under the uppermost leaf or at the fork between the petioles of the upper two leaves. *Pedicels* 1.5–3.5 cm long, pendulous, with variable white pilose indumentum of

Fig. 107. ***Podophyllum difforme.*** A variant with unusual foliage, originating from China. Photograph: Jim McClements.

Fig. 108. ***Podophyllum difforme.*** Cultivated plant from a Japanese nursery. Photograph: Bleddyn Wynn-Jones.

simple hairs. *Sepals* 3–6, lanceolate-obovate, 1.5 cm long, membranous, outer 3 hirsute on outer surface, fugaceous. *Petals* 6, salmon-pink to dark red-purple, oblong-obovate with a rounded apex, 1.5–2(–3.5) cm long. *Stamens* 6, $^1/_3$ to $^1/_2$ as long as the petals, *c.* 8–9 mm, filament and anthers of equal length, *c.* 4 mm, with an apical mucro, 1.6–3 mm long. *Ovary* spherical, glabrous, 3–4 mm diameter with a pronounced style 2–3 mm long; stigma peltate. *Fruit* a globose to ellipsoidal berry, 1–2 cm diameter, surface with scattered tubercles in some cultivated accessions. *Seeds* and cotyledons unknown.

DISTRIBUTION. China, western Hubei, Sichuan and Hunan provinces (Icon. Cormoph. Sinic., 1980), also reported from Guangxi and Guizhou provinces by Ying (1979). Scattered and rare.

ECOLOGY. Growing amongst thick forests (Icon. Cormoph. Sinic., 1980), valley or hillside forests, 1200–1800 m (Fu, 1976).

8. PODOPHYLLUM MAJOENSE

The application of this name has been problematical although it was unambiguously described with adequate herbarium material by François Gagnepain at Paris. The epithet *majoense* is derived from the Chinese town of Majo, near the type locality in Yunnan. However it appears to have been misread as relating to size and becomes 'majorensis' in Ying's (1979) review, where it is applied to a variety of *P. delavayi* with very long petals from Emei Shan (Mt. Omei), since described as var. *longipetalum*. Later Zhuang *et al.* (1993) applied the name to another different plant from Emei Shan, representing a then undescribed taxon (see 11. *P. trilobulus* below).

Separable from *P. delavayi* by the inflorescence positioned just below the leaf on the petiole, the reduced anther mucro and the much shorter petals with a rounded apex. *Podophyllum majoense* may represent hybrids between *P. delavayi* and *P. versipelle* in Guangxi as it occurs mainly towards the south-eastern limits of the range of *P. delavayi*. The type locality is adjacent to colonies of *P. delavayi*, which means that the two species with the longest and shortest petals in the genus are likely to be intimately linked. This raises the possiblity of some interesting genetics on the inheritance and control of petal length.

Podophyllum majoense Gagnep. in Bull. Soc. Bot. France 85: 167 (1938); Rix in Plantsman, 4(1): 10 (1982). Type: China, 'Kouy-tcheou [= Kweichow, but actually Yunnan, see Yeo, (1992)], environs de Majo', 2 May 1908, *Cavalerie s.n.* (holotype P!, isotype P!).

Dysosma majoense (Gagnep.) Hiroe in Pl. Basho's & Buson's Hokku Lit. 8(3): 328 (1973).

Dysosma majorensis (Gagnep.)Ying in Acta Phytotax. Sin. 17(1): 18 (1979), *pro parte excl.* f.2(3) & *Jiang & Zhuang* 30984, *comb. superfl.*

Dysosma majoense (Gagn.) Hsaio & Chen in Yin *et al*. in Acta Bot. Sin. 32(1): 45 (1990), *comb. inval. sine basionym.*

Dysosma lichuanensis Z.Cheng in Fu, Fl. Hupehensis, 1: 402, t.568 (1976) ined.; Su & Liu in J. Wushan Bot. Res. 10(4): 385 (1992); 11(1): 91–93 (1993); 12(2): 111–116 (1994); 12(3): 217–219 (1994), *nom. nud.*

Dysosma lichuanensis Z.Zheng & Y.J.Su in Acta Sci. Nat. Univ. Sunyatseni 36(2): 125–126 (1997). Type: Hubei, Lichuan, Xianfeng, Enshi, 'sub sylvis ad valles', 1500 m, *Li Hongjun* 11048 (holotype WHBI).

ILLUSTRATIONS. Fl. Hupeh. 1: t.568 (1976); Acta Sci. Nat. Univ. Sunyatseni 36(2): 125, f.1 (1997); New Plantsman 7(3): 149, pl.2C. (2000).

DESCRIPTION. *Perennial herb*, 15–50 cm; aerial stem striated, finely pubescent with cataphylls at base; rhizome thick, fleshy with short internodes. *Mature leaves* 2, peltate, dark green or purplish pigmented above, finely hairy beneath, especially on main veins; lobes 7–10 cm long, narrowed at base, deeply trilobulate; apical lobule enlarged; sinuses deep rounded, penetrating to over $^3/_4$ of radius; petioles to 20 cm; lower leaves to 20 × 15 cm with (4–)5–6 lobes; upper leaves to 15 × 10 cm with 4–5 lobes, the lobing severely retarded on adaxial side so that it appears truncate with an almost straight margin. *Inflorescence* of 2–5, usually 3, flowers with pedicels 8–10 mm, elongating in fruit, densely greyish-white pilose with simple hairs to 2 mm. *Buds* ovate, small, purplish. *Sepals* 4–6, very unequal, 10–12 × 6–7 mm, pale green, glabrous, elliptic. *Petals* reddish-purple, 6–7, elliptic-oblong, 11–15(–17) × 3–6 mm, apex obtuse, slightly irregular. *Stamens* 5–7, *c*. 12 mm long; filament short 2–3 mm; anther 8–10 mm with a conspicuous apical mucro longer than the width of the anther, 2–3 mm. *Ovary* elongated, ellipsiod, *c*. 5 mm long, constricted at base; stigma peltate, 1.5 mm wide. *Berry* ovoid, apex somewhat acute with remains of stigma, red when ripe.

DISTRIBUTION. China, sporadic occurrence in eastern Yunnan, Guizhou, Hubei and Guangxi provinces.

ECOLOGY. Grows at 1300–1650m, in dense forest (Ying, 1979). In Hupeh, reported to occur in forest along valleys, around 1500m (Fu, 1976).

9. PODOPHYLLUM GUANGXIENSIS

An evident interest in herbal plants led Wang Yu-shen to describe this *Podophyllum* as a new species, while working at Guilin Medical college in 1984, based on a collection made back in 1958. It has not been collected very much, in fact the only other specimen known to me is even earlier. Dating from 1933 is a collection made by W.T.Tsang (no. 22885, GH, SYS) from Guangxi adjacent to the Guangdong boarder on Tong shan near Sap-luk Po village in Waitsap district. The field notes read, "Fairly common; steep slopes, silt, sandy soil, swamp, scattered shrubs, meadow; 1.2 ft, used as medicine. 25. x. 1933".

One feature that quickly distinguishes this species from all other *Podophyllum* is the parallel sided leaf lobes and sharply angled sinuses. It may represent a hybrid between *P. versipelle* and a member of the *P. delavayi* group. It is not known to be in cultivation.

The epithet *guangxiensis*, refers to Guangxi, the Chinese province in which this plant was found.

Podophyllum guangxiensis (Y.S.Wang) J.M.H.Shaw in New Plantsman 6(3): 163 (1999) [as *guangxiense*]. Type: China, Guangxi, Debao Xian, Yangdong Xiang, Qinja Cun, 1000 m, in woods, 28 November 1958, *Li Zhong-ti* 602053 (holotype IBK, not seen).
Dysosma guangxiensis Y.S.Wang in Guihaia 4(1): 43–44, t. (1984).
Dysosma hispidum sensu S. von R.Altschul, Drugs & Foods from little known Plants 70, no. 980 (1973), *non* Hao.

ILLUSTRATION. Guihaia 4(1): 44 (1984); New Plantsman 7(3): 149, pl.2 B, (2000).
DESCRIPTION. *Perennial herb* to 40 cm; aerial stem with brown, lanceolate, 2.5 × 1.2 cm cataphylls at base, slightly pubescent towards apex; rhizome thick, internodes short. *Juvenile leaves* unknown. *Mature leaves* 2, thick, leathery, peltate, suborbicular, 16–25 cm across, glabrous; lower leaf with (4–)6–7 lobes; upper leaf with 5–6 lobes; lobes oblong, parallel sided from sinus to lobules, longest lobe to 15 cm from apex to petiole, occurring in the centre of the abaxial side of lower leaf; apex trilobulate; midribs conspicuous on underside; margin with sparse mucronate, 1 mm teeth, 5–10 mm apart; petioles to 15 cm long, glabrous. *Inflorescence* with 5–6 flowers; pedicels densely pubescent, 3–4 cm. *Sepals* greenish, obovate. *Petals* 6, reddish-purple, spathulate to obovate, 2.5 × 1–1.2 cm. *Stamens c.* 2 cm; filaments 4–5 mm; anther 10–12 mm with apical mucro to 2 mm. *Ovary* ellipsoidal to inverted pear-shaped, glabrous; style narrow; stigma large, peltate. *Berry* not known.
DISTRIBUTION. China, Guangxi province, moist woodland. A rare local endemic.

Series B. DELAVAYAE

(Species 10–12)

Series *Delavayae* J.M.H.Shaw **ser. nov.** Petala longa (ad 10 cm longa) et antherae connectivo in mucrones amplae producto. Type species: *Podophyllum delavayi* Franch.

The plants of this series are distinct with their unusually long petals and anther connectives extended to form well developed mucros.

10. PODOPHYLLUM DELAVAYI

This species, with the most spectacular flowers in the genus, is now settling down in cultivation, and there is the likelihood of several more introductions. Some forms have good colouring in the new leaves, with the juvenile leaves making for particularly interesting foliage. Plants now in cultivation represent several accessions. A good foliage form collected on Emei Shan (Mt. Omei) by Dan Hinkley has since been propagated by tissue culture at Heronswood Nursery in North America. *Podophyllum delavayi* appears to introgress with *P. hexandrum* in the north-west of its range, with *P. mairei* in Yunnan and *P. versipelle* subsp. *boreale* in Sichuan.

The specific epithet *delavayi*, commemorates Jean Marie Delavay (1834–1895) a French missionary and botanist who collected mainly in north-western Yunnan province from 1882 to 1889.

Podophyllum delavayi Franch. in Bull. Mus. Natl. Hist. Nat. 1: 63 (1895); Diels in Bot. Jahrb. Syst. 29(2–5): 336 (1901); Rix in Plantsman 4(1): 10 (1982). Type: W.China, Yunnan province, 'in silvis circa Long-Ki', *Delavay* 5087 (holotype P!, K photo.!, isotype P!, K!, HUH!).

P. veitchii Hemsl. & E.H.Wilson in Bull. Misc. Inform., Kew 1906: 152 (1906); Dunn,
    Suppl. List Chinese Fl. Pl. 1904–1910, in J. Linn. Soc., Bot. 39: 477 (1911); Rix in
    Plantsman, 4(1): 10 (1982). Type: China, W. Sichuan, in woods and forests, 2500 m,
    Wilson 3170 (holotype K!, isotype K!, BM!, P photo.!).
[*Begonia* ?sp. J.C.Williams, Some plants, shrubs and trees found by Mr. Forrest in 1924
    (1925)].
[*Dysosma pleiantha* (Hance) Woodson in Ann. Missouri Bot. Gard. 15: 339 (1928) *pro
    parte, quoad P. veitchii* in syn.]
Dysosma delavayi (Franch.) Hu in Bull. Fan Mem. Inst., Bot. Ser. 8: 37 (1937).
[? *Podophyllum peltatum sensu* S.Hu *non* L., in J. W. China Border Res. Soc. 15, ser. B:121
    (1945)].
Dysosma veitchii (Hemsl. & E.H.Wilson) Fu [in Icon. Cormoph. Sinic. 1: 759, t.1518
    (1972) *comb. inval. sine basionym*] validated by Ying in Acta Phytotax. Sin. 17(1): 20
    (1979); Icon. Cormoph. Sinic. ed.2, 1: 759, t.1518 (1980); Wu, Wild Fl. Yunnan 2:
    308–9, t.67 (1987); Chen, Ji-Sheng Zheng, Chinese Toxic Pl.: 135, t.2.4 (1987).

10. PODOPHYLLUM DELAVAYI

Fig. 109. **Podophyllum delavayi.** A form originating from the Chinese nursery of Kaichen. Photograph: Jim McClements.

[*Sinopodophyllum emodi* (Wall. ex Honigs.) Ying in Acta Phytotax. Sin. 17(1): 16 (1979) pro parte, quoad *P. delavayi* in syn.]

ILLUSTRATIONS. Icon. Cormoph. Sinic. ed. 2, 1: t.1518 (1980); Fl. Yunnanica 7: 6, t.2 (1–5) (1997); Hinkley, The Explorer's Garden: 118 (1999); New Plantsman 7(3): 149, pl.2, E1–8, (2000).

DESCRIPTION. *Perennial rhizomatous herb*; rhizome short with short internodes, scaly, with many adventitious roots; terminal bud with scales. *Stems* 10–20(–35) cm, grooved, occasionally with scattered hairs, especially near the inflorescence. *Leaves* 1–2, often mottled dark green and dark brown-purple, characteristically deeply lobed with noticeably trilobulate apices. *Juvenile leaves* borne singly on a 5–10 cm stem from the rhizome, peltate, *c.* 10–14 cm diameter, with 4–5 lobes, often subsquarose; lobes broadly triangular or obovate, undivided with an acute apex; sinus rounded, rather shallow, 1–4 cm deep, or reduced to a slight concave curve in the lamina edge; leaf margin mostly smooth with a few mucronulate teeth spaced *c.* 5 mm apart. *Adult leaves* peltate, bilaterally symmetrical, glabrous above, glabrous or subglabrous below, with 5–8 deep lobes narrowed towards the base, sinuses rounded, keyhole shaped, rarely acute, penetrating from $2/3$ to $3/4$ of the radius; lobes usually divided into three lobules at the apex; central lobule from $1/4$ to $1/2$ the length of the lobe; lateral lobules usually smaller, with small usually rounded sinuses between the lobules; first leaf larger with (6–)7(–8) lobes, lobules often small and irregular on the adaxial lobes, or absent, the adaxial lobes then triangular-ovate; second leaf smaller with (3–)4–5 lobes, the adaxial edge of the lamina usually reduced to an almost straight edge with irregularly spaced small teeth. *Inflorescence* of 1–3(-6) flowers, in the fork of the petioles, flower size rather variable. *Pedicels* pendulous, 1–2(–4) cm long, pubescent to pilose, often densely so. *Sepals* 6, membranous, fugacious, oblong-ovate, 1–2.5 cm long, the outer 3 narrower, shorter and pubescent on the outer surface with stiff white hairs, the inner sepals occasionally petaloid. *Petals* 6(–9) spathulate, or lanceolate tapering to a fine point, dark purple or deep pink, 1.5–10 cm long, up to 9 mm wide. *Stamens* (5–)6(–7), 12–20 mm long, thick and rigid, filaments short *c.* 4 mm long, flattened at the base; anther thecae 10–13 mm long with the connective produced into a mucro 5 mm or more long. *Ovary* glabrous, slender oblong-ellipsoidal, *c.* 6 mm long with a long slender neck up to 6 mm; stigma peltate. *Fruit* globose-elliptic, up to 3.5 cm long and 2.5 cm wide, distinctly stipitate in some collections, with remains of stylar neck at apex, probably dark red when ripe. *Seeds* and cotyledons unknown. 2n = 12 (Ma & Hu, 1996).

DISTRIBUTION. Western China, widely distributed in Sichuan, Yunnan, southern Shaanxi and Guizhou provinces.

ECOLOGY. Growing in mountains amongst thick forest (Icon. Cormoph. Sinic., 1980). Riverine forest around 1500 m (Fu, 1976). Common in places, forming colonies in woodland.

Key to varieties of *P. delavayi*

Petals 3.5–7 cm long, 3–5 mm wide .... var. **delavayi**
Petals 8–10 cm long, *c.* 9 mm wide .... var. **longipetalum**

var. **delavayi**

Typical populations are easily distinguished from any other *Podophyllum* by the long narrowly lanceolate to ligulate petals, remarkable stamens with a long, thin mucro, pilose pedicels and leaves with deep, trilobulate lobes. The type collection is widely distibuted in herbaria. It consists of small plants only about 10–15 cm high with leaves 10–12 cm across and ligulate dark purple-red petals *c.* 3.5 cm long.

var. **longipetalum** J.L.Wu & P.Zhuang ex J.M.H.Shaw in New Plantsman 6(3): 163 (1999). Type: China, Sichuan, Emei Shan, *Jiang & Zhang* 30984 (holotype PE).
[*Dysosma majorensis sensu* Ying in Acta Phytotax. Sin. 17(1):21 f.2 (3), (1979)].
Dysosma veitchii var. *longipetalis* J.L.Wu & P.Zhuang in J. Wuhan Bot. Res. 11(1): 41–46 (1993) *nom. nud.*

ILLUSTRATION. Acta Phytotax. Sin. 17(1): 21, f.2 (3) (1979).

This variety differs from typical *P. delavayi* by its longer linear-lanceolate petals, 8–8.5(–10) cm long × *c.* 9 mm wide, shorter (10 mm), curved anthers with shorter apical mucro, and a typically red ovary. The inflorescence is inserted in the fork of the petioles.

DISTRIBUTION. Apparently endemic to Emei Shan around 2200 m.

Informal list of variants of *P. delavayi*

There appear to be two groups of variants, one with narrow ligulate, relatively long petals and a second with broadly lanceolate, relatively short petals. Wide variation in petal colour has been seen in both groups. It is not known to what extent the groups intergrade and further study is necessary before they can be taken up and formally described.

A. A white flowered form with long ligulate petals is known from Cang Shan, Dali, Yunnan and illustrated in Kaiyun, G., *Highland flowers of Yunnan*: 37, pl. 52 (1998).

B. Plants with an entirely glabrous inflorescence are known from the east slope of Gongga Shan, Sichuan, in the Dafengding Natural Reserve. The petals are pink, relatively short, broadly lanceolate and the anther mucro is rather short for this group. A colour illustration appears in Tangjun, Jian, *The rare plants and flowers of Western Sichuan*: 69, t.56 (1984). It may be the plant treated as *D. veitchii* in Liu, Zhaoguang, *Flora of Gongga mountain*: 194 (1983).

C. A white flowered form with broad lanceolate petals is known from cultivation and distributed in North America by Barry Yinger who obtained his plants from a source in Japan. It appears to be different from var. A (above) but, as the petals had not fully expanded when examined, there is a measure of uncertainty.

D. Pink and red flowered forms with broad, lanceolate petals. Many collections display flowers in varying shades of pink to dark purple-red. A collection from Yunnan is illustrated in Wu, *Wild Fl. Yunnan* 2: 308–9, t.67 (1987).

Other forms are known from incomplete collections of this variable species. Most collections in herbaria are incomplete, consisting of leaves only, and consequently it is not possible to assess accurately the significance of the variation encountered at present.

10. × 7. PODOPHYLLUM DELAVAYI × P. DIFFORME

Plants which are thought to represent this hybrid appeared amongst a batch of *P. difforme* imported from a Japanese nursery by Bleddyn Wynn-Jones of Crûg Farm Nursery, North Wales. It has many features in common with *Podophyllum* sp. A (below), but lacks the compound inflorescence caused by branched pedicels and the deep lobing and lobules of the leaf. So far it is known only from cultivated plants but these probably originated from wild collected stock.

P. delavayi × P. difforme

DESCRIPTION. *Aerial stem* green, hirsute with linear hairs 1–2 mm long. *Juvenile leaves* radially symmetric, with 5–7 very shallow broadly triangular lobes, pigmented with

10. × 7. PODOPHYLLUM DELAVAYI × P. DIFFORME

Fig. 110. *Podophyllum delavayi* × *P. difforme*. Photograph: Bleddyn Wynn-Jones.

mottled greens and browns. *Mature leaves* 2–3, with 4–6 shallow triangular lobes, the lobing retarded on adaxial side especially on the upper leaf; margin with few small teeth 5–10 mm apart; lobules absent. *Inflorescence* with 5–9 flowers, densely clustered beneath a small third leaf or held on an elongated peduncle *c*. 15 cm long above the upper (second) leaf; pedicels 2–5 cm, hirsute. *Sepals* 6 with scattered hairs on the exterior; 3 outer sepals narrowly linear-lanceolate, shorter, about $1/2$–$1/3$ as long as the bud; 3 inner sepals broadly lanceolate, equalling length of bud. *Petals* 6, rich maroon-red to wine-coloured, 6–7 cm long, narrowly lanceolate, forming an elongated funnel-shaped chamber. *Stamens* 6, about $1/3$ as long as petals, *c*. 2–3 cm; filaments fleshy, deep maroon; connective lanceolate with elongated thecae; apex projected into a mucro *c*. 4–5 mm long. *Ovary* spherical, green with some reddening at apex; style deep red; stigma brownish-orange corrugated, almost as wide as ovary. *Berry* probably spherical, seeds not seen.

DISTRIBUTION. Known only from cultivated plants, probably propagated in Japan from wild collected material. Possibly collected in mainland China.

11. PODOPHYLLUM TRILOBULUS

Putative hybrid, perhaps involving *P. versipelle*, with an appearance similar to *P. delavayi*, but differing in the following characters: mature leaves with pronounced tri-lobulated lobe apices; inflorescence inserted at or above middle of upper leaf petiole. Although the holotype at Edinburgh lacks flowers there are additional, more complete specimens in other herbaria which help establish this taxon, including *S.Y.Yu* 49384 (SZ!).

The epithet *trilobulus*, Latin, *tri-* (three), *lobulus* Latin (a small lobe), refers to the shape of the leaf lobes.

Podophylum trilobulus J.M.H.Shaw in New Plantsman 7(3): 158–159 (2000). Type: China, Sichuan province, Nanchuan-hsien, road side, 8–9000 ft., 31 May 1928, *W.P.Fang* 1165 (E).
[*Dysosma majorensis sensu* J.L.Wu & P.Zhuang in J. Wuhan Bot. Res. 11(1): 41–46 (1993) *non sensu* Ying, 1979, *non* Gagnepain, 1938].
P. sp. C. J.M.H.Shaw in New Plantsman 6(3): 158, 165 (1999).

ILLUSTRATION. New Plantsman 7(3): 149, pl.2 F, (2000).

DESCRIPTION. *Plant of delicate appearance with annual stems* 20–25(–40) cm high, stem and petioles covered with fine, short linear hairs. *Leaves* two, with trilobulate lobes; marginal teeth fine, 5–9 mm apart where present on upper parts of lobes. *Lower leaf* 18(–28) cm across with 7 lobes; sinuses penetrating to $^2/_3$ radius, rounded and key-hole shaped. *Upper leaf* 15 cm wide with 5 lobes, lobes strongly retarded on adaxial margin, reduced to a few large teeth. *Inflorescence* with (2–)3–4(–5) flowers inserted at or above midpoint on petiole of upper leaf. *Pedicels* 2.2–2.5 cm long, dilated at apex, with brown pilose, linear hairs. *Petals* purple-red, ovate-lanceolate, tapering gradually to a point, 4–5 cm long, 8–10 mm wide. *Anthers c.* 6 mm.

DISTRIBUTION. China, Sichuan province. Known from Emei Shan, 1600 m. Similar plants have been collected from Jingfu Shan (Zhuang *et al.*, 1993) and other sites. Rather local and rare.

ECOLOGY. Details of ecological requirements and pollen fertility are given by Zhuang *et al.*, (1993).

12. PODOPHYLLUM SP. A

Emei Shan (Mt. Omei), a large mountain system in Sichuan province, China, has long been regarded as sacred and consequently has retained its natural vegetation with little disturbance. A tortuous pathway containing many steps leads to a summit on one of the peaks. It is to the margin of this ancient causeway that most botanical visitors have confined their collecting due to the dangerous nature of the terrain. In recent years efforts to collect from other parts of the mountain have yielded many new species. A recent study by Zhuang *et al.* (1993) indicates the existence of *Podophyllum* populations, some of which appear to be new taxa. In response to my enquiry, Ping Zhuang of the Natural Resources Institute of Sichuan at Chengdu, kindly supplied a line drawing of this taxon along with a key to *Podophyllum* on Emei Shan, in Chinese, which was kindly translated by Donglin Zhang and appeared in *The New Plantsman* 6(3): 164–165 (1999). It seems likely that, along with *P. trilobulus*, this taxon represents populations of stable nothomorphs derived from *P. versipelle* × *P. delavayi*. A number of living collections are now in cultivation from Emei Shan and require further investigation.

Podophyllum sp. A.

Dysosma emiensis J.L.Wu & P.Zhuang in J. Wuhan Bot. Res. 11(1): 41–46 (1993) *nom. nud.* *P. sp. B.*, J.M.H.Shaw in New Plantsman 7(3): 158 (2000).

DESCRIPTION. *Perennial herb* to 40–50 cm with thick, creeping rhizome; internodes short; cataphylls present at aerial stem base; stem, petioles and pedicels pubescent. *Juvenile foliage* unknown. *Mature leaves* 2; lower leaf 6–7 lobed; lobes narrow, shallowly trilobulate; sinus penetrating to $^2/_3$ of radius; main veins prominent on underside, pubescent. *Upper leaf* 5–6 lobed; some lobes elobulate, lobing retarded on adaxial margin; margin with fine hairs and small irregular teeth *c.* 5–10 mm apart. *Inflorescence* of 10–13 flowers, emerging at or above middle of upper leaf petiole, apparently subtended by a small bract, compound umbellate with some flowers on solitary pedicels and others branching from peduncles. *Sepals* 10–16 mm, narrowly elliptic, fugacious. *Petals* ovate-lanceolate, 6.5–7.5(–8.5) cm long, obtuse at apex, irregularly curving and spreading at the tips. *Stamens* 6; filaments 10–14 mm; anther *c.* 8 mm with apical mucro slightly curved at the tip, 3–4 mm. *Ovary* globose to ellipsoidal. *Mature berry* and seeds unknown.

DISTRIBUTION: China, Sichuan, known only from Emei Shan, 1800 m.

ECOLOGY. Details of ecology and pollen viability are given in Zhuang *et al.* (1993).

Section iii. PARADYSOSMA

(Species 13)

Section *Paradysosma* J.M.H.Shaw in New Plantsman 6(3): 163 (1999).

Stems reddish-orange. *Leaves* with golden-brown hairs on both surfaces, lobes elobulate; margins with several sizes of mucronate teeth and simple hairs. *Flowers* white, creamy-yellow or rose-pink, or flesh coloured but not dark reddish-purple; petals becoming reflexed with age. *Stamens* with or without a filament; mucro absent or minute. *Pollen* in monads. *Ovary* ellipsoidal to spherical; fruit spherical to pear-shaped. One variable species from Sikkim to southern Yunnan.

This section combines the leaf shape of section *Dysosma* with many of the floral characters of section *Hexandra*. This intermediate position requires the retention of both section *Dysosma* and *P. hexandrum* within *Podophyllum*. The section name *Paradysosma*, Greek *para-* (alongside-), *dysosma* (the name of another section in the genus) is derived from its perceived systematic position.

Podophyllum aurantiocaule is possibly the rarest *Podophyllum* in cultivation while *P. hexandrum* is the most widely grown and is best known for its large red fruits in late summer. The relationship between the two species remains a matter of conjecture with *P. aurantiocaule* being so little known. There are intriguing similarities between the two which require further work to understand.

13. PODOPHYLLUM AURANTIOCAULE

Several famous plant collectors, including Kingdon Ward, George Forrest and Reginald Farrer have provided herbarium specimens of this species, but alas, the introduction of living material had to wait until recently. Mature fruits of *Kingdon Ward* 8265 were reported as "hanging clusters of large pear-shaped scarlet fruits" (Kingdon Ward, 1930a). Kingdon Ward also writes "the large scarlet balloons ... were conspicuous", which evidently refers to the fruits and not the flowers since this was seen by him in autumn (September–October) (Kingdon Ward, 1930b). Seeds of this species, as with most others of the genus with straw-coloured, reniform seeds, are notoriously susceptible to desiccation; even if they had been collected by Kingdon Ward it is unlikely that they would have remained viable.

Chris Sanders of Bridgemere, Cheshire, collected live plants from the south side of Doshang La, [29°29'N, 94°29'E] in south-eastern Tibet at 3048 m (10,000 feet); these have since flowered in cultivation. Plants introduced the previous season and cultivated at Bridgemere perished during a hard frost although in a frame. I have an unconfirmed report of its successful cultivation in Sweden. Seed purported to be this species is sometimes offered in lists, but invariably turns out to be *P. hexandrum*.

The type specimen of *P. sikkimensis* at CAL is in a poor state of preservation, which has led some workers to conclude that it represents *P. hexandrum*. However the isotype at BM, which has been examined thoroughly both by W.T.Stearn and myself, leaves no doubt that it is conspecific with *P. aurantiocaule*. *Podophyllum sikkimensis* is based on a collection by one of Dr. King's anonymous collectors from the Chumbi valley region. Plants from this area and SE Tibet exhibit deeper leaf lobes than their relatives from Arunachal Pradesh, which can cause them superficially to resemble *P. hexandrum*, particularly when the leaf is only partly unfurled, as is the case with the type of *P. sikkimensis*.

The length of stamen parts varies considerably both within and particularly between the subspecies. Subspecies *aurantiocaule* has the shortest filaments which in some

Fig. 111. ***Podophyllum aurantiocaule*** subsp. ***aurantiocaule*** soon after emerging. Photographed in south-eastern Tibet by Chris Sanders.

collections (*L & S* 1660, *L, S & E* 13572) are either very short (<0.5 mm) or apparently absent. There does not appear to be much of a cline, with filaments in most collections in the range 1.5–2 mm long and an average anther length of 6.2 mm. A few collections (*King's collector s.n.*, one plant on *L, S & E* 13572, *Handel-Mazzetti* 9242) have the inflorescence inserted along the petiole, instead of in the fork. They are included in subspecies *aurantiocaule* in view of other concurrent factors including the leaf lobing, anther length and geographical location. In the case of *Handel-Mazzetti* 9242 (type of *P. aurantiocaule*), the inflorescence is very near the fork. *L,S & E* 13572 is of interest since one plant on the sheet has the inflorescence in the typical position. Probably this character is unstable in some populations, as it is in *P. hexandrum*, with which this species is partly

Fig. 112. ***Podophyllum aurantiocaule*** subsp. *aurantiocaule* photographed in its natural habitat in south-eastern Tibet by Chris Sanders.

sympatric and may well have introgressed. In subspecies *furfuraceum* the anther length is shorter than subspecies *aurantiocaule*, with an average value of 4 mm (as opposed to 6.2 mm). Some anthers on the type (*Yu* 15977) appear unusually short; this is due to damage and is probably the reason for the extremely short anthers depicted in the illustration accompanying the original description. Dissection of a mature bud revealed anthers of normal size (3.8 mm).

There seems to be some confusion over the flower colour of subspecies *furfuraceum*. The type collection (*Yu* 15977) from Yunnan, bears on the data label "flowers dark purple", which is followed in the published description "flores purpurei", whereas both isotypes examined from Harvard have white or at most pale pink flowers; these dry a dirty white-brown, while dark purple flowers retain much of their colour. Another Yunnan collection, *1984 Sino-Amer. Bot. Exped.* 320, is in fruit. On both herbarium sheets examined (E, HUH) the fruits are immature but have clearly been developing for some weeks and bear no trace of the petals and yet the label reads "flowers red" which appears to be guesswork. (Compare colour illustration in Kaiyun 1998, listed below). The available evidence indicates that dark purple-red flowers are unknown in this species.

Subspecies *furfuraceum* may introgress with *P. mairei* and be the influence for the open flower with short petals.

There has been much confusion of this species with *P. hexandrum*. For example the key in *Flora of India* (Rao & Hajra, 1993) uses characters which are not mutually exclusive:

"Scales of rootstock membraneous; pedicels erect; flowers usually solitary .... *P. hexandrum*
Scales of rootstock chartaceous; pedicels drooping; flowers 2 .... *P. sikkimensis* [*P. aurantiocaule*]"

All the character states in this couplet apply to *P. hexandrum* except 'Flowers 2'. Presumably there are only the two type sheets of *P. sikkimensis* at CAL, which are very old and in poor condition. A better separation could be obtained using characters from leaf shape, marginal teeth, indumentum and pollen.

Most Chinese publications apply the name *P. aurantiocaule* to plants from central and southern Yunnan with shallow lobes and the inflorescence inserted along the petiole, which are here treated as subsp. *furfuraceum*. Being in fruit, the type specimen of *P. aurantiocaule* from north-western Yunnan is without floral parts but has deeply lobed leaves and the inflorescence inserted near the petiole fork; as such it appears to relate to other specimens from northern. Yunnan, south-eastern Tibet and Bhutan with similar characteristics which have for some time been known as *P. sikkimensis* in western herbaria and were more recently described as *Dysosma tsayuensis* in China. It appears that in some Chinese publications which recognise all three species, the name *P. aurantiocaule* may be applied to specimens from northen Yunnan which are intermediate in character between the two subspecies.

The epithet *aurantiocaule*, Latin *aurantiacus* (orange), *caulis* (stem), recalls the colour of the stem; while *furfuraceum*, Latin *furfuraceus* (scurfy, literally 'like bran') recalls the dense covering of short stubby trichomes on the main veins of the leaf undersurface.

Podophyllum aurantiocaule Hand.-Mazz. in Anz. Akad. Wiss. Wien Math.-Naturwiss. Kl. 61: 163–164 (1924). Type: China, N.W. Yunnan province, 'Prope fines Tibeto-Birmanicas inter fluvios Lu-djiang (Salwin) et Djiou-djiang (Irrawadi), in bambusetis temperatis lateris orient. jugi Tschiangschel, 27.52 N, s.microschistaceo', 3275 m, 3 July 1916, *Handel-Mazzetti* 9242 (holotype W!).

DESCRIPTION. *Perennial rhizomatous herb* with annual stems up to 90 cm tall; rhizome short and thick, *c.* 2 cm in diameter, scaly, with short internodes, chestnut-brown, with circular annual stem scars, *c.* 5 mm in diameter present in a row along upper side; cataphyllary scales present at base of stalk. *Stem* glabrous, grooved and orange-brown colour in the dried state. *Leaves* 1–3, peltate, 15–30 cm diameter, depressed at the centre with 5–7(–9) lobes, lamina with sparse or dense indumentum of golden-brownish hairs, 0.8–1 mm long on both surfaces, usually more dense above; veins prominent on underside with short, dense indumentum; leaf margin with regularly spaced, variable, mucronate teeth, 2–8 mm apart, 0.5–5 mm long, and dense golden-brown hairs; lamina thin and papery when dry, usually dark olive green on upperside, pale green beneath, described as bright green or green mottled with dark purple when young. *Juvenile leaves* borne singly on stems from the rhizome, *c.* 9 cm in diameter, peltate, depressed at the centre, with 4–6 triangular lobes, more or less radially symmetrical. *Adult leaves* 2–3, on petioles of subequal length, usually similar in size (15–)25–30 cm in diameter; lobes (3–)5–7(–9), broadly triangular to ovate or obovate to oblong-lanceolate with an acute to acuminate apex, elobulate, with occasionally one pair of teeth each side of the apex slightly enlarged; sinuses narrow, acute to narrowly keyhole-shaped or broad and triangular, penetrating from $1/3$ to $2/3$ of the radius. *Inflorescence* of 1–9 flowers, sometimes borne on a short, thick peduncle <1 cm long, cither in the fork between the petioles or along the petiole of the second leaf. *Pedicels* 2–11 cm long, thin, glabrous, erect at anthesis if this occurs before the leaves expand (as in *P. hexandrum*), or deflexed later as leaves expand. *Sepals* 6, light green, 10–12 × 3–4 mm, narrowly ovate to oblong, glabrous, fugaceous. *Petals* 6, white, creamy-yellow or cream faintly tinged pink at the base or rose-pink throughout, spathulate to ovate or obovate with a rounded apex, 15–20 × 10–12 mm at widest point. *Stamens* 6, filaments short or almost absent, to 4 mm, flat with midrib visible; anther 3–9 mm long, yellow or pale blue; mucro usually absent, occasionally present as a minute projection of the connective, 0.2–0.4 mm long; pollen released in monads. *Ovary* spherical to ellipsoidal, green or purplish, 6–12 × 3–5 mm with a narrow, rose-pink style 1.5 mm wide, <1 mm long; stigma large, peltate, 3–4 mm broad. *Fruit* a pear-shaped to spherical berry, reddish-orange when ripe, 2–3 cm diameter. *Seeds* many, mature seed and cotyledons not known.

Key to subspecies of *P. aurantiocaule*

Leaf lobes long, obovate to lanceolate; sinuses deep, penetrating $1/2$ to $2/3$ of radius; flowers usually in petiole fork; anthers 5–7.5(–9) mm long .... subsp. **aurantiocaule**
Leaf lobes short, triangular, sinuses shallow penetrating $1/4$ to $1/3$ of radius; flowers borne on petiole; anthers 3.5–4.5 mm .... subsp. **furfuraceum**

subsp. **aurantiocaule**

P. aurantiocaule Hand.-Mazz. in Anz. Akad. Wiss. Wien Math.-Naturwiss. Kl. 61: 163–164 (1924) & Symb. Sin. 7(2): 323 (1931) *pro parte*, excl. Forrest 11897, 24232; Fischer in Bull. Misc. Inform., Kew 1937: 474 (1938); Naithani, Fl. Pl. India, Nepal, Bhutan: 29 (1990); Sharma & Balakrishnan, *Fl. India* 1: 416 (1993).

[*P. versipelle sensu auct. non* Hance: Spare & Fischer, Bull. Misc. Inform., Kew 1929: 249 (1929); Fischer, Bull. Misc. Inform., Kew 1937: 474 (1937); Kingdon Ward, Pl. Hunting on the Edge of the World: 327 (1930) & in Gard. Chron. 1934: 387, f.153 (1934); Rix, Plantsman 4(1): 9 (1982) *pro parte*; Naithani, Fl. Pl. India, Nepal, Bhutan: 29 (1990) [as *versipella*]; Sharma & Balakrishnan, Fl. India 1: 416 (1993)].

Dysosma aurantiocaule (Hand.-Mazz.) Hu, [as *aurantiocaula*] in Bull. Fan. Mem. Inst., Bot. Ser., 8: 37 (1937); Ying in Acta Phytotax. Sin. 17: 19 (1979) *pro parte*, excl. *P. mairei*.

P. sikkimensis Chatterjee & Mukerjee [in Econ. Bot. 6(4): 344, t.7 (1952), *nom. nud.*] & Rec. Bot. Surv. India 16(2): 48, t.1 (1953); Grierson & Long, Fl. Bhutan 1(2): 329 (1984) [as *sikkimense*]; Agarwal & Ghosh, Drug Pl. India Root Drugs: 206 (1985); Naithani, Fl. Pl. India, Nepal, Bhutan: 29 (1990); Sharma & Balakrishnan, Fl. India 1: 415 (1993). Type: Sikkim, Shu-lam-bee, which is a little above Ling-too, about 10,000 ft., 22 June 1882, *King's collector s.n.* (holotype CAL, isotype BM!).

P. sikkimensis var. *major* Chatterjee & Mukerjee [in Econ. Bot. 6(4): 344, t.8 (1952), *nom. nud.*] & Rec. Bot. Surv. India 16(2): 48–49 (1953). Type: Chumbi, Kungaloo, 8 July 1884, *King's collector s.n.* (holotype CAL).

[*P. hexandrum sensu* Hara, Fl. Eastern Himalaya: 34 (1971) *pro parte, quoad P. sikkimensis* in syn.; *non* Royle]

[*Sinopodophyllum emodi* (Wall.) T.S.Ying in Acta Phytotax. Sin. 17(1): 16 (1979) *pro parte, quoad P. sikkimensis* in syn.]

Dysosma tsayuensis T.S.Ying in Acta Phytotax. Sin. 17(1): 20, t.1 (1979); Fl. Xizangica 2: 120 (1985). Type: Tibet, Cha-Yu, *Tsinghai-Tibet Exped.* 73–619, 73–986 (PE).

P. sikkimensis var. *emodi*, an unpublished combination, is listed as a synonym of *P. sikkimensis* by Sharma & Balakrishnan, Fl. India 1: 415 (1993). This appears to be an error, *P. sikkimensis* var. *major* Chatterjee & Mukerjee was evidently intended.

[*P. aurantiacum* hort. An orthographic error for *P. aurantiocaule* which appears in some seed lists.]

ILLUSTRATIONS. Gard. Chron. 1934: 387, f.153 (1934); New Plantsman 7(1): 38; 7(3): 155; 7(4): 223, fig.1, 1–2. (2000); Cox (ed.), F. Kingdon Ward's Riddle of the Tsangpo Gorges: 159, 160 (2001).

DESCRIPTION. *Leaf lobes* obovate to lanceolate, apex acute, with deep rounded sinuses penetrating $1/2$–$2/3$ of radius. *Inflorescence* of 1–7(–16) flowers, usually in the axil of the petioles. *Stamens* with short filaments (0–)0.5–2(–4) mm long, pressed together around style; anthers (4.5–)5–7.5(–9) mm long.

DISTRIBUTION. From Sikkim to northern Yunnan, including Chumbi valley, Bhutan, Tsangpo gorge, centered on S.E. Tibet. Endemic to Sino-Himalayan region.

ECOLOGY. Deciduous and mixed forest, often in wet areas beneath *Rhododendron* and conifers, 2734–3353 m (7–11,000 feet); open stoney alpine pasture amongst dwarf scrub and margins of bamboo thickets, 3353–3962 m (11–13,000 feet), said to be locally common in S.E.Tibet. An extensive population scattered over 200 km along the Dibang valley has recently been located in Arunachal Pradesh by Mithilesh Pathak of the Indian Botanical Survey. He reports that Black wasps and Indian Honey Bees, both of which produce honey have been observed visiting the flowers. The growing season is only about three months. Plants appear above ground from early March, flower during late March and become dormant by late June. Plants seen at Aliny, 1200 m altitude, bore mature red fruits in April while those at Myodia pass, 2700 m altitude, were beginning to flower. By late June there was no trace above ground at Alinye, while the Myodia Pass colony was in fruit.

This is predominantly a wetland species. When growing on slopes it occurs along watercourses usually in full sun or light shade. It has been found in deep shaded forest at Chigupani and also persists well after woodland is removed for pasture. However the aerial stem and leaves are toxic to cattle. Elsewhere in Arunachal Pradesh plants have been collected from Jabrang to Perila, Kameng and the Delei valley, Lohit.

Two varieties of subsp. *aurantiocaule* can be distinguished as follows:
Flowers several .... var. **aurantiocaule**
Flowers solitary .... var. **uniflorum**

var. **aurantiocaule**

Distinguished by multiple flowers in each inflorescence. Throughout the range of the subspecies.

var. **uniflorum** J.M.H.Shaw in New Plantsman 6(3): 163–164 (1999). Type: Bhutan, *Ludlow & Sherriff* 3098 (holotype BM, isotype BM, E photo.)
P. sikkimensis (*pro parte*) Grierson & Long, Flora of Bhutan 1(2): 329 (1984).

DESCRIPTION. *Inflorescence* differs from that of var. *aurantiocaule* in producing only a solitary flower on each sympodial unit, which is usually inserted along the petiole near the fork. *Leaves* very shiny green on upper surface. *Petals* rich pink towards base, gradually becoming pale pink towards apex. *Style* deep pink.

Floral presentation is very similar to *P. hexandrum* with the flower opening above the partly furled leaf and appearing to be apical. This population may be introgressed or represent an intermediate stage in its derivation. It is possible that a single clone has been named here.

DISTRIBUTION. Known only from an isolated population in Central Bhutan, Tongsa district, Phobsikha, 3353 m (11,000 feet). Growing in wet patches under *Rhododendron* and conifers.

The name *uniflorum*, latin *uni-* (one) *flos, floris* (flower), recalls the solitary flower on each shoot.

subsp. **furfuraceum** (S.Y.Bao) J.M.H.Shaw in New Plantsman 6(3): 164 (1999).
Dysosma furfuracea S.Y.Bao in Acta Phytotax. Sin. 25(2): 155, t.5 (1987). Type: China, Yunnan province, Feng qing (Shunning), Snow Range, 3000 m, under thickets, 26 May 1938, *T.T.Yu* 15977 (holotype YUN, isotype A, 2 sheets!).
"polygon-leafed *Podophyllum*" Kingdon Ward in Gard. Chron. 1929: 247 (1929).
P. aurantiocaule Hand.-Mazz., Symb. Sin. 7(2): 323 (1931) *pro parte, quoad* Forrest 11897, 24232.
[*Dysosma aurantiocaule sensu auctt. non* (Hand.-Mazz.) Hu; Ying in Acta Phytotax. Sin. 17(1): 19 (1979) [as *aurantiocaulis*] *pro parte*, excl. *P. mairei* Gagnep. & *Handel-Mazzetti* 9242; Icon. Cormoph. Sinic. 1: 761, t.1521 (1980); Chen Ji-Sheng Zheng, Chinese Toxic Plants: 135 (1987)].
P. aurantiocaule subsp. *multiflorum* Shaw in sched. (BM).
[*P. hispidum sensu* Nowicke & Skvarla in Smithsonian Contr. Bot. 50: 18, 33 (1981), *non* Hao].

ILLUSTRATIONS. Icon. Cormoph. Sinic. 1: t.1521 (1980) [as *D. aurantiocaulis*]; Acta Phytotax. Sin. 25(2): 156 (1987); Kaiyun, Highland Flowers of Yunnan: 37, t.51 (1998) [as *D. aurantiocaulis*]; New Plantsman 7(4): 223, fig. 1, 3–5 (2000).

DESCRIPTION. *Leaf lobes* broadly triangular to elliptic, apex acute or acuminate. *Stamens* each with a total length of 5–8 mm; filaments short, flattened, 2–5 mm long; anthers (3–)3.5–4.5(–5.6) mm long, the ratio of anther to filament variable with the overall length fairly constant.

DISTRIBUTION. North-eastern Burma and western Yunnan (Gongshan, Fenqin, Weixi).
ECOLOGY. Broad-leaved evergreen forest, in small colonies under bamboos or scattered amongst the forest undergrowth, 2438–2743 m (8–9,000 feet); open grassy slopes, moorland meadows amongst scrub and margins of thickets, 3048–3353 m (10–11,000 feet). Collected by Reginald Farrer, on his ill fated last journey, who recorded it growing with *Nomocharis pardanthina* in N. Burma (*Farrer* 1017, E). Described as common in parts of W.Yunnan at 2800 m. In flower May–June; fruit September–October.

Section iv. HEXANDRA

(Species 14)
Section *Hexandra* Seliv.-Gor. in Nov. Sist. Vyssh. Rast. 12: 209 (1975).
Sinopodophyllum Ying in Acta Phytotax. Sin. 17(1): 15–16 (1979).

DESCRIPTION. *Rhizome* with short internodes. *Stems* reddish upon first emerging above ground. *Leaves* basically 3 or 5 lobed, with a large cleft at the base often penetrating to the petiole apex. *Flowers* solitary, white or pink. *Pollen* released in tetrads. Usually self-compatible. One species, W.Himalaya to N.W.China. Type species: *Podophyllum hexandrum* Royle.

This section has been accorded generic status by Ying (1979), which seems to be based largely on the pollen being released in tetrads, a character which is useful at specific and perhaps sectional level, but hardly a character on which to base generic segregation. Other instances are known in which closely related species, differing in pollen organisation, are otherwise retained congenerically. These include *Salpiglossis* (Solanaceae), with the well known *S. sinuata* producing pollen in tetrads and *S. spinescens* producing monads. A similar situation exists in *Reyesia* (Solanaceae) in which only one species, *R. parviflora*, out of the four in the genus produces pollen in tetrads and this is correlated with the presence of hairs on the filaments suggesting that the pollination syndrome is an important factor. In *P. hexandrum* the atypical pollen stucture (see SEM in Nowicke & Skvarla, 1981) and organisation may relate to the unusual pollination system recently elaborated by Xu *et al.* (1997).

14. PODOPHYLLUM HEXANDRUM

The plant now known as *P. hexandrum* was first named *P. emodi*, an appropriate name derived from the Latin, *Emodi Montes* (Himalayan Mountains), coined by the Danish botanist Nathaniel Wallich in 1824. It first appears without description or illustration, along with a var. *royleana*, in the lithographed *Numerical List of Plants in the East India Company's Museum, collected under the superintendence of Dr. Wallich* in 1829. There are no descriptions in this list, but at the time listing a name was accepted as valid publication. In these early days there was no code of plant nomenclature, and names often appeared in lists with no certain way of knowing the author's intention. Subsequently the name *P. emodi* was taken up and first associated with an illustration in Honigsburger's *Thirty-five years in the East* vol. 2: 329, t.20 in 1852, which from a modern viewpoint first validated the name *P. emodi*. However, prior to this in 1834 John Forbes Royle validly published the name *P. hexandrum* in his *Illustrations of the Botany and other Branches of the Natural history of the Himalayan Mountains*. Then in 1844 Cambessedes published a detailed description and plate of *P. hexandrum* in Jacquemont, *Voyage dans l'Inde* vol.2: 10, t.9. As there are these two publications of *P. hexandrum* antedating the first unambiguous publication of *P. emodi* in 1852, the name *P. hexandrum* Royle has priority over *P. emodi* Wall. ex Honigsburger. Additionally, it seems that Honigsburger's 1852 illustration of *P. emodi* was based on the earlier 1844 plate of *P. hexandrum* by Cambessedes. William T. Stearn was the first to recognise the significance of this in establishing the correct name for this taxon and brought it to the attention of F.W.Martin of Harvard in an undated letter from the mid 1950s. Further discussion and additional references are provided by Soejarto *et al.* (1979).

Royle had evidently noticed the variability of the leaf shape in this species — a source of consternation and confusion ever since — and intended to recognise several taxa, for on the type herbarium sheet at Liverpool Museum, consisting of two individuals, is written *Podophyllum acutifolium*, later inked to read *Podophyllum acutum* (Soejarto *et al.*, 1981). Additionally, in the library at Kew, is the manuscript copy of Wallich's *Catalogue* that belonged to Royle. On the reverse of folio 29, opposite entry 814 (*P. emodi*) is written "*P. acutum* Rl." — a further indication of Royle's intention. *Podophyllum acutum* was however never published.

Podophyllum hexandrum is a widespread species with an unusual degree of morphological variation. The variation appears to be somewhat clinal, with broad-lobed, slightly dissected leaves and white to pale pink flowers most frequent in the west. Plants with deep pink flowers and deeply dissected leaves are most common in China, which may be a reflection of introgression from *P. delavayi*. Other characters linking with *P. delavayi* include markedly trilobulate leaf lobes, and the occurrence of long hairs on the pedicels (*Polunin, Sykes & Williams* 954, E) and outer sepals on some collections. On the other hand, characters such as a spherical ovary in some collections from Yunnan, the presence of short stubby trichomes (cf. *P. aurantiocaule* subsp. *furfuraceum*), and the solitary flowers (cf. *P. a.* var. *uniflorum*) might suggest the influence of *P. aurantiocaule* with which it is partly sympatric. Chris Sanders observed large mixed populations of *P. hexandrum* and *P. aurantiocaule* in S.E. Tibet, but did not note any obvious hybrids.

Varieties based on foliar characters
All seedlings begin life with a three-lobed leaf, in some this is retained throughout adulthood, but other individuals produce more dissected foliage as they approach maturity. The var. *jaeschkei* is based on a collection of newly emerged stems in which the leaves have not yet expanded. An isotype sheet at Edinburgh has a stem with fully expanded leaves which are very similar to those of var. *axillaris*. Ideally, the foliar variation requires critical study, perhaps along the lines of Martin (1958), before any of the proposed varieties are taken up. However, an interesting summary of this variation by W.T.Stearn appears in a letter (copy in sched.

Podophyllum hexandrum   MATILDA SMITH

BM) written to F.W.Martin in which Stearn comments on the existence of regional variation. He notes that plants in the dry region of the western Himalaya are similar to Royle's *P. hexandrum*, and could be designated var. *hexandrum* (equivalent to *P. emodi* var. *royleana* Wall.). In Nepal, with its wetter climate, there appears a form with broader, less divided leaf segments, which remains of uncertain name. Finally in the east there is var. *chinense* Sprague, with deep rose flowers and a marbled, deeply dissected leaf.

Varieties based on position of the flower
Chatterjee & Mukerjee (in Chatterjee, 1952, 1953) described several varieties based largely on the position of the inflorescence, but this is very prone to change. Martin (1958) measured the position of the flower in 88 herbarium specimens, in which the position varied from the petiole fork (9.1 per cent) to immediately below the lamina (1.1 per cent). The most frequent position (19.3 per cent) was $1/5$ of the way up the petiole. In 87.5 per cent of the specimens, the flower was on the lower half of the petiole. Cultivation of plants has revealed that on a single plant some shoots have the flower in or very near to the petiole fork while others produce the flower just below the lamina. The cause of this variation is linked with the sympodial nature of shoot organization, which has been shown to be of some taxonomic value in other groups such as the Solanaceae. Its stability, and hence taxonomic value, differs between species; in *P. hexandrum* it is too changeable to be of value in discriminating varieties.

Variation of anther length
Anther length is a very useful character in *Podophyllum*, for example, it is helpful in distinguishing the two subspecies of *P. aurantiocaule*. In *P. hexandrum* it proves to be, like everything else, very variable and ranging from 2.5 to 9.5 mm, with most anthers between 4 and 7 mm. There might be a bimodal distribution with 44 per cent falling within the range 4 to 5 mm, and 34 per cent within 5.8 to 7 mm, with only 2 per cent in between these two ranges. All the specimens at BM, E, and K with intact flowers were measured and the anther lengths plotted on a map of geographical distribution which was then used to produce a three dimensional graph, reproduced in The New Plantsman 7(4): 230 (2000), to afford easier visualisation of any distribution pattern. The values do not fall into any clearly discernible geographical pattern, rather there seem to be pockets of extreme variation, notably from Kumaon to western Nepal in the west of the range, and from Sikkim to Bhutan in the east. Significantly, this may correlate with Stearn's observation regarding foliar variation noted above, since these areas of very unstable anther length appear broadly to coincide with the boundaries between the leaf forms. Crosses with cultivated plants from morphologically different inbred lines, such as would be included within different foliar varieties (e.g. *Kingdon-Ward* KWA 20 × 'Major'), produce plants bearing very short anthers, indicating that a degree of incompatibility exists between some inbred races of *P. hexandrum* and that anther length is genetically controlled. Interestingly, similar plants with very short anthers are known from the wild (they appear as deep troughs on the graph). Their presence, in the midst of pockets of extreme variation, likely indicates zones of interaction between different wild inbreeding

units within *P. hexandrum*, possibly providing a biological basis for the recognition of three somewhat vague varieties within *P. hexandrum*. At present it is very difficult to achieve a satisfactory separation between these three biological units since they are difficult to define and composed of extremely variable individuals. There are also many intermediates between the more typical individuals so that to assign most specimens to a formal variety would be rather arbitrary. It would be interesting to know how anther length correlates with leaf dissection and other variables including flower colour and karyotype, which has also been shown to vary considerably with geographic location (Hara, 1971; Siddique *et al.*, 1990). A collection from Nepal, *Polunin, Sykes & Williams* 2029 (BM) has what appear to be carpellate stamens or unexpanded petaloid organs, previously reported only from *P. peltatum*.

In cultivation it is a very variable species with many clones and inbred seed lines. Horticulturally significant plants are best designated by cultivar names or details of origin when known (Shaw, 1999). The Royal Botanic Garden at Edinburgh maintains a good collection of different examples (Walter *et al.*, 1995). One clone of historical interest maintained in this collection is 'Leichtlinii', named after the German nurseryman Max Leichtlin (1831–1910), which provided material for one of the earliest chromosome counts of *P. hexandrum* published by Langlet (1928). There are several very distinct forms marketed under the name 'Major' by different nurseries and at present it is not possible to say which inbred line originally bore the name. Chen Yi of Kaichen nursery, China reportedly has a variegated clone in cultivation. The most desirable forms for the garden are those from Yunnan with marbled foliage and deep rose pink flowers, usually marketed under the name var. *chinense*. This variety does not appear to have been formally transferred to *P. hexandrum* so the most appropriate designation, in view of the continuous variation, appears for the present to be cultivar status as 'Chinense'.

Experiments with commercial cultivation, which have met with limited success, are related in Krishnamurthy *et al.* (1965), Troup (1915), Nautiyal (1995) and Prasal (2001). As a result the increasing demand for *P. hexandrum* rhizome for medicinal use and more recently for the semisynthesis of chemotherapeutic anti-cancer drugs has led to over collection from the wild. This has resulted in a once common Himalayan plant disappearing from many of its former localities to the point of becoming an endangered species with a CITES listing (Gupta & Sethi, 1983). Recent conservation studies include Kala (2000) and Rai *et al.* (2000). A history of the commercial exploitation of the species is related by Chatterjee (1952). By contrast, medicinal use of the plant by indigenous people was largely confined to the fruit which is edible when ripe and was thought to aid conception (Kapahi, 1990). The letter that sealed the fate of wild *P. hexandrum* in India is still extant, attached to a voucher specimen in the Edinburgh herbarium. Dated 26 December 1888, its neat copperplate writing displays the hand of William Dymock, author of an encyclopaedic work on Indian medicinal plants:

> " My Dear Dr. Watt, You will be pleased to hear the *Podophyllum emodi* is a great success. Hooper got 12 per cent. of resins from it, where as the yield from *P. peltatum* is about 4 per cent. only, medicinally the resins act just in the same way as the official drug. As *Podophyllum* is by far the most popular purgative at the current time, your Department should take steps to collect some and put it on the market. A good supply would drive the American drug out of the European market".

A foresight that proved regrettably accurate.

The epithet *hexandrum*, from the Greek *hexa-* (six-), *andros* (man, male) refers to the six stamens typically found in each flower.

Podophyllum hexandrum Royle, Ill. Bot. Himal. Mts.: 64 (1834); Cambess. in Jacquem., Voy. Inde 4 (Bot.): 10, t.9 (1844); Hara, Photo-album Pl. E. Himalaya: 216 (1968); Seliv.-Gor. in Bot. Zhurn. (Moscow & Leningrad) 54(10): 1604 (1969); Hara, Fl. E. Himalaya 2: 34 (1971) *pro parte*, excl. *P. sikkimense*; Bedi, Herbal Wealth Bhutan: 109 (1972); Browicz, Fl. Iranica 101: 2 (1973); Malla, Fl. Langtang: 47 (1976); Hara & Williams, Enum. Fl. Pl. Nepal 2: 31 (1979); Soejarto *et al.* in Taxon 28: 549 (1979) & 30: 652 (1981); Sharma & Kachroo, Fl. Jammu 1: 91 (1981); Rix in Plantsman 4(1): 8 (1982); Dhar & Kachroo, Alpine Fl. Kashmir: 72 (1983); Chadwell, Kashmir Bot. Exped. 1983 Rep.: 76 (1984); Chowdhery & Wadhwa, Fl. Himachal Pradesh 1: 42–43 (1984); Polunin & Stainton, Flowers of Himalaya 23, t.11 (1984); Grierson & Long, Fl. Bhutan 1(2): 328, 325, t.26h-i (1984); Singh & Kachroo, Forest Fl. Srinagar: 163 (1987); Gupta, The Living Himalayas 2: 19–20 (1989); Stearn in Walters (ed.), Eur. Gard. Fl. 3: 396 (1989); Phillips & Rix, Perennials 1: 38–39 (1991); Sharma & Balakrishnan, Fl. India 1: 415, t.19 (1993); Clement & Foster, Alien Plants of the British Isles 24 (1994). Type: N.W.India; Kedarkanta, 3658 m (12,000 feet), in a moist and shady situation, May 1828, *Royle s.n.* (lectotype designated by Soejarto *et al.* in Taxon 30: 652 (1981): LIV-Herb. Royle, specimen on rhs of sheet 13/1, photo.!; isolectotype: LIV-Herb. Royle, specimen on lhs of sheet 13/1, photo.!, K!).

P. emodi Wall., Cat. no.814 (1829) *nom. nud.*; Falconer ex Royle, Ill. Bot. Himal. Mts.: 379 (1839) ex Honigsburger, Thirty-five years in the East 2: 235, t.20 (1852); Hook.f. & Thomson, Fl. Ind. 1: 232 (1855) *nom. illegit.*; Decne. in Fl. Serres 16: 95, t.1659–1660 (1865–1867); Aitchison in J. Linn. Soc., Bot. 18: 32 (1880); Collet, Simla Fl. 1: 22 (1902); Strachey, Cat. Pl. Kumaon: 8 (1906); Smith & Cave, Veg. Zemu & Llonakh Valleys Sikkim: 171 (1911); Coventry, Wild Fl. Kashmir: 21–22, t.11 (1923); Blatter, Beautiful Fl. Kashmir 1: 27, t.7 (1927); Pampanini, Fl. Caracorum: 114 (1930); Chatterjee & Mukerjee in Econ. Bot. 6: 342 (1952) & Rec. Bot. Surv. India 16(2): 43 (1953); Kihara in F. & Fl. Nepal Himal. 1: 133 (1955); Ahrendt in Candollea 15: 155 (1956); Chaudhri in Pakistan J. Sci. 8(5): 230 (1956); Kitamura, Fl. Afghanistan: 133 (1960); Bernardi in Candollea 18: 254 (1963); Stewart, Ann. Cat. Vasc. Pl. W. Pakistan & Kashmir: 282 (1972); Subramanyam in Rec. Bot. Surv. India 20(2): 32 (1973); Jafri, Fl. W. Pakistan, Podophyllaceae 57: 1–4, t.1 (1974); Icon. Cormophyt. Sinic. 1: 758 (1972, 1980); Kachroo, Sapru & Dhar, Fl. Ladakh: 48 (1977); Nair, Fl. Bashahr Himalaya: 21 (1977). Type: Nepal, Gossain Than, *Wallich* 814 (holotype: K-Herb. Wallich).

P. emodi var. *royleana* Wall., Cat. no. 814C (1829) *nom. nud.*; Sprague in Curtis's Bot. Mag. 146: t.8850 (1920). Type: 'Mons Choor' (holotype K).

P. acutifolium Royle *in sched.* (LIV-Herb. Royle); Soejarto *et al.* in Taxon 30: 652 (1981), *nom. nud.*

P. acutum Royle *in sched.* (LIV-Herb. Royle); Wallich, Cat., Royle's ms. copy, back of folio 29, entry n.814, Kew Library; Soejarto *et al.* in Taxon 30: 652 (1981), *nom. nud.*

P. emodi var. *chinense* Sprague in Curtis's Bot. Mag. 146: t.8850 (1920); Hand.-Mazz., Symb. Sin. 7(2): 322 (1931); Icon. Cormoph. Sinic. 1: 758, t.1516 (1972, 1980); Fu, Fl. Hupehensis 1: 400, t.565 (1976); Rix in Plantsman 4(1): 8 (1982), [as var. *chinense* Wall.].

P. sinense nom. nud. in sched. (*H.J.Elwes s.n.* 1921 K).

P. leichtlinii Langlet in Svensk Bot. Tidskr. 22: 176 (1928); Miyaji, Planta 11: 53 (1930); Darlington, Chromosome Atl. Fl. Pl.: 28 (1955); *vide* Curtis's Bot. Mag. 146: *sub* t.8850 (1920), *nom. nud.*

P. hexandrum var. *chinense* (Sprague) W.T.Stearn in MS (1933); & letter to F.W.Martin, WTS/EB/1326 (BM); Phillips & Rix, Perennials 1: 38–39 (1991) [as var. *chinense* Wall.]. This combination does not appear to have been validly published.

P. emodi var. *hexandrum* (Royle) Chatterjee & Mukerjee [in Econ. Bot. 6(4): 344, t.3 (1952), *comb. inval. sine basionym*] & Rec. Bot. Surv. India 16(2): 45 (1953) *nom. illegit.*

P. emodi var. *axillaris* Chatterjee & Mukerjee [in Econ. Bot. 6(4): 344, t.4 (1952), *nom. nud.*] & Rec. Bot. Surv. India 16(2): 46 (1953); Bernardi in Candollea 18: 255 (1963); Sharma & Balakrishnan, Fl. India 1: 416 (1993). Type: 'Teesta valley, above Tangu', 4115 m (13,500 feet), 6 July 1903, *Younghusband s.n.* (holotype CAL).

P. emodi var. *bhootanensis* Chatterjee & Mukerjee [in Econ. Bot. 6(4): 344, t.5 (1952), *nom. nud.*] & Rec. Bot. Surv. India 16(2): 46–47 (1953). Type: 'Bhootan, Taloong', 2 August 1884, *Dungboo* 289 (holotype CAL).

P. emodi var. *jaeschkei* Chatterjee & Mukerjee [in Econ. Bot. 6(4): 344, t.6 (1952), *nom. nud.*] & Rec. Bot. Surv. India 16(2): 47 (1953); Sharma & Balakrishnan, Fl. India 1: 416 (1993). Type: Lahul, *Jaeschke s.n.* (holotype CAL, isotypes K!, E!).

P. indica Chopra, Indigenous drugs of India, ed. 2: 226 (1958), *nom. nud.* (this pharmaceutical name occurs in an italic list of Latin binomials, as if a botanical name).

P. hexandrum var. *emodi* (Falconer ex Royle) Seliv.-Gor. in Bot. Zhurn. (Moscow & Leningrad) 54(10): 1605 (1969).

Dysosma emodi (Falconer ex Royle) Hiroe in Pl. Basho's & Buson's Hokku Lit. 8(3): 328 (1973), *nom. illegit.*

P. hexandrum var. *axillare* (Chatterjee & Mukerjee) Browicz in Fl. Iranica, *Podophyllaceae* 101: 2 *in nota* (1973).

P. hexandrum var. *bhootanense* (Chatterjee & Mukerjee) Browicz in Fl. Iranica, *Podophyllaceae* 101: 2 *in nota* (1973).

P. hexandrum var. *jaeschkei* (Chatterjee & Mukerjee) Browicz in Fl. Iranica, *Podophyllaceae* 101: 2 *in nota* (1973); Chowdhery & Wadhwa, Fl. Himachal Pradesh 1: 43 (1984).

Sinopodophyllum emodi (Falconer ex Royle) T.S.Ying in Acta Phytotax. Sin. 17(1): 16 (1979) *pro parte,* excl. *P. sikkimensis, P. delavayi*; Chen, Chinese toxic plants: 134 (1987); Naithani, Fl. Pl. India, Nepal & Bhutan: 29 (1990), *nom. illegit.*

P. hexandrum var. *majus* hort.; Rix in Plantsman 4(1): 8 (1982).

[*P. versipelle* hort. *sensu auct. non* Hance; Rix in Plantsman 4(1): 9 (1982), *pro parte*; Phillips & Rix, Perennials 1: 39 (1991), *pro parte, quoad* plant "hardy at Kew".].

Sinopodophyllum hexandrum (Royle) T.S.Ying in Fl. Xizangica 2: 119 (1985), *pro parte*, excl. *P. sikkimensis*.

P. emodi cv. Major hort.; Reader's Digest, Successful Gardening, A-Z of Perennials: 132 (1992).

P. pentaphyllum [*nom. nud. in syn.*] Deno, Seed germination theory and practice (1993).

P. hexandrum cv. Majus hort.; Dewick & Shaw in Garden 113(5): 235, 236 (1988).

[*P. heterophyllum sensu* Gehenio *non* Raf., North American Rock Garden Society seed distribution list, 1996].

P. hexandrum subsp. *substerilis* hort. Used by some European nurseries for lines of *P. hexandrum* with low self compatibility.

P. hendersonii hort. *nom. nud.* Appeared in RHS Plant Finder, 1997–98.

[*P. aurantiacum* hort. An error for *P. aurantiocaule* which appears in some seed lists, however in all cases thus far the identity of the seed is *P. hexandrum*.]

ILLUSTRATIONS. Flore des Serres 16: t.1659, 1660 (1866); Curtis's Bot. Mag. 146: t.8850 (1920); Blatter, Beautiful flowers of Kashmir t.7 (1928); Everard & Morley, Wild Flowers of the World t.91C (1970); Grierson & Long, Fl. Bhutan 1(2): 325, f.26 h-i (1984); Polunin & Stainton, Flowers of the Himalayas: pl. 11 (1984); Garden 113(5): 233, 236, 237 (1988); Phillips & Rix, Perennials 1: 39 (1991); Roberts (ed.), Wild Flowers of Pakistan: pl. 5 (1995); Fl. Yunnanica 7: 3, t.1,1–6 (1997); Kaiyun, Highland flowers of Yunnan: 38 (1998); Hinkley, The Explorer's Garden: 115 (1999); New Plantsman 7(1): 5, 35, 39; 7(3): 151; 7(4): 226, 231, fig.2, (2000).

DESCRIPTION. *Perennial, rhizomatous herb* with annual aerial stems; rhizome thick, scaly, with dense fibrous roots, internodes short; cataphyllary scales enveloping the dormant bud, persisting at the stem base well into the growing season. *Stems* 15–60 cm, smooth, glabrous, often with a waxy bloom, reddish-orange upon emerging above ground, with 1–3 leaves. *Juvenile leaves* peltate, bilaterally symmetrical, palmately divided, with three lobes, the central lobe larger than the lateral lobes; sinuses narrow and acute, except for the adaxial sinus between the two smaller lobes, which is broadly triangular, often penetrating the lamina nearly to or actually reaching the petiole apex; on some plants the juvenile foliage is retained at anthesis, these may be neotenic. *Adult leaves* palmate, 7–25 cm across, with 3–5 lobes joined slightly at the base; lobes narrowly or broadly ovate-lanceolate, bi- or tri-lobulate or becoming laciniate, divided up to $^3/_4$ of their length, with serrate margins, apex acute-acuminate; indumentum variable, linear white hairs usually present on the leaf underside, especially on the prominent main veins, the upper surface with scattered hairs, usually glabrous. *Inflorescence* a single flower borne on a short pedicel from the fork of the two upper petioles, or along the petiole of the upper leaf. *Pedicel* erect at anthesis, pendant and elongating slightly in fruit, occasionally with scattered spreading hairs, which become more dense towards the calyx. *Sepals* 6, ovate, green, often with spreading hairs on the outermost 3, not usually opening but loosely coalescing and falling off as a cap as the petals expand. *Petals* (4–)6, obovate-oblong, 1.5–3 cm long, 1–1.5 cm wide, white or pale pink or rich rose pink. *Stamens* 6(–9);

filaments flattened towards the base, 4–10 mm long; anthers (2.5–)4–6.5(–9) mm long, occasionally with the connective extended to form a minute mucro 0.2 mm long. *Ovary* elliptic, rarely subglobose, *c*. 1 cm, with a large corrugated, peltate stigma. *Fruit* an ovoid to elliptic many-seeded berry with the remains of the stigma at the apex, 2–6 cm long, red, often with a waxy bloom, edible when ripe. *Seeds* obovoid, maroon, 3–4 mm long. *Cotyledons* 2, united along the adaxial edge, margins smooth. 2n = 12 (Langlet, 1928; Hara, 1971; Siddique *et al.*, 1990). Usually self-compatible.

DISTRIBUTION. Along the Himalayan range, from Afghanistan to northern Yunnan, extending northwards into Kansu and Sichuan provinces, China. Recorded by Fu (1976) from Hubei province. Endemic to the Sino-Himalayan region. Cultivated as an ornamental in North America and Europe and occasionally reported as an alien garden escape (Shaw, 1996). A dot map is provided in *New Plantsman* 7(4): 231 (2000).

ECOLOGY. Open mountain slopes and shady coniferous or *Quercus* forests, *Juniperus* – *Rhododendron* scrub or occasionally in alpine meadow in moist places, 1800–4500m. In the western Himalaya the maximum numbers of plants are often associated with *Viburnum nervosum* shrubs (Chaudhri, 1956). Ecological studies have been published by Ma & Hu, 1996; Ma *et al.*, 1997; Xu *et al.*, 1997; Airi *et al.*, 1997).

DOUBTFUL AND EXCLUDED NAMES

Podophyllum diphyllum L., Sp. Pl. ed. 1, 505, (1753). = *Jeffersonia diphylla* (L.) Pers. (*Berberidaceae*).

Podophyllum japonicum T. Ito in Maxim., Mélanges Biol. Bull. Phys.-Math. Acad. Imp. Sci. Saint-Pétersbourg 12: 417 (1886); J. Linn. Soc., 22: 434 (1887). = *Ranzania japonica* T. Ito (*Berberidaceae*).

Podophyllum cavaleriei H.Lév. in Bull. Acad. Int. Géogr. Bot. 24: 142 (1914); Hu in Bull. Fan Mem. Inst., Bot. Ser. 8: 36 (1937); Rix in Plantsman 4(1): 10 (1982) [as *cavalieri*] = *Pilea peperomioides* Diels (*Urticaceae*); see Lauener, A. in Notes Roy. Bot. Gard. Edinburgh 24(1): 75 (1962) & Plantsman 4(2): 127 (1982). For the extraordinary story of this plant see Kew Mag. 1(1): 14–19, t.5 (1984) & 2(3): 334–336 (1985).

Podophyllum himalayense Le Maout & Decne., Traite Gen. Bot.: 377 (1868) *nom. nud*. It seems probable that *P. hexandrum* was intended.

ACKNOWLEDGMENTS

I would like to express my gratitude to the very many people who have helped with this study in many different ways, particularly: J.Anderson, C.D.Brickell, Y.-S.Chen, P.M.Dewick, G.Dunlop, E.V.Eyre, T.-W.Hu, M.Chambers, J.McClements, M.K.Pathak, C.Sanders, J.Sargent, C.Shaw, W.T.Stearn, B.Wynn-Jones, P.Zhuang and my family. Also staff at the following herbaria: A, B, BM, CAL, E, GH, GZU, K, P, SZ, W, for helpful discussions and access to plant material.

BIBLIOGRAPHY

A. REFERENCES AND BIBLIOGRAPHY FOR *EPIMEDIUM* AND OTHER GENERA OF HERBACEOUS BERBERIDACEAE

Antevs, E. (1928). The last Glaciation (*American Geographical Society, Research series* 17): 34–36. New York.

Arldt, T. (1919–1922). *Handbuch der Palaeogeographie*. 2 vols. Lepizig.

Bailey, L.H. (several editions, e.g.1939). *The Standard Cyclopedia of Horticulture* 1: 1121–1122.

Baillon, H. (1862). Remarques sur l'organisation des Berberidées. *Adansonia* 2: 268–291.

Baillon, H. (1871). *Histoire des Plantes* 3: 49–76.

Baker, J.G. (1880). A synopsis of the species and forms of Epimedium, *Gardeners' Chronicle* New Ser. 13: 620, 683–684.

Barbey, A. (1934). Une relique de la sapinière mediterranéene; le Mont Babor. *Monographie de l'Abies numidica Lann*. Paris & Gembloux.

Barker, D.G. (1927). *Epimedium and other herbaceous Berberidaceae*. Great Comberton, Worcestershire.

Berg, R.Y. (1972). Dispersal ecology of Vancouveria (Berberidaceae). *American Journal of Botany* 59: 109–122.

Blackmore, S., Stafford, P. & Persson, V. (1995). Palynology and systematics of Ranunculiflorae. In: Jensen, V. & Kadereit, J.W. (eds.), Systematics and Plant Evolution in the Ranunculiflorae. *Plant Systematics and Evolution* [Suppl.] 9: 71–82.

Blanckenhorn, M. (1910). Das Klima der Quartärperiode in Syrien-Palästina und Aegypten. In: *Congrès Géologique International XI Die Veränderungen des Klimas seit dem Maximum der Letzten Eiszeit*: 425–428.

Bretschneider, E. (1895). *Botanicon Sinicum* 3: 54–56. Shanghai.

Bretschneider, E. (1898). History of European botanical discoveries in China. London.

Brett, J.F. & Posluszny, U. (1982). Floral development in Caulophyllum thalictroides (Berberidaceae). *Canadian Journal of Botany* 60: 2133–2141.

Brooks, C.E.P. (1922). *The Evolution of Climate*. London

Browicz, K. (1976). Genus Bongardia and its range. *Fragmenta Floristica et Geobotanica* 22: 435–444.

Butters, F.K. (1909). The seeds and seedlings of Caulophyllum thalictroides. *Minnesota Botanical Studies* 14: 11–32.

Calloni, S. (1887). Nuovo specie di Vancouveria. *Malpighia* 1: 263–272.

Chamberlain, D.F. (1993). In: Jarvis, C. *et al.*, A list of Linnaean generic names and their types. *Regnum Vegetabile* 127: 45.

Chapman, M. (1936). Carpel anatomy of the Berberidaceae. *American Journal of Botany* 23: 340–8.

Christensen, K.I. & Hansen, H.V. (1998). SEM studies of epidermal patterns of petals in the angiosperms. *Opera Botanica* 135.

Citerne, P.E. (1892). *Berbéridées et Erythrospermées*. Thèse. Paris.

Denton, C.W. & Hughes, T.J. (eds.) (1981). *The last great Ice Sheets*. New York & Chichester.

Dermen, H. (1931). Chromosome numbers in Mahonia and Berberis. *Journal of the Arnold Arboretum* 12: 281–287.

Drake del Castillo, M.F. (1900). Notice sur la vie et les travaux de A.Franchet. *Bulletin de la Société Botanique de France* 47: 158–172.

Du Rietz, G.E. (1930). Fundamental units of biological taxonomy. *Svensk Botanisk Tidskrift* 24: 333–428.

Endress, P. (1989). Chaotic floral phyllotaxis and reduced perianth in Achlys (Berberidaceae). *Botanica Acta* 102: 159–163.

Engler, A. & Gilg, E. (1924). *Syllabus der Pflanzenfamilien* ed. 9–10: 206. Berlin.

Ernst, W. (1964). Berberidaceae. *Journal of the Arnold Arboretum* 45: 1–35.

Fischer, F.E.L. von & Meyer, C.A. (1846). *Sertum Petropolitanum*. St. Petersburg.

Fournier, P. (1932). *Voyages et découvertes scientifiques des missionaires naturalistes français*. Paris.

Franchet, A. (1886). Sur les especes du genre Epimedium. *Bulletin de la Société Botanique de France* 33: 38–41, 103–116.

Fukuda, I. (1967). The biosystematics of Achlys. *Taxon* 16: 308–316.

Fukuda, I. & Baker, H.S. (1970). Achlys californica (Berberidaceae) a new species. *Taxon* 19: 341–344.

Giffen, M.H. (1937). Chromosome numbers of Berberis. *Transactions of the Royal Society of South Africa* 24: 203–206.

Gottsberger, G. (1974). The structure and function of the primitive angiosperm flower — a discussion. *Acta Botanica Neerlandica* 23: 461–471.

Greene, E.L. (1890). New or Noteworthy species VIII. *Pittonia* 2: 100–106.

Gregory, J.W. (1911). The geology of Cirenaica. *Quarterly Journal of the Geological Society* 67, 18–22.

Grossheim, A.A. (1927). Vegetation und Flora des Talysch-Gebiets. *Beihefte zum Botanischen Centralblatt* 43(2): 1–33.

Handel-Mazzetti, H. (1927a). *Naturbilder aus Südwest-China*. Vienna & Leipzig.

Handel-Mazzetti, H. (1927b). Das nordost-birmanisch/west-yünnanisch Hochgebirgs-gebiet. In: Karsten, G. & Schenk, H., *Vegetationsbilder* xvii, Heft 7–8. Jena.

Handel-Mazzetti, H. (1931). Das pflanzengeographische Gliederung und Stellung Chinas. *Botanische Jahrbücher für Systematik, Pflanzengeschichte und Pflanzengeographie* 64: 309–23.

Hinkley, D.J. (1999). *The Explorer's Garden: rare and unusual perennials*: 94–123. Portland, Oregon.

Himmelbaur, W. (1913). Die Berberidaceen und ihre Stellung im System. *Denkschriften der Kaiserlichen Akademie der Wissenschaften Mathematisch-Naturwissenschaftliche Klasse* 89: 733–796.

Holm, T. (1889). Podophyllum peltatum, a morphological study. *The Botanical Gazette (Crawfordsville)* 27: 419–33.

Hooker, J.W. (1842). Botanical Information: Russia. *Journal of Botany (Hooker)* 1: 206–208.

Hoot, S.B. & Crane, P.R. (1995). Inter-familial relationships in the Ranunculidae based on molecular systematics. *Plant Systematics and Evolution* [Suppl.] 9: 119–131.

Hutchinson, J. (1920). Jeffersonia and Plagiorhegma. *Bulletin of Miscellaneous Information, Kew* 1920: 242–245.

Ishidoya, T. (1933). *Chinesische Drogen* 1: 35–37. Keijo, Japan.

Ito, Tokutaro. (1887). Berberidearum Japoniae Conspectus. *Journal of the Linnean Society, Botany* 22: 422–437.

Janchen, E. (1949). Die systematische Gliederung der Ranunculaceen und Berberidaceen. *Österreichische Akademie der Wissenschaften Mathematisch.-Naturwissenschaftliche Klasse* 108(44): 1–82.

Jussieu, A.L. de (1789). *Genera Plantarum*: 286–287. Paris.

Kendrew, W.G. (1927). *Climates of the Continents*. Oxford.

Kim, Y.-D. & Jansen, R.K. (1995). Phylogenetic implications of chloroplast DNA variation in the Berberidaceae. In: Jensen, V. & Kadereit, J.W. (eds.), Systematics and Plant Evolution in the Ranunculiflorae. *Plant Systematics and Evolution* (Suppl.) 9: 341–349.

Kim, Y.-D. & Jansen, R.K. (1997). Phylogenetic implications of *rbcl* and ITS sequence variations in the Berberidaceae. *Systematic Botany* 21: 381–396.

Knuth, P. (1908). *Handbook of Flower Pollination* 2: 57–58. Oxford.

Komarov, V.L. (1908). Revisio critica specierum generis Epimedium L. *Trudy Imperatorskago S.-Petersburgskago botanicheskago sada* 29: 125–154.

Konuklugil, B. & Shaw, J. (1996). Alpines and medicines; a look at Podophyllum and Linum. *Quarterly Bulletin of the Alpine Garden Society* 64: 334–340.

Kryshtofovich, A.N. (1929). Evolution of the Tertiary flora in Asia. *New Phytologist* 28: 303–12.

Kryshtofovich, A.N. (1935). A final link between the Tertiary floras of Asia and Europe. *New Phytologist* 34: 339–44.

Kumazawa, M. (1930). Morphology and biology of Glaucidium palmatum Sieb. & Zucc., with notes on affinities to the allied genera Hydrastis, Podophyllum and Diphylleia. *Journal of the Faculty of Science University of Tokyo, Section III, Botany* 2: 345–380.

Kumazawa, M. (1936). Podophyllum pleianthum Hance, a morphological study, with supplementary notes on allied plants. *Botanical Magazine (Tokyo)* 50: 268–276.

Kumazawa, M. (1937a). Ranzania japonica: its morphology, biology and systematic affinities. *Japanese Journal of Botany* 9: 55–70.

Kumazawa, M. (1937b). Comparative studies on the variation in the Ranunculaceae and Berberidaceae. *Journal of Japanese Botany* 13: 573–586, 659–669, 713–726.

Kuznezow, N.I. (1909). Principles of the division of the Caucasus into botanical-geographical provinces. *Zapiski Imperatorskoj Akademii Nauk po Fiziko-matematičeskomu Otděleniju* VIII, 24: no. 1, maps.

Laar, H.J. van der. (1981). Epimedium. *Dendroflora* 18: 5–12.

Langlet, O. (1928). Einige Beobachtungen über die Zytologie der Berberidazeen. *Svensk Botanisk Tidskrift* 22: 169–184.

Liang, Hairui & Yan, Wenmei. (1995). Studies on the leaf surface of the genus Epimedium in China. *Acta Botanica Boreali-Occidentalis Sinica* 12: 142–148.

Lindley, J. (1846). Paeonia wittmanniana. *Botanical Register* 32: t.9.

Loconte, H. (1993). Berberidaceae. In Kubitzki, K. (ed.), *Families & Genera of Vascular Plants* 2: 147–152.

Loconte, H. & Blackwell, W.H. (1985). Intrageneric taxonomy of Caulophyllum (Berberidaceae). *Rhodora* 87: 463–469.

Loconte, H. & Estes, J.R. (1989a). Generic relationships within Leonticeae (Berberidaceae). *Canadian Journal of Botany* 67: 2310–2316.

Loconte, H. & Estes, J.R. (1989b). Phylogenetic systematics of Berberidaceae and Ranunculales (Magnoliidae). *Systematic Botany* 14: 565–579.

Lubbock, J. (1892). *Contribution to our knowledge of seedlings* 1: 108–114. London.

Maekawa, F. (1955). Species problem and phylogenetic appreciation for diagnostic characters, A case of Epimedium. *Journal of Japanese Botany* 30: 353–358. (In Japanese).

Marchand, L. (1864). Sur les fleurs monstrueuses d'Epimedium. *Adansonia* 4: 127–132.

Mattfeld, J. (1925). Die in Europa und in Mittlemeergebiet wildwachsenden Tannen (Abies). *Mitteilungen der Deutschen Dendrologischen Gesellschaft* 35: 1–37.

Meacham, C.A. (1986). Phylogeny of the Berberidaceae with an evaluation of classifications. *Systematic Botany* 5: 149–172.

Miyaji, Y. (1930). Beiträge zur Chromosomenphylogenie der Berberidaceen. *Planta* 11: 650–651.

Morren, C. (1846). Epimedium pinnatum. *Annales de la Société Royale d'Agriculture et de Botanique de Gand* 2: 139–140.

Morren, C. & Decaisne, J. (1834). Observations sur la flore du Japon suivies de la monographie du genre Epimedium. *Annales des Sciences Naturelles* 11, Botanique 2: 347–361.

Nakai, T. (1944). Epimedium grandiflorum et ejus affinitates, vel species sectionis Macroceras in Imperio Nipponico sponte nascentes. *Journal of Japanese Botany* 20: 65–84.

Nickol, M.G. (1995). Phylogeny and inflorescences of Berberidaceae. In: Jensen, V. & Kadereit, J.W. (eds.), Systematics and Plant Evolution in the Ranunculiflorae. *Plant Systematics and Evolution* [Suppl.] 9: 327–394.

Nowicke, J.W. & Skvarla, J.J. (1981). Pollen morphology phylogenetic relationships of the Berberidaceae. *Smithsonian Contributions to Botany* 50: 1–83.

Ogisu, M. (1996). Epimedium campanulatum (Berberidaceae) a new species from Sichuan. *Kew Bulletin* 51: 393–400.

Pampanini, R. (1903). Essai sur la géographie botanique des Alpes. [Thèse ... Fribourg.] *Mémoires de la Fribourgeoise des Sciences Naturelles. Géologie, Géographie* 3, fasc. 1.

Pellmyr, O. (1984). Yellow jackets disperse Vancouveria seeds. *Madroño* 32: 56.

Penck, A. (1906). Die Entwicklung Europas seit der Tertiärzeit. In Lotzy, J.P., *Résultats scientifiques du Congrès internationale de Botanique, Vienna*, 1905: 12–24. Jena.

Prantl, K. (1888). In Berberidaceae. Engler, A. & Prant, K. (Eds.). *Die Naturlichen Pflanzenfamilien* III. 2: 70–77.

Reid, C. & Reid, E.M. (1915). Pliocene floras of the Dutch-Prussian border. *Mededeelingen van de Rijksopsporing van Delfstoffen* no. 6. The Hague.

Reid, E.M. (1920). Comparative review of Pliocene floras, based on the study of fossil seeds. *Quarterly Journal of the Geological Society* 76: 145–61.

Ridgeway, R. (1912). Colour Standards. Washington, DC.

Robson, G.C. & Richards, O.W. (1936). *The Variation of Animals*. London.

Royal Horticultural Society (1938–43). *Horticultural Colour Chart*. Ed. 3 (1995). RHS, London.

Saccardo, P.A. (1882–1931). *Sylloge Fungorum* 1: 500(1882); 2: 798(1883); 3: 39 (1884); 10: 422 (1892); 11: 597 (1895); 13: 430, 1291 (1898); 16: 326 (1902); 25: 741 91931); Padua (Vol. 13 Berlin)..

Sastri, R.L.N. (1969). Floral morphology, embryology, and relationships of the Berberidaceae. *Australian Journal of Botany* 17: 69–79.

Saunders, E.R. (1928). On carpel polymorphism, I. *Annals of Botany* 39: 123–67.

Savatier, L. (1875). *Livres Kwa-Wi*. Paris.

Schwarz, O. (1936). Entwurf zu einen naturlichen System.......der gattung Quercus. *Notizblatt des Königlichen Botanischen Gartens und Museums, Berlin* 13: 1–22.

Seward, A.C. (1931). *Plant Life through the Ages*. Cambridge.

Shaw, J.M.H. (1999). New taxa, combinations and taxonomic notes on Podophyllum. *New Plantsman* 6(3): 158–165.

Shaw, J.M.H. (2000). A taxonomic revision of Podophyllum in the wild and in cultivation. *Plantsman* 7(1): 30–41; 7(2): 103–113; 7(3): 142–159; 7(4): 220–235.

Shimizu, T. & Umbayashi, M. (1995). *Underground Organs of Herbaceous Angiosperms*. Tokyo.

Simpson, G.C. (1930). The climate during the Pleistocene period. *Proceedings of the Royal Society of Edinburgh* 50: 262–96.

Spae, D. (1858). Notice biographique sur A. Donkelaar. *La Belgique Horticole* 8: 273–283.

Stearn, W.T. (1932). Some cultivated Epimediums. *Gardening Illustrated* 54: 31.

Stearn, W.T. (1933). Some Chinese species of Epimedium. *Journal of Botany* (London) 71: 343–347.

Stearn, W.T. (1938). Epimedium & Vancouveria (Berberidaceae), a monograph. *Journal of the Linnean Society, Botany* 51: 409–555.

Stearn, W.T. (1979). A new hybrid Epimedium (E. cantabrigiense). *The Plantsman* 1: 187–190.

Stearn, W.T. (1989). Ranzania, Epimedium, Vancouveria, Jeffersonia, Achlys, Caulophyllum, Leontiae, Bongardia, Diphylleia, Podophyllum. In: Walters, M. *et al.*, *European Garden Flora* 3: 370–371, 389–396.

Stearn, W.T. (1990). Epimedium dolichostemon (Berberidaceae) and other Chinese species of Epimedium. *Kew Bulletin* 45: 682–692.

Stearn, W.T. (1991). A Chinese puzzle ... inconsistencies in the names of plants introduced from China. *The Garden (Journal of the Royal Horticultural Society)* 116: 85–89.

Stearn, W.T. (1993a). New large-flowered Chinese species of Epimedium (Berberidaceae). *Curtis's Botanical Magazine* 10: 178–184.

Stearn, W.T. (1993b). The small-flowered Chinese species of Epimedium (Berberidaceae). *Kew Bulletin* 48: 807–813.

Stearn, W.T. (1995). New Chinese taxa of Epimedium (Berberidaceae) from Sichuan. *Curtis's Botanical Magazine* 12: 15–25.

Stearn, W.T. (1996). Epimedium acuminatum and allied Chinese species. *Kew Bulletin* 51: 393–400.

Stearn, W.T. (1997). Four new Chinese species of Epimedium (Berberidaceae). *Kew Bulletin* 52: 659–671.

Stearn, W.T. (1998). Four more Chinese species of Epimedium(Berberidaceae) *Kew Bulletin* 53: 213–223.

Stefanoff, B. & Jordanoff, D. (1935). Studies upon the Pliocene flora of the plain of Sofia (Bulgaria). *Sbornik na Bŭlgarskata Akademiya na Naukité* 25.

Stojanoff, N. & Stefanoff, B. (1929). Beitrag zu Kenntnis der Plioz"neflora des Ebene von Sofia. *Spisanie na Bŭlgarskoto Geologichesko Druzhestvo (Zeitschrift der Bulgarischen Geologischen Gesellschaft)* 1, fasc. 3.

Studt, W. (1926). Die heutige unf frühere Verbreitung der Koniferen. *Mitteilungen aus dem Institut für Allgemeine Botanik in Hamburg* 6: 167–308.

Suzuki, K. (1978). Biosystematic studies of Japanese Epimedium (Berberidaceae) (1) Variation of the populations of Shikoku. *Journal of Japanese Botany* 53: 203–212, 225–231.

Suzuki, K. (1982). A contribution to the taxonomy of the genus Epimedium (Berberidaceae) in Japan. *Journal of Japanese Botany* 57: 65–69.

Suzuki, K. (1986). Epimedium trifoliatobinatum, a species derived from hybridization between E. grandiflorum and E. diphyllum. In: Iwatsuki *et al.*, *Modern Aspects of biological Species*: 195–209. Tokyo.

Suzuki, K. (1984). Pollination system and its significance on isolation and hybridization in Japanese Epimedium (Berberidaceae). *Botanical Magazine (Tokyo)* 97: 381–396.

Suzuki, K. (1990). *Nippon no Ikariso* [*Epimedium in Japan*]. Tokyo.

Takahashi, C. (1989). Karyomorphological studies on speciation of Epimedium and its allied Vancouveria with special reference to C-banding. *Journal of Science, Hiroshima University*, Series B., Division 2 (Botany) 22: 159–269.

Takeda, H. (1915). On the genus Achlys, a morphological and systematic study. *Botanical Magazine (Tokyo)* 29: 169–184.

Takhtajan, A. (1997). *Diversity and Classification of Flowering Plants*. New York.

Tanaka, R. & Takahashi, C. (1981). Comparative karyotype analysis in Epimedium species by C-banding (1) E. sempervirens var. hypoglaucum and E. perralderianum. *Journal of Japanese Botany* 56: 17–24.

Terabayashi, S. (1977). Studies in the morphology and systematics of Berberidaceae. I. Floral anatomy of Ranzania japonica. *Acta Phytotaxonomica Geobotanica* 28: 45–57.

Terabayashi, S. (1979). Studies in the morphology and systematics of Berberidaceae. III. Floral anatomy of Epimedium grandiflorum ssp. sempervirens and Vancouveria hexandra. *Acta Phytotaxonomica Geobotanica* 30: 153–168.

Terabayashi S. (1981). Studies in the morphology and systematics of Berberidaceae. IV. Floral anatomy of Plagiorhegma dubia, Jeffersonia diphylla, and Achlys triphylla ssp. japonica. *Botanical Magazine (Tokyo)* 94: 141–157.

Terabayashi, S. (1983a). Studies in the morphology and systematics of Berberidaceae. V. Floral anatomy of Caulophyllum, Leontice, Gymnospermium, and Bongardia.

Memoirs of the Faculty of Science Kyoto University, Series Biology 8: 197–217.

Terabayashi S. (1983b). Studies in the morphology and systematics of Berberidaceae. VI. Floral anatomy of Diphylleia, Podophyllum, and Dysosma. *Acta Phytotaxonomica Geobotanica* 34: 27–47.

Terabayashi, S. (1985a). The comparative floral anatomy and systematics of the Berberidaceae. I. Morphology. *Memoirs of the Faculty of Science Kyoto University*, Series Biology 10: 73–90.

Terabayashi, S. (1985b). The comparative floral anatomy and systematics of the Berberidaceae. II. Systematic considerations. *Acta Phytotaxonomica Geobotanica* 36: 1–13.

Terabayashi S. (1987). Seedling morphology of the Berberidaceae. *Acta Phytotaxonomica Geobotanica* 38: 63–74.

Tischler, G. (1902). Die Berberidaceen und Podophyllaceen. *Botanische Jahrbücher für Systematik, Pflanzengeschichte und Pflanzengeographie* 31: 596–727.

Trabut, L. (1889). De Djidjelli aux Babors par les Beni Foughal. *Bulletin de la Société Botanique de France* 36: 56–64.

Vavilov, N.J. (1922). Law of homologous series in variation. *Journal of Genetics* 12: 47–89.

Weaver, R. (1979). In praise of epimediums. *Arnoldia* 39: 51–66.

White, A.R. (1996). Epimedium: Dawning of a new era. *The Garden* 121: 208–214.

Woldstedt, P. (1929). *Das Eiszeitalter: Grundlinien einer Geologie des Diluviums*. Stuttgart.

Woodson, R.E. (1928). Dysosma: a new genus of Berberidaceae. *Annals of the Missouri Botanical Garden* 15: 335–340.

Wright, W.B. (1914). *Quaternary Ice Age*. (ed. 2, 1936). London.

Wu, Z. (1983). On the significance of Pacific intercontinental discontinuity. *Annals of the Missouri Botanical Garden* 70: 577–590.

Ying, T.-S. (1975). On the Chinese species of Epimedium L. *Acta Phytotaxonomica Sinica* 13(1): 49–56.

Ying, T.-S. (1979). On Dysosma Woodson and Sinopodophyllum Ying, gen. nov. of the Berberidaceae. *Acta Phytotaxonomica Sinica* 17: 17–23.

Ying, T.-S., Terabayashi, S. & Boufford, D. (1984). A monograph of Diphylleia. *Journal of the Arnold Arboretum* 65: 57–94.

Zhong, Guo-Yue, Komatsu, K., Mikage, M. & Nambo, T. (1994). Pharmacognostical studies on the Chinese herbs derived from Berberidaceae plants in Sichuan province. On 'Yingyanghuo' derived from Epimedium plants. *Natural Medicines* 48: 141–154.

Zao, Xiwen et al. (1992). *The Paleoclimate of China*. Geological Publishing House, Beijing.

B. REFERENCES AND BIBLIOGRAPHY FOR *PODOPHYLLUM*

Airi, S., Rawal, R.S., Dhar, U. & Purohit, A.N. (1997). Population studies on Podophyllum hexandrum, a dwindling medicinal plant of the Himalaya. *Plant Genetic Resources Newsletter* 110: 29–34.

Adachi, J., Kosuge, K., Denda, T. & Watanabe, K. (1995) Phylogenetic relationships of the Berberidaceae based on partial sequences of the gapA gene. *Plant Systematics and Evolution* S9: 351–353.

Agarwal, V.S. & Ghosh, B. (1985). *Drug plants of India — Root Drugs*. Kulyani Publishers, New Delhi.

Airy Shaw, H.K. (1936). Glaucidium palmatum. *Curtis's Botanical Magazine* 159: t.9432.

Altschul, S. von Reis (1975). *Drugs and foods from little known plants*: 70. Harvard University Press.

American Herbal Pharmacology Delegation (1975). *Herbal pharmacology in the People's Republic of China*: 51. National Academy of Sciences, Washington D.C.

Anon. (1816). Podophyllum peltatum. *Curtis's Botanical Magazine* 43: t.1819.

Arens, H., Fischer, H., Leyck, S., Romer, A. & Ulbrich, B. (1979). Anti-inflammatory compounds from Plagiorhegma dubium cell culture. *Planta Medica (Stuttgart)* 9: 32–36.

Arens, H., Ulbrich, B., Fischer, H., Parnham, M.J., Romer, A. (1986). Novel antiinflammatory flavanoids from Podophyllum versipelle cell culture. *Planta Medica (Stuttgart)* 6: 468–473.

Arroyo, M.T.K. (1981) Breeding systems and pollination biology in Leguminosae. In: Polhill, R.M. & Raven, P.H. (eds.), *Advances in Legume Systematics*: 723–769. Royal Botanic Gardens, Kew.

Arumugam, N. & Bhojwani, S. (1991). Somatic embryogenesis in tissue cultures of Podophyllum hexandrum. *Canadian Journal of Botany* 68: 487–491.

Arumugam, N. & Bhojwani, S. (1994). Chromosome variation in callus and regenerants of Podophyllum hexandrum (Podophyllaceae). *Caryologia* 47(3–4): 249–256.

Aswal, B.S. & Mehrotra, B.N. (1985). Contribution to the flora of Lahul valley (N.W.Himalayas) — V. Phytogeographical aspects. *Journal of Economic Taxonomic Botany* 7(2): 299–305

Badhwar, R.L. & Sharma, B.K. (1963). A note on the germination of Podophyllum seeds. *The Indian Forester* 89: 445–447.

Bao, S.Y. (1987). New taxa of Berberidaceae from S.W.China. *Acta Phytotaxonomica Sinica* 25(2): 150–159.

Benner, B.L. & Watson, M.A. (1989). Developmental ecology of Mayapple: seasonal patterns of resource distribution in sexual and vegetative rhizome systems. *Functional Ecology* 3: 539–547.

Benson, L. (1979). *Plant Classification*, ed. 2: 146. Heath & Co., Lexington.

Bernhardt, P. (1975). *The pollination ecology of four species of local vernal herbaceous angiosperms*. M.S. Thesis, SUC at Brockport, New York (unpublished).

Bigelow, J. (1840). *Florula Bostoniensis* (3rd. ed.): 229. Boston.

Blakeslee, A.F. (1922). Variations in Datura due to changes in chromosome number. *American Naturalist* 56: 16–31.

Blakeslee, A.F. (1931). Extra chromosomes, a source of variation in the Jimson weed. *Smithsonian Report* 1930: 431–450.

Bracchi, R. & Routledge, P. (1996). Chinese herbal therapies: what GPs need to know. *Prescriber* 5th October 1996: 91–94.

Brickell, C. & Sharman, F. (1986). *The vanishing garden*: 182–183. John Murray/RHS, London.

Britton, N.L. & Brown, A. (1913). *An illustrated flora of the Northern United States and Canada*. New York. Dover reprint (1970) 2: 129–130.

Broomhead, J.A. & Dewick, P.M. (1990). Tumour-inhibitory aryltetralin lignans in Podophyllum versipelle, Diphylleia cymosa and Diphylleia grayi. *Phytochemistry* 29(12): 3831–3837.

Broomhead, J.A., Rahman, M.M.A., Dewick, P.M., Jackson, D.E. & Lucas, J.A. (1991). Matairesinol as precursor of Podophyllum lignans. *Phytochemistry* 30(5): 1489–1492.

Browicz, K. (1973). Podophyllaceae. In: Rechinger, K.H. (ed.) *Flora Iranica* 101: 1–2. Akademische Druck, Graz.

Chang, F.C., Chiang, C. & Nambi Aiyar, V. (1975). Isopicropodophyllone from Podophyllum pleianthum. *Phytochemistry* 14: 1440.

Chang, X., Hu, Z. & Zeng, G. (1980). Studies on the chemical constituents of Chinese medicine Wo-Er-Chi (Diphylleia sinensis Li). *Acta Pharmacologica Sinica* 15(3): 158–162.

Chapman, M. (1936). Carpel anatomy of the Berberidaceae. *American Journal of Botany* 23: 340–348.

Chatterjee, R. (1952). Indian Podophyllum. *Economic Botany* 6(4): 342–352.

Chatterjee, R. (1953). Studies in Indian Berberidaceae. *Records of the Botanical Survey of India* 16(2): 43–51.

Chatterjee, R. & Chakravarti, S.C. (1952). Resin sikkimensis I. *Journal of the American Pharmacological Association, Science Edition* 41: 415–419. *Chemical Abstracts* 47: 5920.

Chatterjee, R. & Datta, D.K. (1950). Podophyllum II. *Indian Journal of Physiology* 4: 61–65. *Chemical Abstracts* 45: 7567.

Chaudhri, I.I. (1956). Medicinal plants of West Pakistan: Podophyllum emodi L. *Pakistan Journal of Science* 8(5): 230–233.

Chen, C., Huang, Y.-C. & Pi, C.-P. (1977). Analysis of heteromorphic karyotype in Dysosma pleiantha Hance. *Memoirs of College of Agriculture, National Taiwan University* 17(1): 39–42.

Chen, J-S.Z.(ed.) (1987). *Chinese toxic plants*. Science Press, Beijing.

Chien, S.S. (1936). Vascular plants from Chekiang: Berberidaceae. *Contributions from Biological Laboratory of the Chinese Association for the Advancement of Science, Section Botany*, 10(2): 108–110.

Child, A. & Lester, R. (1991). Life form and branching within the Solanaceae. In: Hawkes, J.G. et al., *Solanaceae III: Taxonomy, Chemistry, Evolution*: 151–159. RBG Kew and Linnean Society of London.

Chopra, R.N., Chopra, I.C., Handa, K.L., Kapur, L.D. (1958). *Chopra's Indigenous drugs of India*, ed. 2: 226– 229. Dhur & Sons, Calcutta.

Chuang, M-J. & Chang, W-C. (1987). Embryoid formation and plant regeneration in callus cultures derived from vegetative tissues of Dysosma pleiantha. *Journal of Plant Physiology* 128: 279–283.

Clark, L. (1923). The embryogeny of Podophyllum peltatum L. *Minnesota Studies in Plant Sciences* 4: 111– 137.

Clausen, K.S. & Hu, S.Y. (1980). Mapping the collecting localities of E.H.Wilson in China. *Arnoldia* 40(3): 139–145.

Clement, E.J. & Foster, M.C. (1994). *Alien plants of the British Isles*. Botanical Society of the British Isles.

Clute, W.N. (1915). A May-apple with multiple fruits. *American Botanist* 21: 92–93.

Clute, W.N. (1918). Podophyllum peltatum polycarpum. *American Botanist* 24: 113.

Collett, H. (1902). *Flora Simlensis*: 22. Thacker & Spink, Calcutta and Simla.

Corner, E.J.H. (1976). *The seeds of Dicotyledons* 1: 27–29, 218–219. Cambridge University Press.

Cox, D.D. (1985). *Common flowering plants of the Northeast — Their natural history and uses*: 45, t.21. State University of New York Press, Albany.

Cox, K. (ed.) (2001). *Frank Kingdon Ward's Riddle of the Tsangpo Gorges*: 159, 160. Antique Collectors Club, Woodbridge.

Dafni, A. (1992). *Pollination ecology — a practical approach*. IRL Press at Oxford University Press, Oxford.

Dana, W.S. (1918). *How to know the wild flowers*. Dover publications (reprinted 1963).

Darlington, C.D. (1936). The analysis of chromosome movements, 1. Podophyllum versipelle. *Cytologia* 7: 242–247.

Darlington, C.D. & Wylie, A.P. (1955). Chromosome atlas of flowering plants: 27–28. Allen & Unwin, London.

Davis, G.L. (1966). *Systematic embryology of the Angiosperms*: 214–215. J.Wiley & Sons, New York.

Davis, P.H. & Heywood, V.H. (1963). *Principles of angiosperm taxonomy*. Oliver & Boyd, Edinburgh and London.

Deam, C.C. (1940). *Flora of Indianapolis* 475–476. Indianapolis.

DeMaggio, A.E. & Wilson, C.L. (1986). Floral structure and organogenesis in Podophyllum peltatum. *American Journal of Botany* 73(1): 21–32.

Deno, N.C. (1993). *Seed germination theory and practice*, ed. 2. Published by author, State College, PA.

Dennis, J. (1997). More on Gibberellic acid. *Bulletin of the Alpine Garden Society* 65(4): 378–381.

Dewick, P.M. (1989). Tumour inhibitors from plants. In: Evans, W.C., *Trease and Evans' Pharmacognosy*, ed. 13: 637–656. Bailliere Tindall, London.

Dewick, P.M. & Jackson, D.E. (1981). Cytotoxic lignans from Podophyllum, and the nomenclature of aryltetralin lignans. *Phytochemistry* 20(9): 2277–2280.

Dewick, P.M. & Shaw, J.M.H. (1988). Podophyllums. *The Garden*. 113(5): 233–238.

Dickson, A. (1882). On the germination of Podophyllum emodi. *Transactions of the Botanical Society of Edinburgh* 16: 129–130.

Diels, L. (1901). Die flora von Central-China. *Botanische Jahrbuch* 25(2–5). Leipzig.

Dormer, K.J. (1972). *Shoot organization in vascular plants*: 196. Chapman & Hall, London.

Dymock, W., Warden, C.J.H. & Hooper, D. (1890). *Pharmacographica Indica*. Keegan Paul, London. Reprinted as *Hamdard* 15(1–2) (Jan.–Mar. 1972). Pakistan.

Ellis, S. & Fell, K.R. (1962). The morphology and anatomy of the leaf of Podophyllum hexandrum Royle. *Journal of Pharmacy and Pharmacology* 14: 573–586.

Ellis, S. & Fell, K.R. (1963). The morphology and anatomy of the leaf of Podophyllum peltatum L. *Journal of Pharmacy and Pharmacology* 15: 251–267.

Erichsen-Brown, C. (1979). *Medicinal and other plant uses of North American plants*. General Publishing Co., Toronto.

Ernst, W.R. (1964). The genera of the Berberidaceae, Lardizabalaceae and Menispermaceae in the southeastern United States. *Journal of the Arnold Arboretum* 45(1): 1–35.

Evans, W.C. (1987). *Trease and Evans' Pharmacognosy* (ed. 13) Bailliere Tindall, London.

Everard, B. & Morley, B.D. (1970). *Wild flowers of the world*: t.91c. Rainbird reference books, London, for Ebury press and Michael Joseph.

Faegri, K. & van der Pijl, L (1979). *The principles of pollination ecology*, ed 3. Pergamon Press, Oxford.

Farnsworth, N.R., Blomster, R.N., Quimby, M.W. & Schermerhorn, J.W. (1974). *The Lynn Index, a bibliography of phytochemistry* Monograph VIII [Berberidaceae pp.70–83]. University of Illinois, Chicago. published by N.R.Farnsworth.

Federov, A.N.A. (1969). *Chromosome numbers of flowering plants*. Acadamy of Sciences, USSR.

Ferguson, I.K. & Skvarla, J.J. (1982). Pollen morphology in relation to pollinators in Papillionoideae (Leguminosae). *Botanical Journal of the Linnean Society* 84: 183–193.

Fischer, M. (1975). In: The Veronica hederifolia group: taxonomy, ecology, and phylogeny. In: Walters, S.M. & King, C.J. (eds.), *European floristic and taxonomic studies*: 48–60. E.W.Classy, Oxford for BSBI.

Foerste, A.F. (1884). The May Apple. *Bulletin of the Torrey Botanical Club* 11: 62–64.

Forbes, E.B & Hemsley, W.B. (1886). Index Florae Sinensis. *Journal of the Linnean Society, Botany* 23: 1–489.

Foster, S. (1989). Phytogeographic and botanical considerations of medicinal plants in eastern Asia and eastern North America. In: Cracker, L.E., Simon, J.E (eds.), *Herbs, Spices and Medicinal Plants; Recent advances in Botany, Horticulture and Pharmacology* 4: 117–144. Oryx Press, Phoenix, Arizona.

Fournier, P. (1932). Voyages et decouvertes scientifiques des missionnaires naturalistes francais a travers le mond (XVe a XXe siecles). *Encyclopedie Biologique* X. P.Lechevalier & Fils, Paris.

Franchet, A.R. (1895). Sur quelques plantes de la Chine occidentale. *Bulletin du Museum Paris* 1: 62–66.

Frye, D.M. (1977). Seasonal changes in the morphology and physiology of the subterranean portions of Podophyllum peltatum L. Ph.D. Thesis, Rutgers University, New Brunswick, N.J. (unpublished).

Fu, S.H. (1976). *Flora Hupehensis* 1: 400–403.

Fu, L. & Jin, J. (eds.) (1992). *China plant red data book*, Vol. 1. Academic Press, Beijing.

Fukuda, I. (1967). The biosystematics of Achlys. *Taxon* 16: 308–316.

Fukuda, I. & Baker, H. (1970). Achlys californica (Berberidaceae) a new species. *Taxon* 19: 341–344.

Gagnepain, F. (1938). Treize species nouvelles d'Extreme-Orient. *Bulletin de la Société Botanique de France* 85: 165–171.

Gagnepain, F. (1939). In: Humbert, J.H. *Supplement Flora Generale de l'Indo-Chine* 1: 145–147, t.13.

Gibbs, R.D. (1974). *Chemotaxonomy of flowering plants* 1: 673–674; 3: 1584. McGill-Queen's University Press, Montreal & London.

Gray, A. (1848). *The genera of the plants of the United States* 1: 87, t.35 & 36. J.Munroe & Co, Boston.

Gray, A. (1868). *A manual of the botany of the northern United States*, ed. 5. Ivison, Blakeman, Taylor & Co., New York.

Gremmen, J. (1987). Notes on the Mayapple and the life history of the fungus Septotinia podophyllina. *Mycotaxon* 22(1): 255–256.

Grierson, A.J.C. & Long, D.G. (1984). *Flora of Bhutan* 1(2): 328–329, t.26h–i. Royal Botanic Gardens, Edinburgh.

Griffiths, M. (1994). *Index of Garden Plants*: 913. Royal Horticultural Society & Macmillan Press, London & Basingstoke.

Griggs, B. (1981). Green Pharmacy, a history of herbal medicine; cited in Steffey, J., Strange relatives: the Barberry family. *American Horticulturalist (Alexandria)* 64(4): 4–9.

Guedes, M. (1977). Le gynecee des Podophyllum (Berberidaceae). *Comptes Rendus Hebdomadaires des séances de l'Academie des Sciences Paris.* D 285: 755–758.

Gupta, R. & Sethi, K.L. (1983). Conservation of medicinal plant resources in the Himalayan region. In: Jain, S.K. & Mehra, K.L. (eds.), *Conservation of Tropical plant resources*: 101–107. Botanical Survey of India, Howrah.

Halsted, B.D. (1894). Pistillodia of Podophyllum stamen. *Bulletin of the Torrey Botanical Club* 21: 269.

Hamel, P.B. & Chiltoskey, M.U. (1975). *Cherokee Plants*. N.C. Herald publishing Co., Sylva.

Hance, H.F. (1883a). Podophyllum, a Formosan genus. *Journal of Botany (London)* 21: 174–175.

Hance, H.F. (1883b). A second new Chinese Podophyllum. *Journal of Botany (London)* 21: 361–363.

Handel-Mazzetti, H. (1924). Planta novae Sinenses. *Anzeiger der Akademie der Wissenschaften in Wien. Mathematische-naturwissenschaftliche Klasse* 61: 163–164.

Handel-Mazzetti, H. (1931). *Symbolae Sinicae* 7(2): 322–323. Wien.

Hao, K.S. (1934) Planta novae Sinicae. *Repertorium Specierum novarum regni vegetabilis* 36: 222–224.

Hara, H. (1971). *The flora of the eastern Himalaya*, second report: 34–35, 355–356. University of Tokyo Press, Tokyo.

Hara, H (1979). Podophyllum. In: Hara, H & Williams, L.H.J.(eds.), *An enumeration of the flowering plants of Nepal.* 2: 31–32. The Natural History Museum, London.

Harris, J.A. (1909). The leaves of Podophyllum. *The Botanical Gazette (Crawfordsville)* 47: 438–444.

Harvey-Gibson, R.J. & Horsman, E. (1919). Contributions towards a knowledge of the lower Dicotyledons. II. The anatomy of the stem of the Berberidaceae. *Transactions of the Royal Society of Edinburgh* 52(3): 501–515.

Hayata, B. (1915). *Icones Plantarum Formosanarum, nec non et contributiones ad floram Formosanam.* 5: 2–4, fig.1. Bureau of Productive Industries, Government of Taiwan, Taihoku.

Hedge, I.C. (1967). The specimens of Paul Dietrich Giseke in the Edinburgh Herbarium. *Notes from the Royal Botanic Garden Edinburgh* 28(1): 73–86.

Hedge, I.C. & Lamond, J.M. (1970). *Index of collectors in the Edinburgh Herbarium.* HMSO, Edinburgh.

Hemsley, W.B. (1906). XXV — Some new Chinese plants. *Bulletin of Miscellaneous Information, Kew* 1906: 147–163.

Heyenga, G. (1989). *Tissue culture of Podophyllum hexandrum and production of anticancer lignans.* Ph.D. thesis. University of Nottingham (unpublished).

Hickey, L.J. (1979). A revised classification of the architecture of dicotyledonous leaves. In: Metcalf, C.R. & Chalk, L. *Anatomy of the Dicotyledons* (ed. 2) 1: 25–39. Clarendon Press, Oxford.

Himmel, W.J. (1927). A contribution to the biophysics of Podophyllum petioles. *Bulletin of the Torrey Botanical Club* 54: 419–451.

Hinkley, D.J. (1999). *The Explorer's Garden*. Timber Press, Oregon.

Hiroe, M. (1973). The plants of Basho's and Buson's hokku literature. In: Hiroe, M. *The plants of classical literature: classical taxonomy (honzo) and Asiatic plants* 8, parts 1–3. Ariake Book Co., Tokyo. [Supplement 5: The Asiatic plants No. 1: pp.319–348. Dysosma p.326–329.]

Holm, T. (1899). Podophyllum peltatum. A morphological study. *Botanical Gazette (Crawfordsville)* 27(6): 419–433.

Hooker, J.D. (1875). *Flora of British India* 1: 112–113. L.Reeve & Co., Ashford.

Hooker, J.D. (1890). Podophyllum pleianthum. *Curtis's Botanical Magazine* 116: t.7098.

Hooker, J.D. (1891). Podophyllum versipelle. *Hooker's Icones Plantarum* 20: t.1996.

Howard, R.A. (1980). E.H.Wilson as a Botanist, Part 1, *Arnoldia* 40(3): 102–138. Part 2, *Arnoldia* 40(4): 154–193.

Hu, H.H. (1937). Notulae systematicae ad florem Sinensium. *Bulletin of the Fan Memorial Institute, Botanical Series* 8: 31–46.

Hu, S-Y. (1945). Medicinal plants of Chengtu herb shops. *Journal of West China Border Research Society* 15, Series B: 95–177.

Hutchinson, J. (1907). Podophyllum versipelle. *Curtis's Botanical Magazine* 133: t.8154.

Hutchinson, J. (1920). Jeffersonia and Plagiorhegma. *Bulletin of Miscellaneous Information, Kew* 5: 242–245.

Iconographia Cormophytorum Sinicorum (1980), ed. 2, 1: 758–761, t.1516–1521. Beijing.

Iriki,Y., Fukuda, Y., Kakizaki, K., Nakazawa, M., Uesugi, Y., Kozu, T. (1984). Phytochemical study of the cell wall of Berberidaceae. *Nippon Nogeikagaku Kaishi*. 54(4): 367–371.

Ito, T. (1887). Berberidearum Japonicae conspectus. *Journal of the Linnean Society, Botany* 22: 422–437.

Jackson, D.E. & Dewick, P.M. (1984a). Biosynthesis of Podophyllum lignans — I. Cinnamic acid precursors of podophyllotoxin in Podophyllum hexandrum. *Phytochemistry* 23(5): 1029–1035.

Jackson, D.E. & Dewick, P.M. (1984b). Biosynthesis of Podophyllum lignans — II. Interconversions of aryltetralin lignans in Podophyllum hexandrum. *Phytochemistry* 23(5): 1037–1042.

Jackson, D.E. & Dewick, P.M. (1984c). Aryltetralin lignans from Podophyllum hexandrum and Podophyllum peltatum. *Phytochemistry* 23(5): 1147–1152.

Jackson, D.E. & Dewick, P.M. (1985). Tumour-inhibitory aryltetralin lignans from Podophyllum pleianthum. *Phytochemistry* 24(10): 2407–2409.

Jafri, S.M.H. (1974). Podophyllaceae In: Nasir, E. & Ali, S.I. (eds.), *Flora of West Pakistan* 57: 1–4, t.1. Stewart Herbarium, Gordon College and University of Karachi; Rawalpindi and Karachi.

Jain, S.K. & DePhillipps, R.A. (1991). *Medicinal plants of India*. Reference Books Publ. Inc., Michigan.

Jansen, U. (1973). The interpretation of comparative serological results. In: Bendz, G. & Santesson, J. (eds.), *Chemistry in Botanical Classification*: 217–227. Nobel Symposium 25. Academic Press, London and New

Jarvis, C.E., Barrie, F.R., Allan, D.M. & Reveal, J.L. (1993). A list of Linnean generic names and their types. *Regnum Vegetabile* 127: 77. IAPT, Koeltz Sci. Books, Koenigstein.

Jiang, Z & Chen, S. (1989). Chemical components of Dysosma veitchii. *Yunnan zhiwu yanjiu* 11(4): 479–481. *Chemical Abstracts* 112: 232531h.

Kaiyun, G. (ed.) (1998). *Highland Flowers of Yunnan*: 37. Yunnan Science & Technology Press, Kunming.

Kala, C.P. (2000). Status and conservation of rare and endangered medicinal plants in the Indian trans-Himalaya. *Biological Conservation* 93: 371–379.

Kamil, W.M. & Dewick, P.M. (1986a). Biosynthetic relationship of aryltetralin lactone lignans to dibenzylbutyrolactone lignans. *Phytochemistry* 25(9): 2093–2102.

Kamil, W.M. & Dewick, P.M. (1986b). Biosynthesis of the lignans α- and ß-peltatin. *Phytochemistry* 25(9): 2089–2092.

Kapahi, B.K. (1990). Ethno-botanical investigation in Lahaul (Himacal Pradesh). *Journal of Economic Taxonomic Botany* 14(1): 49–55.

Kaufmann, B.P. (1926). Chromosome structure and its relation to the chromosome cycle. II. Podophyllum peltatum. *American Journal of Botany* 13: 355–363.

Kawakami, T. (1910). *A list of plants of Formosa*. Government of Formosa, Taihoku.

Kim, K. & Jansen, R.K. (1993). Phylogenetic studies of the Berberidaceae: Integration of chloroplast DNA and morphological data. *American Journal of Botany* 80(6): 157–158.

Kim, Y.D. & Jansen, R.K. (1996). Phylogenetic implications of rbcl and ITS sequence variation in the Berberidaceae. *Systematic Botany* 21(3): 381–396.

Kimura, Y. (1963). The cytological effects of chemicals on tumors, XXI. Notes on the effects of crude extracts from Japanese podophyllaceous plants on transplantable rat and mouse ascites tumors. *Journal of the Faculty of Science Hokkaido University*, Ser. 6, Zoology 15: 264–271.

Kingdon Ward, F. (1930a). *Field notes of Rhododendrons and other plants collected by Kingdon Ward in 1927/28*. Published privately.

Kingdon Ward, F. (1930b). *Plant hunting on the edge of the world*: 327.

Kofod, H. & Jorgensen, C. (1954). Dehydropodophyllotoxin, a new compound isolated from Podophyllum peltatum L. *Acta Chemica Scandinavica* 8(2): 1296–1297.

Kosenko, V.N. (1979). Comparative karyological study of representatives of the family Berberidaceae. *Botaniceskij Zhurnal (Moscow & Leningrad)* 64(11): 1539–1552.

Kosenko, V.N. (1980). Comparative study of the family Berberidaceae.I. Morphology of the genera Diphylleia, Podophyllum. *Botaniceskij Zhurnal (Moscow & Leningrad)* 65(2): 198–205.

Krishnamurthy, T., Karira, G.V., Sharma, B.K. & Bhatia, K. (1965). Cultivation and exploitation of Podophyllum hexandrum Royle. *The Indian Forester* 91(7): 470–475.

Krochmal, A. (1968). Medicinal plants in Appalachia. *Economic Botany* 22: 332–337.

Krochmal, A., Wilkins, L., Van Lear, D. & Chien, M. (1974). *Mayapple (Podophyllum peltatum L.)*. USDA Forest Service Research Paper NE-296.

Kudo, Y. & Masamune, G. (1932). Genera plantarum Formosanarum. *Annual Report of the Taihoku Botanic Garden* No.2, 1: 81.

Kumazawa, M. (1930). Morphology and biology of Glaucidium palmatum Sieb. & Zucc. with notes on affinities the allied genera Hydrastis, Podophyllum and Diphylliea. *Journal of the Faculty of Science, Imperial University of Tokyo* Series B., Botany 2(4): 345–380.

Kumazawa, M. (1936). Podophyllum pleianthum Hance. A morphological study, with supplementary notes on allied plants. *Botanical Magazine (Tokyo)* 50: 268–276.

Kurita, M. (1956). Karyotype studies in Berberidaceae. *Memoirs of Ehime University, Section 2 Natural Sciences, Series B (Biology)*, 2: 247–252.

Kuroki, Y. (1965). Chromosome study in three species of Berberidaceae. *Memoirs of Ehime University, Section 2 Natural Sciences, Series B (Biology)* 5(2): 19–24.

Kuroki, Y. (1967). Chromosome studies in seven species of Berberidaceae. *Memoirs of Ehime University, Section 2 Natural Sciences, Series B (Biology)* 5: 175–181.

Langlet, O. (1928). Einige beobachtungen uber die zytologie der Berberidazeen. *Svensk Botanisk Tidskrift* 22, II (1–2): 169–184.

Lauener, L.A. (1962). Catalogue of the names published by Hector Leveille: II. *Notes from the Royal Botanic Gardens Edinburgh* 24(1): 73–78.

Laverty, T.M. (1992). Plant interactions for pollinator visits: a test for the magnet species effect. *Oecologia* 89(4): 502–508.

Laverty, T.M. & Plowright, R.C. (1988). Fruit and seed set in Mayapple (Podophyllum peltatum): Influence of intraspecific factors and local enhancement near Pedicularis canadensis. *Canadian Journal of Botany* 66: 173–178.

Leveille, H. (1912). Decades plantarum novarum XCIII–C. *Repertorium Specierum Novarum Regni Vegetabilis* 11: 295–307.

Li, H.L. (1952). Floristic relationships between eastern Asia and eastern North America. *Transactions of the American Philosophical Society*, New Series 42(2): 371–429.

Li, H.-L. (1977). Hallucinogenic plants in Chinese herbals. *Botanical Museum Leaflets (Harvard University)* 25(6): 161–181.

Li, L. (1986). A study on the karyotypes and evolution of Dysosma pleiantha with its relatives. *Acta Botanica Yunnanica* 8(4): 451–457.

Ling, C.S. & Yu, D.Q. (1963). Studies on the constituents of rhizomes and roots of Podophyllum versipelle Hance. *Acta Pharmacologica Sinica* 10: 489–495.

Linnaeus, C. (1737). *Critica botanica* (1938 reprint, translated by A.Hort, revised M.L.Green.) Ray Society, London.

Linnaeus, C. (1753). *Species Plantarum* (1957 reprint) Ray Society, London.

Litardière, M.R.De (1921). Remarque au sujet de quelques processus chromosomiques dans les noyaux diploidiques du Podophyllum peltatum. *Comptes Rendus Hebodomadaires des séances de l'Academie de Sciences* 172: 1066–1069.

Liu, T-S. (1976). Podophyllum. In: Li, H-L., Li, T.S., Huang, T-S., Koyama, T., DeVol, C.E. (eds.), *Flora of Taiwan* 2: 520–521, t.396.

Loconte, H. (1993). Berberidaceae. In: Kubitzki, K. Rohwer, J. & Bittrich, V. (eds.), *The families and genera of vascular plants* 2: 147–152. Dicotyledons. Springer-Verlag, Berlin.

Loconte, H. & Blackwell, W.H. (1984). The Berberidaceae of Ohio. *Castanea* 49(1): 39–43.

Loconte, H. & Estes, J.R. (1989). Phylogenetic systematics of Berberidaceae and Ranunculales (Magnoliidae). *Systematic Botany* 14(4): 565–579.

Lublinerowna, K. (1925a). Recherches sur le development de l'ovule et de la graine dans le genre Podophyllum. *Bulletin International de l'Académie Polonaise des Sciences, Classe des sciences mathémathiques et naturelles. Série B,* 1925: 379–402.

Lublinerowna, K. (1925b). Uber die Plastiden in der Eizelle von Podophyllum peltatum. *Acta Societatis Botanicorum Poloniae* 2: 225–227.

Lublinerowna, K. (1925c). Recherches sur le development des teguments ovulaires et seminaux dans le genre Podophyllum. *Acta Societatis Botanicorum Poloniae* 3: 277–282.

Ma, C., Yang, T.S., Luo, S.R. (1993). Lignans from Diphylleia sinensis. *Yaoxue xuebao* 28(9): 690–694. *Chemical Abstracts* 120: 4678w.

Ma, S. & Hu, Z. (1996). Preliminary studies on the distribution pattern and ecological adaptation of Sinopodophyllum hexandrum (Royle) Ying. *Journal of Wuhan Botanical Research* 14(1): 55–57.

Ma, S., Xu, Z. & Hu, Z. (1997). A contribution to the reproductive biology of Sinopodophyllum hexandrum (Royle) Ying (Berberidaceae). *Acta Botanica Boreali-Occidentalia Sinica* 17(1): 49–55.

Ma, S.-B. & Hu, Z.-H. (1996). A karyotypic study on Podophylloideae. *Acta Botanica Yunnanica* 18(3): 325–330, 1 plate at end of issue.

McLean, R.C. & Ivimey-Cook, W.R. (1956). *A Textbook of Theoretical Botany* Vol. 2. Longmans, Green & Co., London.

McMullen, J.M. & Clovis, J.F. (1975). Anatomical variation in Podophyllum peltatum L. due to aspect and elevation. *Proceedings of West Virginia Academy of Science*, Biology Section 46(3–4): 274–280.

Malla, S.B. (ed.) (1976). *Flora of Langtang*. Department of medicinal plants, Kathmandu, Nepal.

Malla, S.B., Bhattarai, S., Gorkhali, M., Saija, H., Kayastha, M. (1981). Chromosome count reports. *Taxon* 30(1): 75.

Martin, F.W. (1958). *Variation and morphology of Podophyllum peltatum*. Ph.D thesis, Washington University (unpublished). *Dissertation abstracts* 19(3): 424–425 (1958).

Mathews, F.S. (1912). *Field book of American wild flowers*. G.P.Putnam's Sons.

Meacham, C.A. (1980). Phylogeny of the Berberidaceae with an evaluation of classifications. *Systematic Botany* 5(2): 149–172.

Meijer, W. (1974). Podophyllum peltatum — May Apple, a potential new cash-crop plant of eastern North America. *Economic Botany* 28: 68–72.

Melville, R. (1983). The affinity of Paeonia and a second genus of Paeoniaceae. *Kew Bulletin* 38(1): 87–105.

Metcalf, C.R. & Chalk, L. (1979). *Anatomy of the Dicotyledons*, ed. 2, Vol. 1. Clarendon Press, Oxford.

Miyaji, Y. (1930). Beitrage zur chromosomenphylogenie der Berberidaceen. *Planta* 11: 650–659.

Miyase, T., Ueno, A., Takizawa, N., Kobayashi, H. & Oguchi, H. (1989). Ionone and lignan glycosides from Epimedium diphyllum. *Phytochemistry* 28(12): 3483–3485.

Moerman, D.F. (1986). Medicinal plants of native America. *University of Michigan, Museum of Anthropology Technical Reports* 19, Vol. 1: 354.

Moore, D.M. (1982). *Flora Europaea check-list and chromosome index*. Cambridge University Press.

Morley, B.D. (1979). Augustine Henry: his botanical activities in China, 1882–1890. *Glasra (Contributions from the National Botanic Garden, Glasnevin)* 3: 21–81.
Morley, B.D. (1980). Augustine Henry. *The Garden* 105(7): 285–289.
Morley, B.D. (1986). Podophyllum versipelle — a letter. *The Garden* 111(8): 395.
Morton, J.F. (1977). *Major Medicinal Plants*. Charles C. Thomas, Springfield, Illinois.
Mottier, D.M. (1897). Beitrage zur Kenntniss der kerntheilung in der Pollenmutterzellen einiger Dikotylen und Monokotylen. *Jahrbücher für Wissenschaftliche Botanik* 30: 169–204.
Mottier, D.M. (1905). The development of the heterotypic chromosomes in pollen mother cells. *Botanical Gazette (Crawfordsville)* 40: 71–177.
Muhling, G.N. & Wilson, G.B. (1961). The chromosomes of Podophyllum peltatum. *Rhodora* 63: 267–275.
Nadkarni, K.M. (revised Nadkarni, A.K.) (1976). *The Indian Materia Medica* Vol.1: 994–995. Bombay.
Naithami, H.B. (1990). *Flowering plants of India, Nepal and Bhutan* (not recorded in Sir. J.D.Hooker's Flora of British India): 29. Surya publications, Dehra Dun.
Nautiyal, M.C. (1988). Germination studies on some high altitude medicinal plants species. In: Kaushik, P. (ed.), *Indigenous medicinal plants symposium*: 107–112. New Delhi.
Nautiyal, M.C. (1995). Agro-technique of some high altitude medicinal herbs. In: Sundriyal, R.C. & Sharma, E. (eds.), *Cultivation of Medicinal Plants and Orchids in Sikkim Himalaya*. Himavikas occasional publication no. 7. Bishen Singh Mahendra Pal Singh, Dehra Dun.
Nautiyal, M.C., Rawat, A.S., Bhadula, S.L. & Purohit, A.N. (1987). Seed germination in Podophyllum hexandrum. *Seed Research* 15(2): 206–209.
Newman, L.J. (1959). Chromosomal aberrations in Podophyllum peltatum. *Evolution* 13: 276–279.
Newman, L.J. (1966). Bridge and fragment aberrations in Podophyllum peltatum. *Genetics* 53: 55–63.
Noda, S. & Fujimura, T. (1970). Karyotypes in root-tip cells and endosperm nucleus of Diphylleia grayi. *Kromosomo* 79–80: 2548–2551.
Nowicke, J.W. & Skvarla, J.J. (1981). Pollen morphology and phylogenetic relationships of the Berberidaceae. *Smithsonian Contributions to Botany* 50: 1–83.
Oganozova, G.G. (1974). Anatomical structure of the leaf in Berberidaceae s.l. related to the taxonomy of the family. *Botaniceskij Zhurnal (Moscow & Leningrad)* 59(12): 1780–1793.
Overton, J.B. (1905). Ueber Reduktionsteilung in den Pollenmutterzellen einiger Dikotylen. *Jahrbücher für Wissenschaftliche Botanik* 42: 121–154.
Overton, J.B. (1909). The organization and reconstruction of the nuclei of Podophyllum peltatum. *Proceedings of the British Association for the Advancement of Science* 1909: 678–679.
Overton, J.B. (1922). The organization of the nuclei in the root tips of Podophyllum peltatum. *Transactions of the Wisconsin Academy of Sciences*, Art. Lett. 20: 275–320.
Parker, A.G. (1910). Iroquois uses of Maize and other food plants. *New York State Museum Bulletin* 1910: 144.
Parker, M.A. (1989). Disease impact and local genetic diversity in the clonal plant Podophyllum peltatum. *Evolution* 43(3): 540–547.
Phillips, R. & Rix, M. (1991). *Perennials* 1: 31–32. Pan Books, London.
Plitt, C.C. (1931). Two abnormalities in Podophyllum peltatum. *Rhodora* 33: 228–229.

Policansky, D. (1983). Patches, clones and self-fertility of Mayapples. *Rhodora* 85(843): 253–256.

Polunin, O. & Stainton, A. (1984). *Flowers of the Himalaya*: 23. T.11. Oxford University Press.

Poole, M.M. (1981). Pollen diversity in Zimmermannia (Euphorbiaceae). *Kew Bulletin* 36(1): 129–138.

Porter, T.C. (1877). Variations in Podophyllum peltatum Linn. *Botanical Gazette (Crawfordsville)* 2(9): 117–118.

Prasad, P. (2001). Impact of cultivation on active constituents of the medicinal plants Podophyllum hexandrum and Aconitum heterophyllum in Sikkim. *Plant Genetic Resources Newsletter* 124: 33–35.

Puri, H.S. & Jain, S.P. (1988). Ainsliaea latifolia: An adulterant of Indian Podophyllum. *Planta Medica* 54(3): 269.

Rafinesque-Schmaltz, C.S. (1817). *Florula ludoviciana — a flora of the state of Louisiana*. (Translated, revised and improved from the French by C.C.Robin). New York.

Rafinesque-Schmaltz, C.S. (1828). *Medical Flora; or manual of the medical botany of the United States of North America* 1: 59. Atkinson, Philadelphia.

Rai, L.K., Prasad, P. & Sharma, E. (2000). Conservation threats to some important medicinal plants of the Sikkim Himalaya. *Biological Conservation* 93: 27–33.

Rao, R.R. & Hajra, P.K. (1993). Podophyllum. In: Sharma, B.D. & Balakrishnan, N.P. *Flora of India* 1: 414– 416. Botanical Survey of India, Calcutta.

Raven, P.H. (1972). Plant species disjunctions: a summary. *Annals of the Missouri Botanical Garden* 59: 234– 246.

Raymond, M. (1948). A red-fruited form of Podophyllum peltatum. *Rhodora* 50: 18.

Richards, A. (1909). Mitosis in the root-tip cells of Podophyllum peltatum. *Kansas University Science Bulletin* 5(6): 87–93.

Rickett, F.C. (1966). *Wildflowers of the United States*. McGraw-Hill Co., New York.

Rix, M. (1982). The herbaceous Berberidaceae. *The Plantsman* 4(1): 1–15.

Royle, J.F. (1834). *Illustrations of the botany and other branches of the natural history of the Himalayan mountains and the Flora of Cachmere* 2(1): 64 (p.379 in footnote, 1839 edition). W.H.Allen & Co., London.

Rust, R.W. & Roth, R.R. (1981). Seed production and seedling establishment in the Mayapple, Podophyllum peltatum. *American Midland Naturalist* 105: 51–60.

Saunders, E.R. (1925). On carpel polymorphism.I. *Annals of Botany* 39: 123–167.

Saunders, E.R. (1928). Illustrations of carpel polymorphism.II. *New Phytologist* 27: 175–214.

Sawyer, M.L. (1926). Carpeloid stamens of Podophyllum peltatum. *Botanical Gazette (Crawfordsville)* 82: 329–332.

Schilling, E.E. & Calie, P.J. (1982). Petal flavonoids of white-flowered spring ephemerals. *Bulletin of the Torrey Botanical Club* 109(1): 7–12.

Scoggan, H.J. (1978). *The Flora of Canada* 3: 762. National Museum of Natural Sciences, Ottowa.

Selivanova-Gorodkova, E.A. (1969). On two Himalayan species of Podophyllum L. *Botaniceskij Zhurnal (Moscow & Leningrad)* 54(10): 1604–1605.

Selivanova-Gorodkova, E.A. (1975). Sectio nova generis Podophyllum L. *Novosti Sistematiki Vyssikh Rastenij* 12: 209–210.

Senior, R.M. (1965). Podophyllum emodi Wall. *Quarterly Bulletin of the Alpine Garden Society* 33(3): 258– 261.

Shao-Bin & Zhi-Hao (1996). A karyotypic study on Podophylloideae (Berberidaceae). *Acta Botanica Yunnanica* 18(3): 325–330.

Shaw, J.M.H. (1996). Notes on Podophyllum as potential British aliens. *BSBI News* 73: 43–45.

Shaw, J.M.H. (1999). Variation in Podophyllum hexandrum and its nomenclatural consequences. In: Andrews, S., Leslie, A.C. & Alexander, C. (eds.), *Taxonomy of cultivated plants: Third International Symposium*: 405–408. Royal Botanic Gardens, Kew.

Shaw, J.M.H. (2000). A taxonomic revision of Podophyllum in the wild and in cultivation. *The New Plantsman* 7(1): 30–41; 7(2): 103–113; 7(3): 142–159; 7(4): 220–235.

Shibata, S., Murata, T. & Fujita, M. (1961). Studies on the constituents of Japanese and Chinese drugs. VI. On the constituents of rhizome and roots of Podophyllum pleianthum Hance. *Journal of the Pharmacology Society of Japan* 82(5): 777.

Siddique, M.A.A., Wafai, B.A. & Dhar, U. (1990). Chromosome complement and nucleolar organization in Podophyllum hexandrum Royle. *Genetica* 82: 59–62.

Silva, C.G., Power, J.B. *et al.* (1998). Plant regeneration from root explants of the medicinal plant Podophyllum hexandrum cultured in liquid medium. *Experimental Botany* (Supplement) 49: 88–89.

Singh, G. & Kachroo, P. (1987). *Forest Flora of Srinagar and Plants of the Neighbourhood*: 163. Periodical Expert Book Agency.

Singh, V., Jain, D.K. & Sharma, M. (1974). Epidermal studies in the Berberidaceae and their taxonomic significance. *Journal of the Indian Botanical Society* 53: 271–276.

Singh, V., Jain, D.K. & Sharma, M. (1978). Leaf architecture in the Berberidaceae and its bearing on the circumscription of the family. *Journal of the Indian Botanical Society* 57(3): 272–280.

Smith, G.H. (1923). *Floral Anatomy of the Ranales* (unpublished thesis). Cornel University, New York..

Smith, H.H. (1923). Ethnobotany of the Menomini Indians. *Bulletin Public Museum Milwaukee* 4: 1–174.

Smith, H.H. (1928) Ethnobotany of the Meskwaki Indians. *Bulletin Public Museum Milwaukee* 4: 175–326.

Soejarto, D.D., Faden, R.B. & Farnsworth, N.R. (1979). Indian podophyllum: Is it Podophyllum emodi or Podophyllum hexandrum? *Taxon* 28(5/6): 549–551.

Soejarto, D.D., Greenwood, B.D., Lauener, L.A. & Farnsworth, N.R. (1981). Typification of Podophyllum hexandrum Royle. *Taxon* 30: 652–656, t.1.

Sohn, J. & Policansky, D. (1977). The cost of reproduction in the Mayapple, Podophyllum peltatum. *Ecology* 58: 1366–1374.

Sprague, T.A. (1920). Podophyllum hexandrum var. chinense. *Curtis's Botanical Magazine* 146: t.8850.

Stace, C.A. (ed.) (1975). *Hybridization and the flora of the British Isles*. Academic Press, London.

Stace, C.A. (1991). *New Flora of the British Isles*: 1020. Cambridge University Press.

Stearn, W.T. (c. 1958, no date). *Letter*, ref: WTS/EB/1326, to F.W.Martin, Northeast Louisiana State College. Kept in the general herbarium, Natural History Museum, London (unpublished).

Stearn, W.T. (1971). In: Smith, A.W., *A Gardener's Dictionary of Plant Names*. Cassell & Co., London.

Stearn, W.T. (1976). Frank Ludlow and the Ludlow-Sherriff expeditions to Bhutan and South-East Tibet of 1933–1950. *Bulletin of the British Museum (Natural History). Botany* 5(5): 243–268.

Stearn, W.T. (1978). Berberidaceae. In: Heywood, V.H.(ed.), *Flowering Plants of the World*: 45–46. Oxford University Press.

Stearn, W.T. (1989). Podophyllum. In: Walters, S.M. (ed.), *The European Garden Flora* 3: 395–396. Cambridge University Press.

Steffey, J. (1985). Strange relatives: the Barberry family. *American Horticulturist* 64(4): 4–9.

Steyermark, J.A. (1952). Color-forms of the may-apple. *Rhodora* 54: 131–134.

Steyermark, J.A. (1963). *Flora of Missouri*: 710–711. Iowa State University Press, Ames.

Strausbaugh, P.D. & Core, E.L. (1979). *Flora of West Virginia* ed. 2: 402–403. Seneca Books, Grantsville.

Su, Y., Chen, L., Gan, G. & Liu, Q. (1993). Studies on the anatomy and pollen morphology of Dysosma lichuananensis. *Journal of Wuhan Botanical Research* 11: 91–93 (two plates follow p. 96).

Su, Y., Chen, L., Gan, G. & Liu, Q. (1994). The anatomical study on Dysosma pleiantha. *Journal of Wuhan Botanical Research* 12: 295–298 (a plate follows p. 398).

Su, Y. & Liu, Q. (1992). Studies on the leaf epidermis of Dysosma in Hubei province. *Journal of Wuhan Botanical Research* 10(4): 385–386, t.1–3.

Su, Y. & Liu, Q. (1994). A study on the peroxidase isoenzymes of Dysosma from Hubei Province. *Journal of Wuhan Botanical Research* 12(1): 44–48.

Su, Y. & Liu, Q. (1994). A study on the pollen morphology of Dysosma from Hubei. *Journal of Wuhan Botanical Research* 12: 217–219.

Su, Y. & Liu, Q. (1994). Anatomical study on three species of Dysosma in Hubei province. *Journal of Wuhan Botanical Research* 12: 111–116.

Suwal, P.N. (ed.) (1970). Medicinal plants of Nepal. *Bulletin of Department of Medicinal Plants, Thapathali* 3: 124 & plate opposite. Government of Nepal, Kathmandu.

Swanson, S.D. & Sohmer, S.H. (1976a). Reproductive biology of Podophyllum peltatum. The comparative fertility of inter- and intra-populational crosses. *Transactions of the Wisconsin Academy of Science and Arts* Letters 64: 109–114.

Swanson, S.D. & Sohmer, S.H. (1976b). The biology of Podophyllum peltatum L., the May apple. II. The transfer of pollen and success of sexual reproduction. *Bulletin of the Torrey Botanical Club* 103(5): 223–226.

Tantaquidgeon, G. (1942). *A study of Delaware Indians medicine practice and folk beliefs*. Historical Commission, Harrisburg.

Terabayashi, S. (1982). Systematic consideration of the Berberidaceae. *Acta Phytotaxonomica Geobotanica* 33: 355–370.

Terabayashi, S. (1983). Studies in the morphology and systematics of Berberidaceae.VI. Floral anatomy of Diphylleia, Podophyllum and Dysosma. *Acta Phytotaxonomica Geobotanica* 34: 27–47.

Terabayashi, S. (1985). The comparative floral anatomy and systematics of the Berberidaceae. I. Morphology. *Memoirs of Faculty of Science Kyoto University*, Series Biology X: 73–90.

Terabayashi, S. (1987). Seedling morphology of the Berberidaceae. *Acta Phytotaxonomica Geobotanica* 36: 63–74.

Tischler, G. (1902). Die Berberidaceen und Podophyllaceen. *Botanische Jahrbucher* 31: 596–727.

Toyokuni, H. & Toyokuni, Y. (1964). Ein neuer anhalt für die teilung der Podophyllaceen in zwei unterfamilien. *Botanical Magazine (Tokyo)* 77: 197.

Torres, R., Delle Monache, F. & Marini-Bettolo, Q.B. (1979). Biogenetic relationships between lignans and alkaloids in Berberis genus. Lignans and berbamine from Berberis chilensis. *Journal of medicinal plant research* 37: 32–36.

Troup, R.S. (1915). A note on the cultivation of Podophyllum emodi. *Indian Forester* 41(10): 361–365.

Tucker, A.O., Duke, J.A., Foster, S. (1989). Botanical nomenclature of medicinal plants. In: Cracker, L.E., Simon, J.E. (eds.), *Herbs, spices and medicinal plants: Recent advances in Botany, Horticulture & Pharmacology* 4: 191–192. Oryx Press, Phoenix.

Uittien, H. (1928). Uber den Zusammenhang zwischen Blattnervatur und sprossverzweigung. *Recueil des Travaux Botaniques Néerlandais* 25: 390–493.

Vogel, S. (1990). *The role of scent glands in pollination.* Amerind pubishing Co., New Delhi.

Wadmond, S.C. (1898). Leaf retardation in Podophyllum peltatum. *Asa Gray Bulletin* 6: 66–67.

Wallis, T.E. & Goldberg, S. (1931). Podophyllum rhizome — American and Indian. *Quarterly Journal of Pharmacy and Pharmacology* 4: 28–32.

Wallis, T.E. & Goldberg, S. (1937a). The histology of Podophyllum. *Quarterly Journal of Pharmacy and Pharmacology* 10: 40–51.

Wallis, T.E. & Goldberg, S. (1937b). The histology of Indian Podophyllum. *Quarterly Journal of Pharmacy and Pharmacology* 10: 311–318.

Walter, K.S. & Gillett, H.J. (1998). *1997 IUCN Red List of Threatened Plants*: 74. IUCN, Cambridge.

Walter, K.S., Chamberlain, D.F., Gardiner, M.F. et al. (1995). *Catalogue of Plants*. Royal Botanic Gardens, Edinburgh.

Wang, Y.S. (1984). A new species of Dysosma from Guangxi. *Guihaia* 4(1): 43–44.

Wherry, E.T., Fogg, Jr., J.M. & Wahl, H.A. (1979). *Atlas of the Flora of Pennsylvania*. Morris Arboretum, Philadelphia.

Whetzel, H.H. (1937). Septotinia, a new genus of the Ciborioideae. *Mycologia* 29: 128–146.

Whisler, S.L. & Snow, A.A. (1992). Potential for the loss of self-incompatability in pollen-limited populations of Mayapple (Podophyllum peltatum). *American Journal of Botany* 79(11): 1273–1278.

Wilkinson, H.P. (1979). The plant surface (mainly leaf), Part 2 Hydathodes. In: Metcalf, C.R. & Chalk, L., *The anatomy of the Dicotyledons* ed 2: 117–124. Clarendon Press, Oxford.

Williams, J.C. (1925). *Some plants, shrubs and trees found by Mr. Forrest in 1924*. Published privately.

Witthoft, J. (1947). An early Cherokee ethnobotanical note. Journal of the Washington Academy of Sciences 37(3): 73–75.

Wood, C.E. (1972). Morphology and phytogeography: the classical approach to the study of disjunctions. *Annals of the Missouri Botanical Garden* 59: 107–124.

Woodson, R.E. (1928). Dysosma, a new genus of Berberidaceae. *Annals of the Missouri Botanical Garden* 15: 335–340.

Wordsell, W.C. (1908). A study of the vascular system in certain orders of the Ranales. *Annals of Botany* 22: 651–682.

Wu, C.Y. (ed.) (1987). *Wild flowers of Yunnan* 2: 308–309, t.67. Japan Broadcast Publishers Ltd.

Wu, S.K. (1974). *Notes on localities of Western botanists' plant collecting in Sichuan, Yunnan and Guizhou Provinces.* Committee for the Flora of China, Beijing.

Xiao, P-G. (1989). Excerpts of the Chinese Pharmacopoeia. In: Cracker, L.E. & Simon, J.E (eds.), *Herbs, Spices and Medicinal plants: Recent advances in Botany, Horticulture and Pharmacology* 4: 45, 87. Orynx Press, Phoenix.

Xu, Z., Ma, S., Hu, C., Yang, C. & Hu, Z. (1997). The floral biology and its evolutionary significance of Sinopodophyllum hexandrum (Royle) Ying. *Journal of Wuhan Botanical Research* 15(3): 223–227.

Yeo, P.F. (1992). A revision of Geranium L. in south-west China. *Edinburgh Journal of Botany* 49(2): 123–211.

Yin, M., Chen, C. & Wang, Q. (1987). [Chemical constituents of Podophyllum emodi var. chinense.] *Zhongcaoyao* 18(12): 535–538.

Yin, M. & Chen, Z. (1989a). Chemical constituents of Dysosma aurantiocaulis (H.-M.) Hu and D. pleiantha Woods. *Zhongguo zhongyao zazhi* 14(7): 420–421. *Chemical Abstracts* 111: 229003m.

Yin, M., Chen, Z., Gu, Z. & Xie, Y. (1989b). Dysoanthraquinone and 2-Demethyldysoanthraquinone from Dysosma majoensis. *Acta Chimica Sinica* (English Edition) 5: 468–470. *Chemical Abstracts* 112: 175628n.

Yin, M., Chen, Z., Gu, Z. & Xie, Y. (1990). Separation and identification of chemical constituents of Dysosma majorense (Gagn.) Hsiao et Y.H.Chen. *Acta Botanica Sinica* 32(1): 45–48.

Ying, T.S. (1979). On Dysosma Woodson and Sinopodophyllum Ying gen. nov. of the Berberidaceae. *Acta Phytotaxonomica Sinica* 17(1): 15–22.

Ying, T.S., Terabayashi, S. & Boufford, D.E. (1984) A monograph of Diphylleia. *Journal of the Arnold Arboretum* 65: 57–94.

Yu, P., Wang, L. & Chen, Z. (1991). A new podophyllotoxin-type lignan from Dysosma versipellis var. tomentosa. Journal of Natural Products 54(5): 1422–1424. *Chemical Abstracts* 116(5): 37939c.

Zhang, D., Shao, J. & Li, D. (1991). A study on the karyotypes of Dysosma versipellis and D. pleiantha endemic to China. *Guihaia* 11(1): 58–62.

Zhuang, P., Wu, H., Wu, J., Liang, K. & Zhou, F. (1993). The study on ecological and biologic characters of the genus Dysosma in Mount Emei. *Journal of Wuhan Botanical Research* 11(1): 41–46.

Zihua, J. & Chen, S. (1989). Chemical components of Dysosma veitchii. *Acta Botanica Yunnanica* 11(4): 479–481.

INDEX OF SCIENTIFIC NAMES

Accepted names in **bold**, synonyms in *italics*. Page numbers for illustrations are in *italics*.

Aceranthus 41
A. diphyllus 152
A. macrophyllus 126
A. sagittatus 126
A. triphyllus 126
Achlys 211
A. californica 213
A. japonica 213
A. triphylla *212*
    subsp. **californica** 213
    subsp. **japonica** 213
    subsp. **triphylla** 211
        var. *japonica* 214, 219
        var. *typica* 213
Anapodophyllum 258
A. canadense 263
Bongardia 223
B. chrysogonum *225, 226,* 227
B. margalla 227
B. olivieri 227
B. rauwolfii 227
× *Bonstedtia* 41
× *B. lilacina* 194
    var. *youngiana* 193
B. youngiana 193
Caulophyllum 214
C. giganteum 215
C. robustum 213, 215
C. thalictroides 215
    var. *giganteum* 215
Diphylleia 215
D. cymosa *216,* 218
    subsp. *grayi* 218
    subsp. *sinensis* 218
D. grayi 218
D. sinensis 218
D. pleiantha 273
Dysosma 269
D. aurantiocaule 303
D. aurantiocaulis 285
D. chengii 273
D. delavayi 292
D. difformis 287
D. emiensis 298
D. emodi 312
D. furfuracea 305
D. guangxiensis 291
D. hispida 280
D. hispidum 291
D. lichuanensis 290
D. mairei 285
D. majoense 290
D. majorense 290
D. majorensis 290, 294, 297
D. pleiantha 273, *279,* 287, 292
D. pysosma 284, 285
D. tonkinense 280
D. tsayuensis 304
D. veitchii 292
    var. *longipetalis* 294
D. versipellis 276, 280, 281
Endoplectris 41
Epimedium 41
    subgenus **Epimedium** 24, 25, 48
    subgenus **Rhizophyllum** 24, 26, 164
    subgenus *Vancouveria* 203
    section *Dimorphophyllum* 164
    section **Diphyllon** 24, 25, 48
    section **Epimedium** 24, 26, 155
    section *Gymnocaulon* 164
    section **Macroceras** 24, 26, 137
    section *Microceras* 155
    section *Phyllocaulon* 48
    section **Polyphyllon** 24, 26, 154
    section *Rhizophyllum* 164
    section *Vancouveria* 203
    subsection *Diphyllon* 154
    subsection *Polyphyllon* 154
    series **Brachycerae** 24, 26, 114
    series **Campanulatae** 24, 25, 48
    series **Davidianae** 24, 25, 55
    series **Dolichocerae** 24, 25, 81
    series *Diphyllon* 48
    series *Elatae* 154

INDEX OF SCIENTIFIC NAMES

series *Elongatae* 81
series *Microcerae* 155
series *Polyphylla* 154
E. acuminatum *18, 20, 38, 61*, 91, *92, 93, 94, 95, 96*
   'Galaxy' 200
   'Ruby Shan' 200
E. alpinum *4, 19, 20,* 155, *157,* 177
E. alpinum
   var. *pubigerum* 162
   var. *rubrum* 180
   f. *normale* 158
   f. *rubrum* 180
E. 'Amanogawa' 199
E. 'Amber Queen' 199
E. angustifolium 197
E. 'Arctic Wings' 199
E. atroroseum 194
E. baiealiguizhouense 136
E. baojingense 113
E. 'Baoxing Mist' 199
E. 'Black Sea' 199
E. borealiguizhouense 136
E. brachyrrhizum *18, 20,* 109, *110, 111*
E. brevicornu *19, 20,* 118, *119, 120*
E. 'Buckland Spider' 199
E. campanulatum *18,* 49, *49,* 50
E. × cantabrigiense *177,* 179
E. cavaleriei 197
E. chlorandrum *18, 20,* 101, *101, 102, 103*
E. chrysanthum 208
E. circinnato-cucullatum 170
E. citrinum 188
E. coactum 135
   var. **coactum** 135
   var. **longtouhum** 136
E. coccineum 180
E. colchicum 169
E. concinnum 194
E. cremeum 142
E. davidii *18, 20,* 55, *56, 57, 58, 59*
   var. *hunanense* 63
E. dioscoridis 158
E. diphyllum *19,* 149, *150, 152*
   subsp. **diphyllum** 151
   subsp. **kitamuranum** 151, 152
   'Nanum' 200
   'Roseum' 200
E. discolor 187
E. dodonaei 158
E. dolichostemon *19, 20,* 131, *132*
E. ecalcaratum *16, 18,* 50, *53,* 54
E. 'Egret' 199

E. elachyphyllum 134
E. elatum *19, 20, 153,* 154, *156*
E. elegans 169
E. elongatum *18, 20,* 81, *82, 83*
E. 'Enchantress' 184, 199
E. enshiense 100
E. epsteinii *18, 20,* 68, *68, 69*
E. fangii *18, 20,* 59, 60, *60, 61,* 62
E. fargesii *16, 19, 20,* 131, *133*
   'Pink Constellation' 200
E. 'Fire Dragon' 199
E. flavum *16, 18, 20,* 64, 65, *65*
E. 'Flowers of Sulphur' 199
E. franchetii *18, 20,* 97, *98, 99*
   'Brimstone Butterfly' 200
E. glandulosopilosum 113
E. grandiflorum *19, 20,* 137, *138, 142*
   subsp. *koreanum* 146
   subsp. *sempervirens* 145
   var. *thunbergianum* 143
   var. *violaceum* 143
   f. **flavescens** 141
   f. **grandiflorum** 141
   f. *normale* 141
   f. **violaceum** *138, 142, 143*
   'Akakage' 200
   'Akebono' 200
   'Album' 200
   'Beni-chidori' 200
   'Crimson Beauty' 200
   'Crimson Queen' 200
   'Elfenkönigin' 200
   'Koji' 200
   'La Rocaille' 200
   'Lilacinum' 201
   'Lilafee' *139,* 201
   'Mount Kitadake' 201
   'Nanum' 201
   'Nanum Freya' 201
   'Purple Prince' 201
   'Queen Esta' 201
   'Rose Queen' *138,* 201
   'Rubinkrone' 201
   'Saturn' 201
   'Saxton Purple' 201
   'Shiho' 201
   'Sirius' 201
   'Sunset' 201
   'Tama no Genpei' 201
   'Violaceum' 201
   'White Beauty' 201
   'White Queen' *140,* 201
   'Yellow Princess' *141,* 201

E. grandiflorum Marnock 152
E. hexandrum 206
   f. *planipetalum* 209
E. 'Honeybee' 199
E. hunanense 63
E. hybridum 194
E. hydaspidis 154
E. ikariso 126
E. ilicifolium *18, 20*, 66, *67*
E. japonicum Siebold ex Miq. 152
E. japonicum Makino 197
E. 'Kaguyahime' 185, 199
E. 'Kew Hybrid' 185
E. kitamuranum 154
E. komarovii 96
E. koreanum *142*, 146
E. latisepalum *20*, 70, *70, 71*
E. leptorrhizum *18, 20*, 107, *107, 108*
   'Mariko' 201
E. 'Lilac Charm' 200
E. lilacinum 194, 197
   var. *concinnum* 194
E. lishihchenii *18, 20*, 88, *89, 90*
E. 'Little Shrimp' 200
E. lobophyllum 136
E. macranthum 140, 141
   var. *hypoglaucum* 145
   var. *musschianum* 198
   var. *niveum* 195
   var. *normale* 141
   var. *roseum* 194
   var. *sulphureum* 188
   var. *thunbergianum* 142
   var. *versicolor* 187
   var. *violaceum* 143
   f. *niveum* 195
   f. *roseum* 194
   f. *sulphureum* 188
   f. *versicolor cupreum* 188
   f. *versicolor* 187
   f. *violaceum* 143
E. macrosepalum 146, *147*
E. membranaceum *18, 20*, 84, *84, 85*
   subsp. *genuinum* 84
   subsp. *orientale* 91
E. mikinorii *18, 20*, 78, *78, 79, 80*
E. musschianum 191, 195, 198
   var. *multifoliolatum* 198
   var. *trifoliolato-binatum* 148
   var. *vulgare* 193
E. myrianthum *19, 20*, 126, *127*
E. niveum 195
E. ochroleucum 188

E. ogisui *18, 20*, 72, *73, 74*
E. × omeiense *61*, 183, *184*
   'Akame' 202
   'Emei Shan' 184, 202
   'Myriad Years' 202
   'Stormcloud' 184, 202
E. orientale, flore albo 162
E. orientale, flore albo flavescente 162
E. parviflorum 209
E. parvifolium 134
E. parvulum 198
E. pauciflorum 75, *76, 77*
E. × perralchicum 186
   'Fröhnleiten' 202
   'Weihenstephan' 202
   'Wisley' *188*, 202
E. perralderianum *19, 20*, 171, *172, 173*
E. 'Pink Elf' 200
E. pinnatum *19, 20*, 164
   subsp. **circinatum** 170
   subsp. **colchicum** *163*, 165, 167, 169
   subsp. *originarum* 166
   subsp. **pinnatum** 166, *168*
   var. *colchicum* 169
   var. *integrifolium* 169
   var. *perralderianum* 173
   var. *sulphureum* 188
   f. *colchicum* 169
   (?f.) *elegans* 169
   f. *integrifolio* 169
E. pinnatum DC. 169
E. planipetalum 209
E. platypetalum *18, 50*, 51, *52*
   var. *tenuis* 75
E. pteroceras Baker 166
E. pteroceras C.Morren 198
E. pubescens *11*, 114, *115*, 117
   subsp. *cavaleriei* 116
   subsp. *primarium* 116
   subsp. **pubescens** 116
   var. *cavaleriei* 116
E. pubigerum *19, 20, 160*, 161, *161, 163*
E. purpureum 180
E. quorundam 158
E. reticulatum 121
E. rhizomatosum *18, 20*, 86, *87, 88*
   'Golden Eagle' 201
E. roseum 194
E. rotundatum 118
E. × rubrum *19, 20, 36*, 177, 180
   Cobblewood Form 202
   'Sweetheart' 202
E. rugosum 146

E. **sagittatum** *16, 19, 20,* 121, *122, 123, 124, 125*
   subsp. *pyramidale* 127
   subsp. *typicum* 126
   var. *pyramidale* 127
E. sagittifolius 126
E. **'Sasaki'** 200
E. **sempervirens** *19, 20,* 144, *144, 145*
   'Aurora' 201
   'Candy Hearts' 201
   var. *hypoglauceum* 145
   var. **rugosum** 146
   var. **sempervirens** 146
E. **shuichengense** 54
E. sieboldianum 166
E. **simplicifolium** 112
E. sinense 126, 194
   var. *pyramidale* 127
E. **'Sohayaki'** 200
E. **'Starlet'** 200
E. **stellulatum** *19, 20,* 128, *129,* 130
   'Wudang Star' *129,* 201
E. sulphurellum 142
E. sulphureum 188
E. **sutchuenense** 100
E. **takhtajanii** 196
E. **trifoliolatobinatum** *19,* 148
   subsp. **maritimum** 149, *149*
   subsp. **trifoliolatobinatum** 149
E. **truncatum** 134
E. versicolor 187
   cupreum 187
   cl. *versicolor* 187
   var. *neo-sulphureum* 190
   var. *sulphureum* 188
E. × **versicolor** 186
   'Cupreum' 187, 202
   'Neosulphureum' *37,* 190, 202
   'Sulphureum' 188, *189,* 202
   'Versicolor' 187, *189,* 202
E. violaceo-diphyllum 194
E. violaceum 140, 142
   var. *grandiflorum* 142
E. × **warleyense** *19, 20,* 181, *182*
   'Orangekönigin' 202
E. **'William Stearn'** 200
E. **wushanense** *18, 20,* 104, *104, 105, 106*
   'Caramel' 201
E. × **youngianum** 190, 195
   'Beni-kujaku' 202
   'Capella' 202
   'Lilacinum' 202
   'Merlin' *192,* 195, 202
   'Niveum' 19, 20, *192, 193,* 195, 202

'Pink Ruffles' 202
'Roseum' *192,* 194, 202
'Shikinomai' 202
'Tamabotan' 202
'Yenomoto' *195,* 202
'Youngianum' *191,* 202
   var. *concinnum* 194
   var. *niveum* 195
   var. *roseum* 194
   var. *typicum* 193
   f. *rubrum* 180
E. **zhushanense** 112
Gymnospermium 227
G. albertii *228, 229,* 230, *232*
G. altaicum 230
   subsp. *odessanum* 230
G. darwasicum 231
G. kiangnanensis 231
G. microrrhynchum 231, *233*
   var. **venosum** *233*
G. odessanum 230
G. scipetarum 230
G. shquipetarum 230
G. silvaticum 231
G. smirnovii 230
G. vitellinum 231
Jeffersonia 219
J. bartonis 222
J. binata 222
J. diphylla *221,* 222
J. dubia *220,* 222, *224*
   'Alba' 224
Leontice 234
L. albertii 231
L. altaica 230
   var. *odessana* 230
L. apiifolia 236
L. armeniaca 236, *236*
L. chrysogonum 227
L. darwasica 231
L. incerta *235,* 236
L. kiangnanensis 231
L. leontopetalum *226,* 234
   subsp. *armeniaca* 236
   subsp. **ewersmannii** 234
   subsp. **leontopetalum** 234
L. microrrhyncha 231
   f. *venosa* 231
L. minor 236
L. odessana 230
L. smirnovii 230
L. sylvatica 231
L. thalictroides 215

L. triphylla 213
L. vesicaria 236
Plagiorhegma 222
P. dubium 223
Podophyllum 239, 258
    section **Dysosma** 253, 269
    section **Hexandra** 254, 306
    section **Paradysosma** 254, 298
    section **Podophyllum** 251, 260
    section **Delavayae** 292
    section **Pleianthae** 270
P. acutifolium 311
P. acutum 311
P. aurantiacum 304
P. aurantiocaule 299
    subsp. **aurantiocaule** *300, 301*, 303
    subsp. **furfuraceum** 305
    subsp. *multiflorum* 305
    var. **aurantiocaule** 304, 305
    var. **uniflorum** 304, 305
P. callicarpum 267
P. cavaleriei 314
P. chengii 273
P. delavayi 292, *293*
    var. **delavayi** 294
    var. **longipetalum** 294
    × **P. difforme** 295, *296*
P. difforme 286, *288*, 289
P. diphyllum 222, 314
P. emodi 311, 313
    var. *axillaris* 312
    var. *bhootanensis* 312
    var. *chinense* 312
    var. *hexandrum* 312
    var. *jaeschkei* 312
    var. *royleana* 311
P. esquirolii 279
P. glaucescens 285
P. guangxiensis 291
P. hemsleyi 283
P. hendersonii 313
P. hexandrum 306, *308*
    subsp. *substerilis* 313
    var. *axillare* 312
    var. *bhootanense* 312
    var. *chinense* 312
    var. *emodi* 312
    var. *jaeschkei* 312
    var. *majus* 312
    cv. Majus 313
P. himalayense 314
P. hispidum 279, 305
P. indica 312

P. × **inexpectatum** 267, *269*
P. japonicum 219, 314
P. leichtlinii 312
P. mairei 284
P. majoense 289
P. montanum 263
    var. *acuminatum* 263
    var. *parviflorum* 263
P. onzoi 273
P. peltatum 260, *261*
    var. **annulare** 267
    var. *elatior* 263
    var. *extraxillare* 263
    var. *grandiflorum* 263
    var. *heterophyllum* 263
    var. *odoratum* 263
    var. *oligodon* 263
    var. **peltatum** 265
    var. *pumilum* 263
    var. *triphyllum* 263
    f. *aphyllum* 263
    f. **biltmoreanum** 266
    f. **callicarpum** 267
    f. **deamii** 266
    f. *polycarpum* 263
    f. *roseum* 266
P. pentaphyllum 313
P. pleianthum 271, *272, 274*
    var. **album** 276
    var. **pleianthum** 276
    × **P. versipelle** 276
P. pleianthum sensu Diels 281
P. podophyllum 263
P. sikkimensis 303, 305
    var. *emodi* 304
    var. *major* 303, 304
P. sinense 312
P. sp. **A** 297
P. sp. *B.* 298
P. sp. *C.* 297
P. tonkinense 280
P. triangulum 287
P. trilobulus 297
P. veitchii 292
P. versipelle 276
    subsp. **boreale** 277, 280
    subsp. **versipelle** 279
    var. **boreale** 281
    var. **sichuanense** 282
    var. *tomentosum* 282
P. versipelle Hance, *sensu* auctt. 279, 280
Ranzania 219
R. japonica *217*, 219

Sculeria 203
S. geminata 206
Sinopodophyllum 258, 306
S. emodi 293, 312
S. hexandrum 313
Vancouveria 203
V. brevicula 206
V. chrysantha *204, 207*, 208
V. concolor 210
V. crispa 210
V. hexandra *204*, 205, *207*
   var. *aurea* 208
   var. *chrysantha* 208
V. parviflora 209
V. parvifolia 206
V. picta 206
V. planipetala *204, 207*, 209
V. vaseyi 210
Vindicta 41
V. begonifolia 152
Yatabea japonica 219